Human Variability in Response to Chemical Exposures

Measures, Modeling, and Risk Assessment

Edited by

David A. Neumann
Risk Science Institute
International Life Sciences Institute
Washington, D.C.

Carole A. Kimmel
U.S. Environmental Protection Agency
National Center for Environmental Assessment
Washington, D.C.

CRC Press
Taylor & Francis Group
Boca Raton London New York

CRC Press is an imprint of the
Taylor & Francis Group, an **informa** business

CRC Press
Taylor & Francis Group
2385 NW Executive Center Drive, Suite 320, Boca Raton FL 33431

and by CRC Press
4 Park Square, Milton Park, Abingdon, Oxon, OX14 4RN

CRC Press is an imprint of Taylor & Francis Group, LLC

© 1998 Taylor & Francis Group, LLC

First issued in paperback 2019

Reasonable efforts have been made to publish reliable data and information, but the author and publisher cannot assume responsibility for the validity of all materials or the consequences of their use. The authors and publishers have attempted to trace the copyright holders of all material reproduced in this publication and apologize to copyright holders if permission to publish in this form has not been obtained. If any copyright material has not been acknowledged please write and let us know so we may rectify in any future reprint.

Except as permitted under U.S. Copyright Law, no part of this book may be reprinted, reproduced, transmitted, or utilized in any form by any electronic, mechanical, or other means, now known or hereafter invented, including photocopying, microfilming, and recording, or in any information storage or retrieval system, without written permission from the publishers.

For permission to photocopy or use material electronically from this work, access www.copyright.com or contact the Copyright Clearance Center, Inc. (CCC), 222 Rosewood Drive, Danvers, MA 01923, 978-750-8400. For works that are not available on CCC please contact mpkbookspermissions@tandf.co.uk

Trademark notice: Product or corporate names may be trademarks or registered trademarks and are used only for identification and explanation without intent to infringe.

ISBN 13: 978-0-367-44776-2 (pbk)
ISBN 13: 978-0-8493-2805-3 (hbk)

Visit the Taylor & Francis Web site at
http://www.taylorandfrancis.com

and the CRC Press Web site at
http://www.crcpress.com

Foreword

Even the most casual observer is quick to recognize the extensive variation manifest in the human population. Human beings differ considerably in size, shape, and color and within each of these and other phenotypic continua, a host of other variable traits is evident to most of us. In addition to this obvious phenotypic variation that reflects underlying genotypic variation, humans dwell in a variety of environments, consume qualitatively and quantitatively different diets, are subject to a host of diseases and ailments, and live in settings dictated by their socioeconomic status.

Despite such familiar examples of interindividual variability as well as the evidence emerging from recent and rapidly accruing advances in molecular genetics, the potential application of such information for risk assessment has been largely neglected. Does the 10-fold uncertainty factor that was established to account for individual-to-individual variability within the human population adequately provide for observed variability in human response and susceptibility? Do the so-called conservative default assumptions used in cancer risk assessment account for interindividual variability relative to human exposure to carcinogens? Importantly, how well does the scientific community understand human variability and the implications of that information for risk assessment?

These and related questions have been raised many times within the field of risk assessment but often have gone unaddressed. This volume is the result of a concerted and systematic attempt to address these issues through a rigorous, multidisciplinary review of the relevant scientific information on interindividual variability within the human population. These eight chapters, reflecting the contributions and insights of scientists from academia, industry, and government, offer a state-of-the-science assessment of how we characterize and quantify human variability and how that information might influence current thinking about risk assessment.

As is typical of such examinations, there is both good news and bad. On the positive side, we can frequently quantitate human variability and often have significant insight into the implications of that variability. On the other hand, the dearth of mechanistic information about toxic and carcinogenic processes limits our ability to understand and use this information to assess the adequacy of current public health protection practices. Nevertheless, this volume offers specific guidance and recommendations that, if pursued, will enhance our understanding of the nature of human variability and will demonstrate how to apply that understanding to risk assessment. I hope that the scientists, policy makers, students, and others who read this book will benefit from its insights and take up the challenge of pursuing the identified gaps in knowledge and understanding.

Alex Malaspina, President
International Life Sciences Institute

Preface

A host of factors contribute to interindividual variability in human responses and susceptibility to the effects of potential toxicants in the workplace, home, and environment. Concern about interindividual variability within the human population is reflected in current risk assessment practices. In the case of cancer risk assessment, allowances for human variability are considered to be encompassed in the conservative assumptions that are routinely used in the risk assessment process. Implicit in these assumptions is recognition that there are a number of issues for which scientific knowledge and understanding are limited and that to protect the public in the face of such uncertainties, risk assessments must err on the side of conservatism. Noncancer risk assessment expressly considers interindividual variability through the use of an uncertainty factor that is intended to reflect the variability in human response and susceptibility. This factor is typically assigned a value of 10, but in practice it may vary depending on the availability of relevant information. Moreover, some have suggested that uncertainty factors that address animal-to-human extrapolation and other considerations for which there is a dearth of information provide a buffer or cushion in terms of accounting for interindividual variability within the human population.

Recent developments in medicine, epidemiology, and toxicology have reinforced the notion that individuals differ considerably in their responses and susceptibility to various chemical and biologic agents. The concepts of sensitive subpopulations, populations at risk, genetic susceptibility factors, and environmental justice have underscored the need to evaluate how human variability is addressed in risk assessment. Are the assumptions and the use of an uncertainty factor adequately protecting the public from cancer or other effects associated with exposure to potentially toxic substances? Indeed, in 1994 the National Research Council* posed a similar question to the U.S. Environmental Protection Agency (EPA): "Has the EPA sufficiently considered the extensive variation among individuals in their exposures to toxic substances and in their susceptibilities to cancer and other health effects?"

Questions surrounding the adequacy of the various assumptions and the application of an uncertainty factor for noncancer risk assessment prompted this effort to understand the current state of scientific knowledge about interindividual variability and its relevance and application to risk assessment. In current risk assessment practice, considerations of interindividual variability may be based on quantitative or qualitative data or both. Obviously, quantitative data are preferable, particularly in the case of noncancer risk assessments, where the possible derivation of an uncertainty factor above or below the default value of 10 could have profound implications on the outcome of the risk assessment. The most valuable data in this context are those that quantitatively describe human responses to defined stimuli where differences in the measured parameter — e.g., a biomarker of susceptibility, exposure, or effect — are predictive of the magnitude or severity of the response. To facilitate evaluation of how information about interindividual variability is used in risk assessment, it is important to know what quantitative measures of response variability are available to risk assessors and whether these measures are predictive of risk. Such information would provide insight into, and lead to development of, models for facilitating the use of human variability data for risk assessment.

The ILSI Risk Science Institute formed a collaborative partnership with the EPA Offices of Research and Development; Water; Pesticide Programs; and Program, Planning, and Evaluation; the American Industrial Health Council; and the National Institute for Occupational Safety and

---

* National Research Council (1994) Science and Judgment in Risk Assessment. National Academy Press, Washington, DC.

Health to address these difficult issues. A steering committee* of scientists from government, industry, and academia recommended that the assessment begin by critically reviewing what is known about the variability of human responses relative to pulmonary toxicity, reproductive and developmental toxicity, neurotoxicity, and cancer. Those reviews, which are included in this volume, explore whether and how age, gender, ethnicity, and socioeconomic or dietary status influence quantitative measures of interindividual variability. These areas were selected because they were considered to be relatively rich in terms of the quality and quantity of available human data.

During preparation of these reviews, a workshop was convened to consider the implications of this information for current and future risk assessment practices. Workshop participants included the authors of the reviews, other scientists with expertise in these areas, and scientists from the risk assessment community. Both the reviews and the workshop were guided by the following questions:

1. Are there suitable quantifiable parameters that allow an assessment of the range of interindividual variability in the human response to carcinogenic processes or to agents that elicit neurotoxic, pulmonary, or reproductive and developmental effects?
2. Can these parameters be used to predict the specificity (target tissue) or severity of the response or process?
3. Are such parameters influenced by age, gender, ethnicity, preexisting disease, diet and nutritional status, socioeconomic status, or other factors?
4. Are the findings and relationships unique to the compound(s), endpoint(s), and process(es) under consideration or are they more broadly applicable?
5. How might this information be used in cancer and noncancer risk assessment?
6. Do the current default methodologies for cancer risk assessment in the United States adequately account for interindividual variability in the human population? Does this information enhance or diminish confidence in the currently used 10-fold uncertainty factor in noncancer risk assessment to account for interindividual variability?
7. Are there specific research needs relative to quantitatively assessing interindividual variability that, if addressed, would improve the practice of risk assessment?

The following reports provide informed and critical insight into inherent variability in human response and susceptibility. Jean Grassman, Carole Kimmel, and David Neumann set the stage by examining current risk assessment practices in the context of variability in human response and susceptibility to chemical exposures. Dale Hattis reviews the distributional aspects of response variability in human populations and describes procedures that can be used to analyze variability relative to both quantal and continuous processes. David Eckerman, John Glowa, and Kent Anger examine variability in the human response to neurotoxicants such as alcohol and lead. The same compounds and others are considered in the context of variability in reproductive and developmental outcomes by Anthony Scialli and Armand Lione. A single compound, ozone, is used by Philip Bromberg to explore interindividual variability in response to a pulmonary toxicant. Lynn Frame, Christine Ambrosone, Fred Kadlubar, and Nicholas Lang explore host-environment interactions that influence human susceptibility to cancer. Neil Caporaso and Nathaniel Rothman use glutathione S-transferase deficiencies and cigarette smoke-associated cancers as examples to examine the relationships between genetic polymorphisms and cancer. David Neumann and Carole Kimmel summarize the workshop and the implications of this information for current and future risk

* Neil Caporaso (National Cancer Institute), George Daston (American Industrial Health Council), Penelope Fenner-Crisp (EPA), Adam Finkel (U.S. Department of Labor), John Groopman (Johns Hopkins University), Bryan Hardin (National Institute for Occupational Safety and Health), Carole Kimmel (EPA), David Neumann (ILSI Risk Science Institute), Edward Ohanian (EPA), Nathaniel Rothman (National Cancer Institute), Stephen Spielberg (Merck Research Laboratories).

assessment practices. The general conclusions and recommendations emanating from these reports are captured in the *Summary and Recommendations* (following this preface).

This examination of the complex issues related to the collection and application of information on human variability in susceptibility and response to human health risk assessment was conducted under cooperative agreements between the ILSI Risk Science Institute and the EPA Office of Water and the EPA Office of Pesticide Programs. Additional funding was provided by the American Industrial Health Council; the National Institute for Occupational Safety and Health; the EPA Offices of Research and Development and Program, Planning, and Evaluation; and the ILSI Risk Science Institute. The contributions by the steering committee, the authors of the accompanying reports, and the workshop participants are greatly appreciated.

The logistical details associated with the workshop were very capably managed by Diane Dalisera of the ILSI Risk Science Institute. Eugenia Macarthy, Andrea Gasper, and Kerry White of the ILSI Risk Science Institute attended to the many administrative details inherent to such an undertaking. Jennifer Allen, also of this Institute, coordinated the transformation of eight manuscripts, multiple electronic files, and innumerable ancillary pieces of paper into this nearly 300 page volume. Jeffery Foran, Executive Director of the ILSI Risk Science Institute, provided continual encouragement and advice throughout the project. Finally, we are appreciative of the interest and enthusiasm that this project has engendered within the scientific community.

David A. Neumann
ILSI Risk Science Institute

Carole A. Kimmel
National Center for Environmental Assessment, EPA

Summary and Recommendations

David A. Neumann and Carole A. Kimmel

The activity described in this volume provided a unique opportunity to address the current and future application of information on interindividual variability within the human population to the risk assessment process. For the purpose of this exercise, interindividual variability refers to variability in susceptibility and response independent of exposure variability, recognizing that the latter is clearly critical to the risk assessment process. The four types of effects that were examined — neurotoxicity, pulmonary toxicity, reproductive and developmental effects, and cancer — reflect areas of investigation that are rich in both human and animal data. The questions posed to both the authors and the workshop participants were intended to frame the discussions and provide insight into the current state of scientific knowledge. Central to this assessment was an interest in determining, to the extent possible, whether present scientific knowledge about variability in human response and susceptibility supports the current assumptions and approaches used in cancer and noncancer risk assessment, which are considered overly conservative by some and not conservative enough by others.

There were no expectations that the answers would be simple or definitive, yet by asking such questions the scientific community, as represented by those involved in this activity, would be better informed about the challenges and opportunities afforded by increasing awareness of and appreciation for human diversity and the outpouring of information from molecular genetics, molecular epidemiology, and human genome project laboratories around the world.

The following points constitute a synthesis of the salient features of the reports in this volume, discussions by over 40 scientists who participated in the workshop, and comments by the steering committee, both before and after the workshop. Specific examples cited in this summary are explored more fully in the various reports.

Suitable quantifiable parameters are available for assessing the range of interindividual variability in response to exposure to pulmonary toxicants (e.g., pulmonary function or inflammatory changes), neurotoxicants (e.g., neurochemical assays, imaging, electrophysiological changes, behavioral parameters), and reproductive and developmental toxicants (e.g., perinatal mortality, incidence of malformations, enzyme activity, sperm quantity or quality). In the case of cancer, although adduct quantitation or other measures of dosimetry may illuminate the issue of interindividual variability, the absence of mechanistic understanding limits the utility of such data. Data available for these and other health effects typically have not been evaluated to specifically characterize response variability. It is critical to identify and quantify parameters (biomarkers) that can be measured repeatedly in the same individuals to distinguish between variability inherent to the test method(s) and intra- and interindividual variability.

Parameters such as those identified above can be used to predict target tissue specificity and severity for certain neurotoxic agents and endpoints — e.g., lead concentrations in blood and the neurotoxic effects of lead. Parameters that measure lung function clearly predict target organ (lung) effects, but their ability to predict severity is unclear. Quantitative data on interindividual variability currently cannot predict the specificity or severity of effects resulting from exposure to reproductive or developmental toxicants. Similarly, the specificity and severity of response elicited by exposure to carcinogens cannot be predicted from parameters identified as potential measures of interindividual variability. For both reproductive and developmental toxicants and carcinogens, the complex and incompletely understood mechanisms and processes associated with the health effects of interest confound and limit the predictive potential of such parameters.

Age, gender, ethnicity, nutritional and socioeconomic status, and other factors are known or are likely to influence variability in human responsiveness to many agents. It is unclear how much and to what extent such factors influence responsiveness and what mechanisms account for these influences. For example, few data are available to assess how such factors influence responses to neurotoxic agents. Conversely, it is well established that children, young adults, and people who exercise are more affected by exposure to the respiratory toxicant ozone than are other members of the population. Factors such as age, parity, ethnicity, and nutritional and socioeconomic status influence the likelihood that fetal alcohol syndrome will occur among alcohol-abusing women. Such factors likely influence the carcinogenic process, but, again, the underlying mechanisms are largely unknown.

Evidence suggests that certain gene polymorphisms (e.g., *BRCA1* for breast cancer, cytochrome P-450 and glutathione *S*-transferases for lung cancer) can influence the outcome of carcinogenic processes. Similar data link specific enzyme polymorphisms with both lead and isoniazid toxicity and neuropathy, indicating that the pharmacokinetic differences between enzyme isoforms are important determinants of interindividual variability. Such polymorphisms likely contribute to variability in response to pulmonary and to reproductive and developmental toxicants, but there currently is little supporting evidence. In the case of developmental toxicants, identification of relevant genetic factors has begun (e.g., transforming growth factor α and cleft palate), but it is confounded by the need to evaluate the influence of genes from both the mother and the offspring. Data from the human genome project likely will increase the number of genes implicated and provide insight into their possible roles in toxicologic and carcinogenic processes, yet they may necessitate development of new approaches for using such information in risk assessment.

The influences of factors such as age, gender, ethnicity, etc., on toxic processes and endpoints appear to be unique to specific compounds and may not be predictive of responses to exposure to other compounds. These limitations reflect the current paucity of data and incomplete understanding of the underlying mechanisms. Individuals characterized as susceptible on the basis of one parameter may not be similarly categorized if another parameter is evaluated. This has significant implications for risk assessment in terms of identifying those individuals or subpopulations that are most likely to have an adverse or severe response to a chemical exposure. Because embryonic and fetal development involves changing patterns and rates of development, susceptibility to developmental toxicants changes rapidly over time. This will influence both the probability of any outcome occurring as well as the type of outcome. Although the responses do not appear to be generalizable, the design of studies to address how such factors influence the response to specific compounds may have broad applicability.

In the case of cancer, the influences of such factors may be more generalizable, although there are too few examples to be definitive. Existing human and animal data on the carcinogenic effects of aflatoxin and arsenic may offer insight into this issue because of the availability of both dosimetry and effects data in multiple species. The critical impediments to understanding how age, gender, and other factors influence carcinogenic processes are the lack of information on target tissue dosimetry and limited understanding of the rate-limiting steps that control those processes.

Information on interindividual variability, where available, is being used in risk assessment. The ability to incorporate such information is limited, in part, by uncertainty about the sources of interindividual variability and the relationship between external exposure and target tissue dose and between target tissue dose and response. These considerations are further confounded by a lack of understanding of the mechanisms responsible for the variability. Dose–response relationships in both animals and humans, particularly the latter, are often poorly characterized with respect to individual-to-individual variability in responsiveness. Animal studies may provide insight into variability in response relative to the specific species and strain being examined but have uncertain relevance for understanding variability within the human population. However, animal studies have the potential to identify fundamental biological processes that contribute to response variability that may be applicable to humans.

An important concern for the risk assessment process is derivation of the value to which the uncertainty factor for interindividual variability is applied. It is often unclear how much confidence risk assessors have in estimates based on no-observed-adverse-effect levels (NOAEL) or lowest-observed-adverse-effect levels (LOAEL) derived from animal studies. There also may be uncertainty about where the estimated value occurs on the animal dose–response curve, particularly as it relates to the variability in response for the test species and strain. Use of modeling approaches, e.g., benchmark dose, may allow estimates of both the risk at given doses and the variability within the species and strain. The application of the uncertainty factor(s) to this estimate provides no insight into where on the human dose–response curve the calculated exposure value falls. Physiologically based pharmacokinetics and chemical-specific modeling will ultimately facilitate estimation of the human dose–response.

It is unclear whether current approaches to addressing interindividual variability in risk assessments are adequate to protect public health. Although there are no unequivocal cases where an uncertainty factor has failed, evaluation of their adequacy often is hampered by a lack of relevant information. Such evaluations are further complicated because the final risk assessment for non-cancer effects reflects the use of multiple uncertainty factors. Thus, any inadequacy associated with the uncertainty factor for human variability may be compensated for by the use of other uncertainty factors. Similarly, the derivation of the value to which the uncertainty factor is applied (see above) will determine, in part, the adequacy of the uncertainty factor. Estimates based on human data are often derived from data obtained from healthy young males or other subject groups that may not fully reflect the range of human responsiveness. One approach to assessing the adequacy of the uncertainty factor might be to determine the portion of the human population that is protected by the uncertainty factor. This would make fuller use of available information and could provide insight into whether the uncertainty factor used is over- or underprotective with respect to variation in response.

Evaluation of the adequacy of the uncertainty factor for human variability is particularly complex in the area of reproductive and developmental toxicology because exposure may be indirect, i.e., occurring prenatally or via lactation, and reflect the pharmacokinetics and pharmacodynamics of both the mother and the offspring. Reproductive and developmental toxicity may be manifested by a variety of effects that are differentially modulated by specific toxicants, doses, exposure times and/or duration, or other factors. This further confounds the evaluation of the adequacy of the uncertainty factor and raises the possibility that, given the dearth of information on the types and severity of possible outcomes, additional variability might need to be factored into the risk assessment process for reproductive and developmental toxicants.

Although the default cancer risk assessment process does not explicitly address interindividual variability, the EPA Cancer Risk Assessment Guidelines make the assumption that the conservative nature of the process provides adequate allowance for individual-to-individual variability in cancer susceptibility. The veracity of these assumptions has been challenged; however, few data are available to either support or refute such assumptions. It is also clear that there are insufficient data available to determine whether variability associated with specific subpopulations of particular concern, e.g., children, is reflected in the risk assessment process. Some of the specific issues that limit such evaluations relative to the cancer risk assessment process include tissue dosimetry and genetic polymorphisms.

Workshop participants were in agreement that the scientific community should and could improve upon the collection and analysis of quantitative information on interindividual variability. This might be accomplished by establishing better sampling strategies for human studies as well as analytical approaches that recognize the possibility that sensitive subpopulations may occur within study populations. Future epidemiologic studies could be expressly designed and evaluated to characterize interindividual variability and to develop appropriate, accessible databases. Current and completed studies could be modified or reevaluated, respectively, to yield more information on human variability. Such approaches would be facilitated by identification of relevant biomarkers of exposure, effect, or susceptibility and by better characterization of sensitive subpopulations.

Better appreciation and understanding of how information on human variability might affect risk assessment could be approached by reexamining previously determined reference doses (RfDs), acceptable daily intakes (ADIs), and risk assessments. This approach would address both the adequacy of the currently used assumptions and default value as well as determining how new and emerging information on interindividual variability can be incorporated into the risk assessment process. Such an approach would be facilitated by research to improve understanding of target tissue dosimetry, the mechanisms contributing to the observed toxicologic and carcinogenic effects, and the underlying genetics of susceptibility and response variability.

Contributors

Christine B. Ambrosone, Ph.D.
National Center for Toxicological Research
Division of Molecular Biology
3900 NCTR Drive
MS HFT-100
Jefferson, AR 72079-9502

W. Kent Anger, Ph.D.
Center for Research on Occupational and
  Environmental Toxicology (L606)
Oregon Health Sciences University
3180 SW Sam Jackson Parkway
Portland, OR 97201

Philip A. Bromberg, M.D.
Center for Environmental Medicine and
  Lung Biology
School of Medicine
University of North Carolina
104 Mason Farm Road
EPA-Human Studies Facility
Chapel Hill, NC 27599

Neil Caporaso, Ph.D.
National Cancer Institute
Genetic Epidemiology Branch
Executive Plaza North, Room 439
6130 Executive Blvd., MSC 7372
Bethesda, MD 20892-7372

David A. Eckerman, Ph.D.
Department of Psychology
University of North Carolina
Davie Hall
Campus Box 3270
Chapel Hill, NC 27599

Lynn T. Frame, Ph.D.
The Institute of Environmental and Human
  Health
Texas Tech University
Reese Center
Lubbock, TX 79416

John R. Glowa, Ph.D.
Departments of Pharmacology, Psychiatry
  and Psychology
Louisiana State University Medical Center
1501 Kings Highway
Shreveport, LA 71130

Jean A. Grassman, Ph.D.
National Institute of Environmental Health
  Sciences
P.O. Box 12233
MC C4-O5
111 Alexander Drive
Research Triangle Park, NC 27709

Dale Hattis, Ph.D.
Marsh Institute Center for Technology
  Environment and Development (CENTED)
Clark University
950 Main Street
Worcester, MA 01610

Frederick F. Kadlubar, Ph.D.
National Center for Toxicological Research
3900 NCTR Drive
MS HFT-100
Jefferson, AR 72079

Carole A. Kimmel, Ph.D.
U.S. Environmental Protection Agency
National Center for Environmental
  Assessment
401 M Street, SW (8623 D)
Washington, DC 20460

Nicholas P. Lang, M.D.
Chief of Surgery
Little Rock VA Hospital
4300 W. 7th
(112-LR)
Little Rock, AR 72205

Armand Lione, Ph.D.
Reproductive Toxicology Center
Columbia Hospital for Women Medical
  Center
2440 M Street, NW
Suite 217
Washington, DC 20037-1404

David A. Neumann, Ph.D.
ILSI Risk Science Institute
1126 16th Street, NW
Washington, DC 20036

Nathaniel Rothman, Ph.D.
National Cancer Institute
Environmental Epidemiology Branch
Executive Plaza North, Room 439
6130 Executive Blvd.
Rockville, MD 20892-7372

Anthony R. Scialli, M.D.
Department of Obstetrics and Gynecology
(3 PHC)
Georgetown University Medical Center
3800 Reservoir Road, NW
Washington, DC 20007-2197

Contents

Foreword

Preface

Summary and Recommendations
David A. Neumann and Carole A. Kimmel

1 Accounting for Variability in Responsiveness in Human Health Risk Assessment

Jean A. Grassman, Carole A. Kimmel, and David A. Neumann

CONTENTS

INTRODUCTION

The health effects produced by smoking cigarettes show that, when exposed to a toxicant, humans differ in their response. Although cigarette smoking is popularly associated with lung cancer (Villeneuve and Mao 1994), there are qualitative differences in the causes of smoking-related mortality in lifetime male smokers. Depending on the individual, smoking can cause death due to lung cancer, cardiovascular disease, chronic obstructive pulmonary disease, or a variety of other cancers (Doll et al. 1994, Bartechhi et al. 1994). When similar levels of smoking are compared, there are quantitative differences in the risk of specific health effects. Lung cancer risk due to smoking appears to be higher than average in females (Zang and Wynder 1996) and in individuals who slowly detoxify the carcinogens in cigarette smoke (Nakachi et al. 1993). In contrast, smokers who consume diets rich in carotenoid-containing vegetables appear to be at lower risk (van Poppel and Goldbohm 1995).

The example of cigarette smoking demonstrates the functional impact of human diversity in biochemical and genetic traits. Many traits show a high degree of variability. The ability to metabolize drugs such as debrisoquine can vary by more than 1000-fold (Nakamura et al. 1985), and the odds of any two people having the same profile of HLA antigens on their cell surfaces are 1 in 25 million (Lewontin 1995).

0-8493-2805-5/99/$0.00+$.50
© 1999 by CRC Press LLC

There are several reasons for the diversity of human responses. Analysis of mitochondrial DNA suggests that the divergence between African and non-African populations occurred approximately 140,000 years ago (Horai et al. 1995); the divergence between European and Japanese populations occurred approximately 70,000 years ago (Horai et al. 1995). During their subsequent dissemination, humans have covered the globe and thrived in conditions ranging from the cold of the Arctic, to the high thin air of Tibet, to the malarial regions surrounding the Mediterranean. Situations arose where populations were founded by a small number of individuals. Founders' effects accompanied by a wide variety of selective pressures have contributed to present day genetic variation.

Genetic variation alone cannot account for the range of differences among individuals because genetically identical individuals do not have identical susceptibilities. Monozygotic twins predisposed to insulin-dependent diabetes often show discordance in development of the disease because their susceptibility is modified by environmental factors (Verge et al. 1995, Bach 1994). Behaviors such as cigarette smoking, high-fat diets, and excessive consumption of alcohol augment biologic diversity by increasing the likelihood of coronary heart disease, stroke, and loss of physical mobility (Higgins and Thom 1993, LaCroix et al. 1993, Stamler et al. 1993).

Upon this background of genetic and acquired human diversity, populations may be exposed to environmental contaminants capable of producing biologic effects. Exposure is defined as "the contact of an organism with a chemical or physical agent" [U.S. Environmental Protection Agency (EPA) 1992]. Evidence of the extent and variety of exposure can be found in bodily tissues. Adipose tissues and bone act as reservoirs for dioxins (Orban et al. 1994), polycyclic aromatic hydrocarbons (Phillips 1992), chlorinated pesticides such as dichlorodiphenlytrichloroethane (DDT) (Phillips 1992), and lead (Silbergeld 1991). Blood contains contaminants as diverse as dioxins (Patterson et al. 1994), lead (Brody et al. 1994), and benzene (Mannino et al. 1995). Volatile organic solvents such as trichloroethylene and benzene are excreted in exhaled breath (Hattemer-Frey et al. 1990); nonpolar metabolites are found in the urine (Sipes and Gandolfi 1986); and polar contaminants such as polychlorinated biphenyls (PCBs), dioxins, and chlorinated pesticides exit through the bile and breast milk (Pluim et al. 1994, Schlaud et al. 1995).

Although exposure to contaminants is virtually universal, the magnitude of exposure varies according to area of residence, personal habits, and employment. Toxic release inventory chemicals are disproportionately dispersed in the southeast (Perlin et al. 1995), whereas the Los Angeles South Coast Air Basin has reported the highest levels of ozone in the country (EPA 1993). People who consume fish caught in the Great Lakes may be highly exposed to mercury and PCBs (Lambert and Hsu 1992). Youngsters learning to walk may be exposed to higher levels of lead than adults in similar environments because crawling places them in contact with contaminated soil and dust. They also have a tendency to ingest larger amounts of lead (Mushak and Crocetti 1990).

A consequence of human diversity is that, in addition to exposure, individual characteristics determine the type and degree of response to environmental contaminants. Because responsiveness to contaminants varies, members of the population differ in their susceptibility, which is determined by the intrinsic or acquired traits that modify their risk of illness. In this context, the concern is with differences in susceptibility that modify risk after exposure to environmental contaminant(s).

Protecting the general population from the detrimental effects of environmental exposures requires characterization of the response variability that results in differences in susceptibility. Response variability has two aspects. The first is the magnitude of the range of response. For example, if a population is uniformly exposed to a contaminant, is the response of the most sensitive person 10 times or 10,000 times the response of the least sensitive person? The magnitude of variability in response associated with most environmentally mediated adverse health effects has not been characterized.

The intrinsic characteristics that determine an individual's response constitute the second dimension of response variability. Many of the factors that describe individual responsiveness fall into the following categories: genetics, gender, ethnicity, developmental stage, the impact of lifestyle, and preexisting illness.

Fluctuations in the prevalence of illnesses with a possible environmental etiology, such as asthma, suggest that there is much about intrinsic response to contaminants that is not well understood. Environmental factors have been implicated in the increase in asthma prevalence and mortality observed over the past decade [Morbidity Monthly Weekly Review (MMWR) 1990, Balmes 1993, Jackson et al. 1988]. Once initiated, asthma figures prominently in subsequent susceptibility to air pollutants (Balmes 1993). Although approximately 10 million people in the United States are affected (Balmes 1993), the factors that predispose individuals to asthma have not been identified. The course of other chronic diseases, such as Alzheimer's disease, may also be influenced by susceptibility to environmental contaminants (Breitner et al. 1991).

Regulatory agencies such as the EPA have developed risk assessments to help determine acceptable environmental levels of both localized and widespread contaminants. Risk assessments can be performed for any chemical and for any health effect. When data are available, quantitative risk assessments use existing dose–response information to estimate the health effects at projected levels of exposure. Although risk management may consider economic, social, and technical factors, a risk assessment of potential health effects underlies most decisions about the permissible level of contaminants in air, water, and soil.

Risk assessments do not generate new data. Instead, data describing the range of health effects and the relationship between dose and response are drawn from the existing toxicologic and epidemiologic literature. Historically, variability in response has been perceived as a source of error rather than a parameter of interest (Hattis 1996). Because few investigations are designed to assess response variability, the opportunities for incorporating variability into risk assessments is limited. In the rare instances when variability can be estimated, it is often difficult to separate response variability from the contribution of variability due to analytical methods or the normal variability within an individual over time.

Although the likelihood of illness depends on both the level of exposure and an individual's responsiveness, risk assessments emphasize the contribution of exposure because quantitative data describing response variability are usually unavailable. Asthmatics and children exposed to legally permissible levels of air particulates respond with increased hospitalizations and reduced pulmonary function (Pope and Dockery 1992, Schwartz et al. 1993). Observations such as these suggest that a failure to fully account for response variability can result in the promulgation of nonprotective standards.

The impact of emphasizing exposure in risk estimates as opposed to including both exposure and responsiveness is shown in the hypothetical dose–response curves in Figure 1. Figure 1a represents a population dose–response curve typical of what would be used in a risk assessment. In Figure 1b, the dose–response curve has been amended to include the responsiveness of all members of the population so that, at any given level of exposure, the members of a population exhibit a range of responses. Risk is determined by the interaction between the magnitude of an individual's exposure and how the individual responds to that exposure. The risk surface illustrated in Figure 1b suggests that a risk assessment based solely on the relationship between exposure and risk will overestimate the risk to some members of the population and underestimate the risk to others.

RESPONSE VARIABILITY AND SUSCEPTIBILITY

Response variability is the variation in the type or magnitude of biologic effect due to intrinsic or acquired differences between individuals under identical conditions of exposure. Response, or interindividual, variability is only one of many contributors to total variability. As shown in Figure 2, total variability can be partitioned into exposure, biologic response, and methodological variability and stochastic processes. Methodological variability is associated with the acquisition of information describing the outcome and includes variability associated with the analytical methods. The role of exposure variability is widely recognized because virtually all investigations examine the

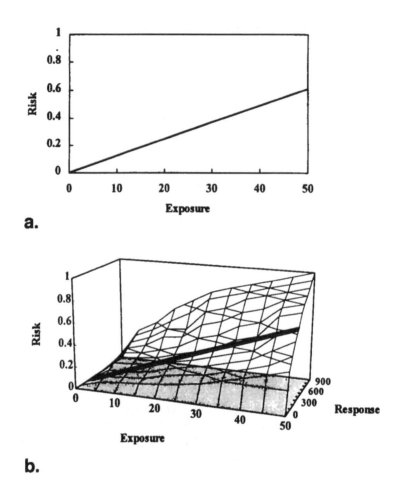

a.

b.

FIGURE 1   Dose–response curves for a hypothetical population where risk is estimated from exposure only (a) and on the basis of both exposure and responsiveness (b).

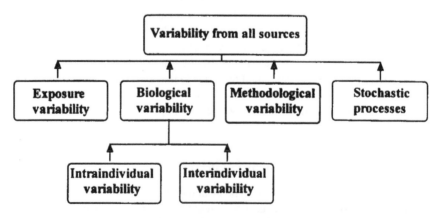

FIGURE 2   Total variability observed after exposure to environmental contaminants has several components, including exposure, biologic, and methodological variability. Biologic variability can be subdivided into intraindividual variability, which is the difference within an individual over time, and interindividual variability, which is the difference between individuals. Stochastic processes can also contribute to total variability.

effects of contaminants as a function of exposure or dose. Biologic variability includes both the differences among individuals (interindividual variability) as well as the variability in an individual's response over time (intraindividual variability). Stochastic processes are events that occur randomly or by chance. A predominantly deterministic process, as opposed to a stochastic process, makes it possible to estimate the magnitude of interindividual variability by subtracting the contribution of other sources of variability.

Although the terms "variability" and "uncertainty" may be used in similar circumstances, they describe different phenomena. Critics of current risk assessment procedures cite the need for better quantitation of variability and uncertainty [National Research Council (NRC) 1994, Bogen and Spear 1987). Any collection of observations describing the response to an environmental agent will include uncertainty and variability from a variety of sources. Uncertainty is due to a lack of knowledge, whereas variability is a characteristic of the data. In risk assessments, sources of uncertainty include the shape of the dose–response curve, the comparability of animal models with human response, and the methods used for data collection. Uncertainty can be reduced by a better understanding of the disease process at low doses and by improvements in the technical methods for measuring exposure and identifying responses.

Quantitative variability in response can be described as the differences in the magnitude of a given response encountered among individuals within an exposed population. Response variability can also be qualitative and described by the variety of responses produced by an exposure. The quantitative aspects of response variability tend to be emphasized because they are easier to measure and summarize. Consider the effects of exposure to organophosphate pesticides, which are cho-linesterase inhibitors. Quantitatively, the response variability associated with a given level of exposure can be described as the percent difference in acetylcholinesterase inhibition across the population. Qualitative variability in response is also evident. When groups of agricultural workers were exposed to relatively uniform levels of organophosphates, some individuals experienced symptoms of nausea, constriction of the pupils, weakness, and diarrhea, whereas others displayed signs of central nervous system disruption such as incoordination and unconsciousness (Richter et al. 1992, Ames et al. 1989, Quinones et al. 1976).

Quantitative response variability can be described with either quantal or continuous outcomes (Hattis 1995). Responses that are quantal can be classified according to their presence or absence, as is done with cancer, birth defects, and asthma. Quantal health effects are usually expressed as a cumulative distribution of the exposure or time required to produce the illness. Because the relevant endpoint is the onset of illness, individual dose–response curves are not possible. For example, Figure 3a shows the concentration at which illness appeared in each individual within a hypothetical population of 1000; for a given individual, there is no dose–response curve, only the level of exposure that produced his or her illness. The population's response variability can be expressed as the range of exposures causing the onset of illness, as indicated by the arrow in Figure 3a.

When a continuous outcome is measured in a population, the magnitude or intensity of response is given as a function of the level of exposure, as shown in Figure 4a. Each square represents an individual positioned according to exposure and response. In this hypothetical example, other sources of variability are assumed to be either controlled or negligible. The horizontal dispersion of the data shows the variability in exposure, whereas the vertical dispersion (marked with a vertical arrow) indicates the variability in response. The response variability can be observed by examining the range of responses at a given level of exposure. Exposure to approximately 70 units produces responses that vary from 20 to 150. Note that as the magnitude of the exposure changes, the variability in response may also change. If a toxicant produces acute and reversible health effects, as is the case with pulmonary irritants such as acids and ozone, it may be possible to construct a dose–response curve for each individual (Kulle et al. 1985, Utell et al. 1983). Figure 4b and 4c

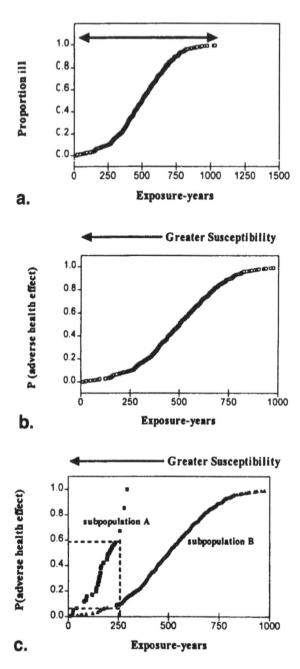

FIGURE 3 Response variability (a), susceptibility (b), and a susceptible subpopulation (c) demonstrated in a hypothetical population where the health effect can be expressed as a quantal response. (a) Response variability is evidenced by the range of exposures causing illness (horizontal arrow). (b) Susceptibility can be inferred based on the exposure, expressed as exposure years, that produces illness. With deterministic processes, highly susceptible people become ill at lower levels of exposure than less susceptible people. (c) Susceptible subpopulations can be identified by classifying individuals. Here most, but not all, individuals in subpopulation A are more susceptible than those in subpopulation B. In addition, the variability within any single subpopulation is less than the variability for the entire population.

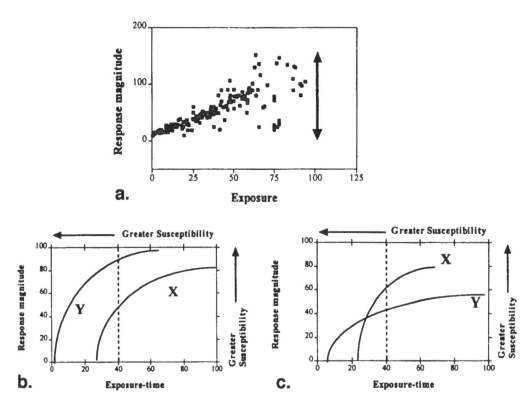

FIGURE 4 Response variability (a) and susceptibility (b and c) demonstrated in hypothetical populations where the health effect can be expressed as a continuous response. (a) Response variability is apparent as the vertical dispersion in a scatter plot where each square corresponds to an individual's response. Vertical arrow represents the variability in response observed within the population. (b) Susceptibility inferred from the dose–response characteristics of two individuals with parallel dose–response curves. Individual Y is more susceptible than individual X because the response occurs at lower concentrations and is always greater. (c) Susceptibility inferred from the dose–response characteristics of two individuals with intersecting dose–response curves. At low concentrations, individual Y is more susceptible because the response is greater. When exposure exceeds 25 units, individual X is more susceptible because the response is greater.

show hypothetical dose–response curves based on the responses of two of the individuals selected from Figure 4a.

Unlike these hypothetical examples, the full range of variability present in a population can rarely be measured, although it can be estimated by statistical and biologic modeling techniques. An alternative to measuring population variability is to examine responsiveness in a population on the basis of identifiable characteristics. Possible comparisons include obese/normal weight, male/female, fast acetylator/slow acetylator, and African/(white) European/Asian ancestry. Differences in response can be used to make inferences about the relative sensitivity and susceptibility to illness after exposure to environmental contaminants.

Sensitivity and susceptibility are defined as an individual's intrinsic or acquired traits that modify the risk of illness. Although highly exposed individuals have been referred to as highly susceptible (EPA 1992), exposure and sensitivity/susceptibility should be considered independent contributors to the risk of illness. Sensitivity and susceptibility are often used interchangeably despite slightly different connotations. Sensitivity refers to the relative level of responses observed or expected in a population, whereas susceptibility also suggests risk, as with individuals with a high degree of susceptibility to cancer. Susceptibility is the preferred term because sensitivity has

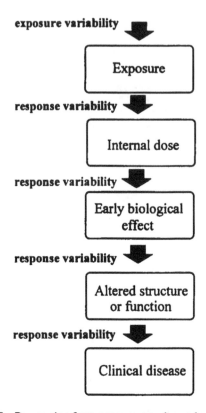

FIGURE 5   Progression from exposure to adverse health effect.

several other usages, including the detection limit for analytical methods and the likelihood of correctly identifying a case in epidemiology.

"Susceptibility" and "response variability" are related but not synonymous terms that are used to describe the different ways individuals respond when exposed to environmental contaminants. Figure 5 shows the progression of biologic changes from exposure to development of adverse health outcomes. In a population, there will be response variability associated with any given step along the progression. Some of the response variability observed after exposure may be due to homeostatic fluctuations and not correlated with susceptibility. For example, induction of enzymes such as cytochrome P450s may occur in the absence of adverse health effects (Nakamura et al. 1985). Susceptibility refers to the overall process, with an emphasis on the likelihood of an adverse health outcome.

In the model of contaminant-induced illness shown in Figure 5, changes occurring early in the process can be used as biomarkers, which are defined as "indicators of events in biological systems" (NRC 1989). When the risk encountered after exposure is being estimated, the most useful biomarkers are those that are highly correlated with either exposure or susceptibility. Biomarkers may aid in evaluating response variability or susceptibility when health outcome data are unavailable. For instance, a propensity to form DNA adducts may indicate susceptibility to carcinogens. Smokers who are lung cancer patients have higher levels of lymphocyte benzo[a]pyrene DNA adducts than are found in smokers without lung cancer (Perera et al. 1989).

Differences in susceptibility can be inferred when health outcome data are examined. With quantal data, as shown in Figure 3b, susceptibility increases as the dose at which the disease occurs decreases. Expressed more conventionally, highly susceptible people are those who become ill at lower concentrations. There is a caveat to this generality. When an illness has a large random or stochastic component, as may be the case with cancer, some unlucky individuals may become ill at low levels of exposure even though they are not highly susceptible. Dale Hattis (this volume) provides a comprehensive discussion of stochastic processes.

Continuous responses to exposure can be used to compare susceptibility. Figure 4b and 4c presents the hypothetical continuous dose–response curves for two individuals. In the parallel curves, individual Y always shows greater susceptibility to the health effect being measured than individual X. Individual Y responds at lower concentrations, and, at a given concentration, the response is always greater. However, the situation may not always be so straightforward, as shown in Figure 4c where the dose–response curves cross. If the criterion for susceptibility is the lowest exposure that produces a response, individual Y is more susceptible. However, if the criterion is the level of response at a given exposure, individual X is more susceptible when exposure exceeds approximately 25 units.

The variability in adverse health effects within a population is manifested by differential susceptibility, which is due to nonhomogeneity of response. The arrows labeled "greater suscepti-bility" in Figures 3 and 4 also indicate differential susceptibility. Differential susceptibility is distinct from differential risk, which is due to nonhomogeneity of either response or exposure.

Susceptible subpopulations are often identified by comparing the risk in those with and without a trait. Figure 3c shows hypothetical subpopulation A selected on the basis of a trait. The trait could be directly associated with susceptibility, as with age, gender, or genetic predisposition. Alterna-tively, subpopulation A may be identified on the basis of a surrogate marker for susceptibility such as socioeconomic status or county of residence.

Characterization of risk on the basis of membership in a susceptible subpopulation can lead to erroneous conclusions about individual risk. As a group, membership in subpopulation A appears to confer greater susceptibility than membership in subpopulation B. For example, when the risks at 250 exposure units are compared, the risk to a member of subpopulation A is approximately 0.6, whereas for a member of subpopulation B it is less than 0.1. However, membership in the more susceptible subpopulation A does not guarantee greater susceptibility. A few members of subpop-ulation B, which is considered to be less susceptible, are affected at lower levels of exposure than some of the members of subpopulation A.

When health effects are continuous, susceptible subpopulations can be identified by comparing their responses after uniform exposures, as shown in Figure 6. This is similar to the approach that was originally used to identify the acetylator enzyme polymorphism. The ability of volunteers to acetylate isoniazid was assessed by measuring the concentration of the unmetabolized drug in their plasma several hours after administration. The resultant frequency distribution of plasma concen-trations was distinctly bimodal (Nebert 1991). Figure 6a shows a hypothetical population where the response distribution is bimodally segregated into two distinct subpopulations. The bimodality could be used to classify the population into subpopulations on the basis of differences in their response. However, a bimodal distribution is not a prerequisite for establishing the existence of a susceptible subpopulation, as shown in Figure 6b where the population exhibits a unimodal response. The members of the susceptible subpopulation would be those whose responsiveness places them at the upper tail of the Gaussian distribution.

There is a tendency to dichotomize a population into susceptible and nonsusceptible individuals. In most instances, all members of the population are at risk of adverse health effects after exposure, although the risk for some members may be extremely low and for others it may be extremely high. Therefore, it is more accurate to speak in terms of greater and lesser susceptibility.

Inequities in income, access to health care, education, and political power exist within our society. In recent years, concerns have been raised that poorer and minority communities may be subjected to more than their share of environmental contamination. Environmental equity refers to "the provision of adequate protection from environmental toxicants for all people, regardless of age, ethnicity, gender, health status, social class or race" (Sexton and Anderson 1993). Although

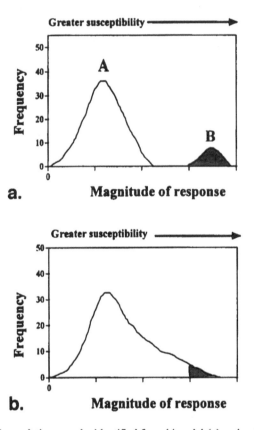

FIGURE 6 Susceptible subpopulations can be identified from bimodal (a) and unimodal (b) response distributions. Here all individuals within both hypothetical populations are assumed to have uniform exposures. (a) Susceptible subpopulation B forms a distinct peak consisting of highly responsive individuals. (b) Susceptible subpopulation (shaded area) consists of individuals with the highest magnitude of response.

most concerns about environmental equity have focused on situations in which communities were at risk because of their exposure, differences in responses to contaminants may also play a role. Greater susceptibility may be attributable to genetic factors or to the overall age of the population. Alternatively, differences may be due to acquired factors such as health care or quality of diet.

These hypothetical populations represent a situation in which the response of each member of the population is known, there is no uncertainty, and the variability due to factors other than response and exposure is nonexistent. Review of the literature will show that studies with these characteristics do not exist. When information on variability or susceptibility is found, it is often accompanied by the following limitations. First, any measure of variability is made from a sampling of the population. The variability in the population that serves as the source of dose–response information is not necessarily comparable to that found in the population at large (Finkel 1994). Second, a failure to measure exposure precisely and accurately limits the ability to measure variability due to response. Third, measurements of only response variability do not provide insight into the reasons for differences in susceptibility. The bottom line is that comprehensive quantitative and qualitative measures of population response variability are rare.

Therefore, it is informative to study examples where an assessment of response variability is possible. David Eckerman, Kent Anger, and John Glowa (this volume) examine human response variability after exposure to neurotoxicants and lead. Phillip Bromberg (this volume), using the example of ozone, explores interindividual variability to a pulmonary toxicant.

TABLE 1
Categories of Susceptibility Factors

Category	Susceptibility factor	Exposure	Health effect
Constitutive susceptibility factors			
Genetic	Slow acetylators	Arylamines[a]	Bladder cancer
Ethnicity	European	Sunlight[b]	Melanoma
Gender	Male	DBCP[c]	Sterility
Developmental stage	12 to 16 weeks gestation	Methyl mercury[d]	Neurological retardation
Acquired susceptibility factors			
Preexisting illness	Asthma	Sulfur dioxide[e]	Bronchoconstriction
Quality of life	Cigarette smoking	Asbestos[f]	Lung cancer

Note: DBCP, dibromochloropropane.

[a] Kadlubar et al. (1992).
[b] Reintgen et al. (1982).
[c] Whorton et al. (1979).
[d] Mortensen (1992).
[e] Jaeger et al. (1979).
[f] Coultas and Samet (1992).

FACTORS CONTRIBUTING TO RESPONSE VARIABILITY

The responses produced by exposure to a contaminant are sometimes associated with specific factors. Populations exposed to methyl mercury in Iraq (Mortensen 1992, Marsh et al. 1981) and to PCBs and furans in Taiwan (Rogan et al. 1988) provide tragic examples of the effect of developmental stage on susceptibility. These two incidents permit comparison of the health effects sustained by mothers and their unborn children after short-lived episodes of very high exposure by ingestion. In both instances, the symptoms in the mothers were transient, consisting of nausea, paresthesias, and anxiety after exposure (Rogan et al. 1988, Marsh et al. 1981). In contrast, the children exposed *in utero* developed serious and irreversible defects. Children exposed to PCBs had defective tooth enamel, abnormally pigmented skin, and deformed nails (Rogan et al. 1988). Both groups of exposed children displayed neurologic symptoms that included profound deficits in motor and cognitive development (Chen et al.1992, 1994; Marsh et al. 1981). The effects of mercury were the most severe when exposure occurred during the second trimester of pregnancy (Marsh et al. 1981). These examples indicate that response is influenced by developmental stage and demonstrate that the differences are both qualitative and quantitative in their expression.

Factors associated with response variability and susceptibility can be grouped into the categories shown in Table 1. These factors are effect modifiers because they "produce different exposure-response relationships at different levels of that factor" (EPA 1996b). Aggregate effect modifiers such as socioeconomic status and level of education are not included because they are surrogates for unidentified factors that can be related to exposure as well as responsiveness.

The effect modifiers are also grouped according to whether they describe constitutive (intrinsic) or acquired traits. Constitutive factors are fixed traits that modify susceptibility and are grouped according to gender, genetics, developmental stage, and ethnicity. Within a population, these factors do not change unpredictably except by birth, death, and migration. Acquired susceptibility factors are related to preexisting illness and quality of life, which include susceptibility factors related to lifestyle and environmental influences. The effects of diet, cigarette smoking, and quality of medical care are among the factors that determine acquired susceptibility. In contrast to constitutive susceptibility factors, public health interventions can actually alter the prevalence of many acquired susceptibility factors.

GENETICS

An individual's ability to metabolize environmental contaminants is determined by an array of enzymes, many of which occur in multiple forms that are inherited in a Mendelian fashion. Genes are considered to be polymorphic when there are multiple alleles with a frequency greater than 1% in the population. In many instances, specific genetic polymorphisms correspond to phenotypic differences in enzyme function. Functional enzyme may be absent, as is the case with individuals homozygous for the nonfunctional allele of glutathione S-transferase (Seidergård et al. 1988). Alternatively, enzyme activity may differ across the population by more than 3 orders of magnitude, as with cytochrome P4502D6 (Nakamura et al. 1985). Because of these differences, the ability to metabolize agents as diverse as polycyclic aromatic hydrocarbons, dioxins, heterocyclic amines, alcohols, and organophosphate pesticides is at least partially heritable. Polymorphic enzymes that have been linked to differential risk include several cytochrome P450s (Butler et al. 1992, Nakachi et al. 1991, Caporaso et al. 1989), glutathione S-transferases (Bell et al. 1993b), N-acetyltransferases (Cartwright et al. 1982), alcohol dehydrogenase (Agarwal and Goedde 1992), and paraoxonase (Furlong et al. 1988). Neil Caporaso and Nathaniel Rothman (this volume) examine the relationship between genetic polymorphisms and cancer by using glutathione S-transferase deficiencies and cigarette smoke-induced cancers as examples.

Several categories of genetic traits have been linked with differences in the capacity to respond to environmental contaminants. Defects in the enzymes responsible for oxidative integrity may result in increased levels of tissue damage after exposure. Examples include glucose-6-phosphate dehydrogenase, which protects hemoglobin against the oxidative effects of ozone and nitrogen dioxide (Amoruso et al. 1986), and α_1-antitrypsin, which protects the lungs from cigarette smoke (Ashford et al. 1990). Major histocompatibility genes may influence responsiveness to contaminants such as polyvinyl chloride (Black et al. 1986) and beryllium (Richeldi et al. 1993). People with ataxia telangiectasia (AT), Bloom's syndrome, or xeroderma pigmentosum are unable to effectively repair the DNA damage caused by ultraviolet light or ionizing radiation (Steingrimsdottir et al. 1995, Sullivan and Willis 1992, Swift et al. 1990). Although homozygosity for these syndromes is rare and debilitating, people who are heterozygous lead normal lives but may be at an increased risk of cancer because of their reduced DNA repair capabilities. The potential impact on cancer cases is enormous because it is estimated that over 1% of the population in the United States carries the gene for AT (Swift et al. 1990).

Susceptibility factors related to ethnicity are often difficult to identify because of confounding due to differences in socioeconomic factors. Confounders are factors that are unequally distributed among the comparison groups and related to the outcome under examination. Clearly, differences in the rates of cancer (Gorey and Vena 1994), asthma (MMWR 1990), and perinatal mortality (Collins and David 1993) are found when populations of African-Americans, Hispanic-Americans, and white Americans are compared. However, the role of differential responses to environmental contaminants in these outcomes is far from clear.

The prevalence of polymorphic genes known to be associated with differences in risk varies substantially among different ethnic groups. For instance, the slow acetylator phenotype is associated with a higher risk of bladder cancer after exposure to heterocyclic amines. The prevalence of slow acetylators is greater than 90% among North Africans but less than 5% among Canadian Eskimos. The prevalence of slow acetylators among Asians and Europeans is intermediate (Bell et al. 1993a, Lin et al. 1993).

The higher rates of cancer mortality in African-Americans compared with white Americans is due primarily to occurrences at three sites: stomach, lung, and cervix (Gorey and Vena 1994). This difference is not entirely explained by adjusting for socioeconomic status (Polednak 1990), which implies that unidentified risk factors are responsible. Specifically, African-Americans and white Americans of similar socioeconomic status may differ in their exposure or responsiveness (Gorey and Vena 1994). In addition, it is possible that exogenous exposures are not related to the excess cancers.

Perinatal mortality and asthma likewise show patterns of occurrence where the underlying mechanisms responsible for the differences among ethnic groups cannot be identified readily and where the role of environmental exposures is somewhat speculative. In the late 1980s, a worldwide increase in asthma prevalence and mortality from asthma were reported (Jackson et al. 1988). Those most affected in the United States are African-Americans and people over 65 years old (MMWR 1990). African-American children are more likely than white children to have asthma (Gold et al. 1993), to succumb from respiratory insufficiency (MMWR 1990), and to be admitted to emergency rooms during episodes of high ozone concentrations (White et al. 1994). Again, the role of differences in intrinsic response, exposure, and access to effective medical care cannot be determined.

Perinatal mortality varies according to race but shows an unexpected pattern in biracial infants (Collins and David 1993). Compared with white children, children of African-American mothers and white fathers have higher mortality rates, but children of white mothers and African-American fathers are not at an increased risk of perinatal mortality. This circumstantial association suggests that environmental factors related to socioeconomic status instead of intrinsic factors are responsible for the differences (Collins and David 1993). Genetic imprinting, which involves selective expression of either maternal or paternal genes (Reik 1996), may also contribute to these differences.

Gender can contribute to response variability by several distinct mechanisms. Sex-related differences in body type can influence responsiveness to contaminants. Females typically have a higher percentage of body fat than males. As a result, females often have higher body burdens of lipid-soluble contaminants. This difference may be somewhat offset by periods of lactation, a route of excretion not available to males. Lactation has been found to reduce the body burden of dioxins, polycyclic aromatic hydrocarbons, and chlorinated hydrocarbons (Becher et al. 1995, Schlaud et al. 1995, Wolff 1983).

The presence of sensitive sex-specific tissues or cells can result in gender-specific vulnerability to contaminants. In males, spermatogonia are highly sensitive to the fumigant dibromochloropropane, and exposure can result in permanent sterility (Whorton et al. 1979).

Hormonal status can alter bioavailability and influence pharmacodynamics. The changes in hormonal status that accompany menopause and lead to bone loss may also result in the release of lead into the bloodstream (Silbergeld et al. 1988). Pregnancy may also cause changes in the bioavailability of lead, although this hypothesis has not been evaluated (Silbergeld, 1991). Animal models provide evidence that hormones play a role in dioxin-mediated toxicity. Female rats with intact ovarian hormones show more oxidative damage (Tritscher et al. 1996) and increased incidence of liver tumors than either male or ovariectomized female rats (Kociba et al. 1978, Huff et al. 1994). Lynn Frame and colleagues (this volume) examine how age, gender, and ethnicity modify cancer risk.

Developmental stage encompasses all age-related effects from conception to death and as such includes developmental effects that are the consequence of exposures sustained *in utero* or postnatally.

Embryos and fetuses develop quickly and the timing of exposure can be an extremely important determinant in disruption of normal development. For example, exposure during organogenesis can result in premature death, structural abnormalities, growth retardation, or functional deficits. This is the time during which alcohol or certain drugs and chemicals may induce craniofacial or neural tube defects. Exposure during later fetal and neonatal development is not likely to cause structural abnormalities but may result in other types of alterations as well as in carcinogenesis. In some cases the effects are apparent at birth or shortly after exposure, but in many cases the alterations are latent and may not become apparent until much later in life (Kimmel and Kimmel 1994). The potential for later expression of developmental effects is demonstrated with lead and PCBs, where early exposure has been linked to learning and behavioral deficits in primary school (Bellinger et al. 1990, Jacobson and Jacobson 1996, Needleman et al. 1996).

After birth, susceptibility may be related to the immature state of some organ systems (e.g., nervous, skeletal, and endocrine systems). Children often have greater uptake of environmental agents because of differences in physiologic permeability, absorption, and respiration. For example, children absorb ingested lead through their gastrointestinal tracts more efficiently than adults (Preuss 1993). Infants breathe faster and have more surface area per body weight (Kacew 1992) and so will have a higher dose per body weight than an adult with a similar exposure. Young children can be sensitive because of their immature metabolism (e.g., neonates lack the enzymes necessary for glucuronidation) (Viccellio 1993). For a more extensive review, the reader is referred to NRC (1993) and Guzelian et al. (1992).

During aging, susceptibility to the health effects elicited by environmental contaminants may increase because of a decline in metabolic, pulmonary, immune, and renal function. Compared with childhood, less is known about the impact of aging on susceptibility to the health effects produced by environmental contaminants. Older people may be more susceptible to carcinogens because their DNA repair capabilities decrease as the number of cancer initiating events accumulate (Liu et al. 1994). Older people, however, can be less responsive to pulmonary irritants such as ozone (Drechsler-Parks et al. 1989; Bromberg, this volume).

The variability in reproductive and developmental outcomes produced by exposure to toxicants is reviewed by Anthony Scialli and Armand Lione (this volume).

Preexisting illness, whether due to an infectious agent or a disease process, can influence the response to contaminants. For instance, hepatitis B is an infectious agent that appears to modify susceptibility to liver cancer. People exposed to both hepatitis B and aflatoxin are 10 times more likely to develop liver cancer than those exposed to only hepatitis B (Wu-Williams et al. 1992). Asthma is another preexisting illness that can alter susceptibility to air pollutants, particularly sulfur dioxide (Jaeger et al. 1979).

Preexisting illness can be the consequence of earlier exposure. Exposure can sensitize individuals to toluene diisocyanate (Dean et al. 1986) and red cedar (Marabini et al. 1993). After sensitization, individuals are extremely susceptible to pulmonary effects when exposed to the same agents.

QUALITY OF LIFE

Health is modified by the conditions people are subjected to and the lifestyles they adopt. Longevity and superior health have been associated with adequate health care, low-fat diets, regular exercise, avoidance of cigarette smoke, and moderate consumption of alcohol. Although lifestyle practices certainly alter physiology, their influence on the response to environmental contaminants is rarely examined. Because genetically conferred low levels of detoxifying enzymes have been associated with lung (London et al. 1995, Nazar-Stewart et al. 1993) and bladder (Bell et al. 1993b) cancer, it is reasonable to speculate that factors that influence glutathione and antioxidant levels may similarly affect cancer risk. There is limited evidence that vegetarians have higher glutathione levels (Flagg et al. 1993) and that vegetable consumption is correlated with a lower risk of gastrointestinal cancer (Steinmetz and Potter 1991). Thus, it is possible that diet may modulate the response to environmental contaminants, particularly those that require glutathione for detoxification. Quality-of-life factors probably modulate the effect of environmental contaminants in the entire population but their effect is often subtle and difficult to measure.

The categories of factors discussed above alter susceptibility because they are predictably associated with specific capacities for uptake, pharmacodynamics, or pharmacokinetics (Hattis

1995). Differences in uptake, because of permeability of the skin and respiratory rates, can influence the internal dose. Once uptake has occurred, metabolism can be described through pharmacokinetic and pharmacodynamic processes. Pharmacokinetics is defined as the rates of absorption, distribution, metabolism, and excretion of a chemical throughout the body (Benet et al. 1996). Physical characteristics such as body composition can influence pharmacokinetics. Aging results in altered pharmacokinetics because the proportion of body fat increases and the proportion of muscle decreases (Viccellio 1993). Pharmacodynamics is defined as "the magnitude and type of effect induced by a given concentration or amount of a chemical" (O'Flaherty and Clark 1994). Examples of pharmacodynamic effects include changes in heart rate and pulmonary function. Receptor variants and the efficiency of signal transduction can contribute to interindividual differences in pharmacodynamics (Ross, 1996).

In summary, an individual's susceptibility to illness after exposure to environmental contaminants is determined by his or her unique profile of factors. These factors are determined by normal processes such as growth and aging, individual characteristics such as gender and genetics, and the accumulated physiologic impact of lifestyle and circumstances.

EVALUATION OF RESPONSE VARIABILITY IN RISK ASSESSMENTS

Virtually all the regulatory limits governing permissible levels of environmental contaminants are based on some form of risk assessment. A notable exception occurs under the Clean Air Act where risk assessments evaluate residual health risks after the installation of pollution controls. The National Academy of Sciences describes risk assessment as the "evaluation of scientific information on the hazardous properties of environmental agents and the extent of human exposure to those agents. The product of the evaluation is a statement regarding the probability that populations so exposed will be harmed and to what degree" (NRC 1994). What risk assessments seek to accomplish is not conceptually difficult to understand. They estimate risk based on the anticipated level of exposure and the hazard posed by the chemical in question. The existence of response variability means that the hazard is not uniform for everyone.

Risk assessments are performed by using variations of the steps shown in Figure 7. In the toxicity assessment, hazard identification evaluates the quality of the data and identifies the potential health effects of the contaminant. This information is used in the dose–response assessment, which determines toxicity as a function of dose. The exposure assessment estimates the level of exposure in human populations. Risk characterization integrates the human exposure estimates with the hazard and dose–response assessments and summarizes the results. Equally important, risk characterization summarizes data deficits, assumptions, and sources of uncertainty and recommends actions to minimize the risk.

Ideally, a quantitative estimate of the magnitude of response variability would be incorporated into the toxicity assessment. The variability in response combined with the variability in exposure would be used to estimate the range of risks for the entire population. This ideal situation is complicated by the need to extend the dose–response curve from the *observed data* into the *range of extrapolation* relevant for most human environmental exposures. The challenge facing the risk assessor is how to use a limited amount of data to estimate response variability within the extrapolated range. To cope with this, risk assessment guidelines recommend a variety of nonquantitative procedures that are intended to compensate for the lack of quantitative estimates of human variability.

INCLUDING RESPONSE VARIABILITY: NONQUANTITATIVE CONSIDERATIONS

Risk assessment guidelines usually contain language stating that response variability should be incorporated into the process. For instance, the *Guidelines for Exposure Assessment* (EPA 1992) direct risk assessors to identify "subpopulations with heightened susceptibility (either because of

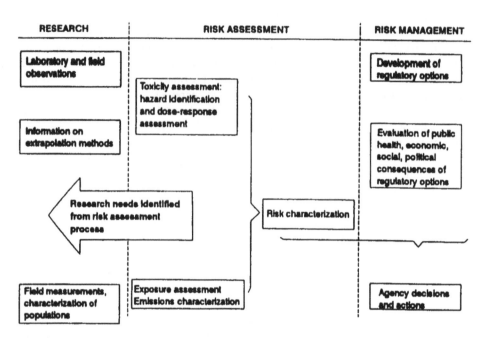

FIGURE 7   The NRC risk assessment/management paradigm showing the relationship between the health-based aspects of risk assessment: toxicity assessment (including hazard identification), dose–response assessment, exposure assessment, and risk characterization in the context of research and risk management. (Used with permission from *Science and Judgment in Risk Assessment.* Copyright 1994 by the National Academy of Sciences. Courtesy of the National Academy Press, Washington, D.C.)

exposure or predisposition)." EPA's *Policy for Risk Characterization* suggests calculating the risk to susceptible individuals and populations as data permit (EPA 1995b). It also states that, in addition to recognizing the different dose–response characteristics of susceptible populations, the magnitude of exposure and the size of highly susceptible population(s) also should be considered. The risk characterization step of the recently developed *Guidelines for Reproductive Toxicity Risk Assessment* (EPA 1996a) and the proposed *Guidelines for Neurotoxicity Risk Assessment* (EPA 1995a) address response variability in terms of identifiable susceptible populations. These guidelines emphasize that differential risk may occur under uniform exposure conditions because of interindividual differences in absorption or metabolism, nutritional status, age, lactation, or genetic polymorphisms.

Dose–Response Data from the Population Presumed to Be Most Susceptible

The dose–response data used for risk assessment may be based on a population that is judged to be highly susceptible without any direct analysis of response variability. This approach may be justified when a segment of the population undergoes health effects that are qualitatively different from the possible health effects in the general population. An example is development of methemoglobinemia in very young infants consuming nitrate-contaminated drinking water. Only infants with intestinal biota capable of reducing nitrates to nitrites develop methemoglobinemia, which limits the oxygen-carrying capacity of their blood (Walton 1951). Because intestinal reduction of nitrates is virtually unknown in adults, the regulatory standard was based on case studies of infants (IRIS 1990). Similarly, safe levels of fluoride in drinking water are based on the risk of dental fluorosis, which occurs primarily in children (IRIS 1989).

Reliance on the responses of a segment of the population judged to be highly sensitive may be convenient in terms of data availability but may result in risk estimates that fail to protect the most sensitive members of the population. This is because, in the absence of comparative data, it may be difficult to determine whether the perceived susceptible subpopulation is the most highly responsive. For example, children are almost invariably presumed to be more vulnerable to toxicants than adults (ILSI 1996). However, compared with adults, children are highly tolerant of acetaminophen-induced hepatotoxicity (Kauffman 1992) and radiation-induced lung cancer (Boice 1996). Thus, a risk assessment based only on the responses of children would underestimate the risk for much of the population.

Even if there is evidence that a group is relatively sensitive to the toxicant, as with infants consuming nitrates, the variability in responsiveness throughout the population needs to be considered. As shown in Figure 3c, assessment of response within a subpopulation will underestimate the variability in responses for the entire population. The importance of evaluating the total population variability is evident when considering the risk posed by broadly dispersed contaminants such as the criteria air pollutants (EPA 1986a). Risk assessments need to provide estimates of risk that protect the most susceptible individuals among the more than 265 million people in the United States. Such a value cannot rely on a single group, such as the elderly or asthmatics, but needs to be based on the best estimate of a population distribution.

Population-Based Risk Assessment Guidelines

Risk assessment guidelines have been developed for populations with response characteristics that may cause them to be highly susceptible. Specifically, the *Guidelines for Developmental Toxicity Risk Assessment* (EPA 1991) and the *Guidelines for Reproductive Toxicity Risk Assessment* (EPA 1996a) are directed toward populations with susceptibilities due to factors related to either age or gender. These guidelines augment the assessment of risk by guiding the interpretation of the specialized endpoints used to evaluate reproductive and developmental toxicity. Otherwise, their approach to the quantitative determination of permissible levels does not differ substantially from that used for other nongenotoxic, noncarcinogenic contaminants. Other endpoint-specific risk assessment guidelines focus on carcinogenesis [EPA 1986b, 1996a (proposed)], mutagenesis (EPA 1986c), and neurotoxicity [EPA 1995a (proposed)].

Assumptions and Defaults

Defaults in risk assessment can refer to either specific values, such as the 10-fold default uncertainty factor (UF), or specific approaches, such as the use of linearized multistage extrapolations for low dose response. A combination of defaults and conservative assumptions is required because the database for even extensively studied contaminants is limited. Several of the most commonly used defaults and assumptions address the deficit of knowledge about the variability in human response. Although data gaps are unavoidable, it is questionable whether the default values adequately substitute for variability data and whether conservative assumptions are, in fact, conservative of human health (NRC 1994). An area of contention about the use of defaults is the cancer risk assessment procedure, which historically has not used UFs to account for response variability. UFs were argued to be unnecessary because the assumptions used throughout the process were considered to be adequately conservative to protect the most sensitive members of the population (EPA 1996b).

The cancer risk assessment guidelines recommend that any agent found to elicit cancer in one human population should be assumed to be a carcinogen in all human populations. The explanation provided by the guidelines is that most human carcinogens were identified in healthy populations with occupational exposures where individuals with higher than average susceptibility may be underrepresented. As a result, the detection of cancer in people who are able to work implies that other, possibly more susceptible populations, will also be affected (EPA 1996b).

Many defaults address procedures to be used in the range of extrapolation. Risk assessments for noncancer health effects account for interindividual variation in humans with a UF that has a default value up to 10. This UF is one of several used for computing the reference dose (RfD) or reference concentration (RfC). The RfD/RfC is "an estimate of a daily exposure (with uncertainty spanning perhaps an order of magnitude) to the human population (including sensitive subgroups) that is likely to be without an appreciable risk of deleterious effects during a lifetime" (Barnes and Dourson 1988). RfD/RfCs are used when a nonlinear dose–response relationship is assumed based on homeostatic and compensatory processes at lower dose levels.

Strictly speaking, low-dose extrapolations are not a part of the calculation of the RfD or RfC. However, like low-dose extrapolations, RfDs and RfCs are by definition below the range of the observable data. When adequate data are available, the RfD/RfC values are based on the no-observed-adverse-effect level (NOAEL) divided by a series of UFs, which include one to account for human variability. In instances where no NOAEL is available, the lowest-observed-adverse-effect level (LOAEL) is used with an additional UF to account for the missing NOAEL. Although a NOAEL is generally interpreted as a biologic phenomenon, the magnitude of the NOAEL is greatly influenced by experimental or epidemiologic study conditions, such as selection of doses or exposure levels and sample size (EPA 1996a).

Using the NOAEL or LOAEL for calculating the RfD/RfC has the effect of basing risk estimates on a single value. This approach ignores the shape of the dose–response curve (EPA 1991) and fails to make use of information on dose–response variability even when it is available. For instance, in the scenario presented in Figure 4c, subpopulation X would be considered to be less susceptible under all circumstances despite the greater response at exposures exceeding 25 units. If this phenomenon occurs at levels of exposure near the RfD, then a slight exceedance could adversely affect a portion of the population.

An alternative to the NOAEL is the benchmark dose where a curve is fitted to the observed data and used to determine "the statistical lower bound on a dose corresponding to a specified level of risk" (Allen et al. 1994). The lower confidence limit on a calculated incidence of 5 to 10% has been proposed for use as the benchmark dose (EPA 1996a, 1996b). A UF with a default value up to 10 is applied to account for response variability. Whether based on the NOAEL, the LOAEL, or the benchmark dose, the UF for response variability may deviate from the default value of 10 if data are available.

The *Proposed Guidelines for Carcinogen Risk Assessment* (EPA 1996b) specify several assumptions for low-dose extrapolations. The assumptions are based on the evidence regarding the mechanism of action as well as uncertainties associated with carcinogenesis. Assumptions that may affect the degree to which highly sensitive individuals are protected include the point of departure for the extrapolation and whether the risk estimate is based on a low-dose extrapolation or margin of exposure.

The EPA proposes using the lower 95% confidence limit on the dose associated with a 10% response as a point of departure for the dose–response curve (EPA 1996b). Although this point of departure is intended to account for experimental rather than response variability, it is often argued to be conservative of human health as well. Earlier guidelines recommended extrapolating the dose response directly from the data rather than a confidence interval (EPA 1986b). The current recommendations for using the lower confidence interval will result in greater estimates of risk for a given dose but, like all indirect estimates of response variability, the degree to which this assumption protects highly susceptible individuals cannot be gauged.

EPA risk assessments for carcinogens use a linear extrapolation when there is evidence that the carcinogen acts through modes of action anticipated to be linear, such as genotoxic mechanisms. The use of a linear extrapolation without a UF is assumed to be more likely to overestimate than to underestimate risk. The assumption that the linear extrapolation is adequately conservative could be misplaced if animal models do not duplicate human response or if human response to a particular chemical is highly variable. Furthermore, linear extrapolations may greatly underestimate risk if the true dose response is supralinear. Of the 315 chemicals examined in rodent models by the National Toxicology Program, a quadratic curve (sublinear) gave the best fit for 47%, a linear curve gave the best fit for 20%, and a square-root curve (supralinear) gave the best fit for 33% (Hoel and Portier 1994). The significance of the fit was determined by standard goodness-of-fit testing at the = 0.10 level. Although the optimal fits indicate that a linear dose–response extrapolation most often overestimates the risk, these data also support the contention that the risk may be underestimated for a substantial number of contaminants. Bailar and collaborators (1988) also found that linear models could underestimate the cancer risk in rodents based on fitting the data collected from the 1212 bioassays listed in the Carcinogenesis Bioassay Database System.

A default factor of at least 10 has been proposed to account for human variability when calculating a margin of exposure (MOE) for carcinogens (EPA 1996b). A MOE is calculated for carcinogens for which there is no evidence of linearity and sufficient evidence to support a nonlinear mode of action (EPA 1996b). The margin of exposure is the "point of departure divided by the environmental exposure of interest" (EPA 1996b).

The conservative nature of these assumptions and alternatives to them were discussed in detail in *Science and Judgment in Risk Assessment* (NRC 1994). In this report, the NRC recommended that the EPA reexamine the assumptions used to address response variability to determine whether they adequately protect susceptible individuals from the adverse effects of environmental contaminants. This volume is, in part, the first of a series of steps in that process.

INCLUDING RESPONSE VARIABILITY: QUANTITATIVE CONSIDERATIONS

Whenever possible, the dose–response step of a risk assessment should be based on an accurate summarization of possible health effects and complete data describing the relationship between exposure and the health effects for a large and diverse population. For scaled health effects, such as the pulmonary measure of forced expiratory volume (FEV), the data might include individual dose–response curves. Complete data for quantal health effects such as lung cancer would consist of knowledge of the magnitude of all the exposures sufficient to cause the illness. However, data describing variability in either responsiveness or susceptibility to health effects are usually scarce, even with well-studied chemicals such as ozone and carbon monoxide. For new industrial chemicals regulated under Section 5 of the Toxic Substances Control Act, human or animal toxicity data are usually unavailable and qualitative dose–response assessments must rely on structure-activity relationships (DeVito 1994).

Characterization of human data within the observed range indicates that variability can be considerable. For example, a conservative estimate of variability can be obtained by comparing the responses of individuals in the 5th and 95th percentiles. For this segment of the population, the quantity of cigarette smoke producing a 1% decrease in FEV (a measure of pulmonary function) varies by more than 8-fold. The total variability would be much larger because this estimate does not include the most responsive 5% of the population (Hattis 1995).

Human dose–response data can be obtained from case reports, epidemiologic studies, and controlled exposure studies. The epidemiologic study designs most likely to contain useful dose–response data are either case control studies where the study population is chosen on the basis of illness or cohort studies where the population is chosen on the basis of exposure. The utility of case reports is usually limited by the small number of subjects, a lack of controls, and inadequate exposure measurements. Controlled exposure studies can be performed when the contaminant produces reversible short-term effects. This approach has been used successfully to characterize the dose–response characteristics of air pollutants such as ozone, nitrogen dioxide, and acids (McDonnell et al. 1993, Drechsler-Parks et al. 1989, Koenig et al. 1993).

Although epidemiologic studies generally do not directly measure population variability, they often estimate the effect of risk factors associated with differences in responsiveness. Epidemiologic evaluation of the effect of air pollutants in those with and without asthma (Balmes 1993) and assessment of blood lead levels in pre- and postmenopausal women (Silbergeld et al. 1988) provide data that can be used to modify the noncarcinogen default UF for human variability. The impact of risk factors on cancer susceptibility, such as functional glutathione S-transferase and lung cancer risk, have also been evaluated epidemiologically (Bell et al. 1993b).

Modeling

Mathematical models of the dose–response curve can estimate the responses of the most sensitive members of the population. Biologically based models can use physiologically based pharmacokinetic (PBPK) or pharmacodynamic/mechanistic approaches. In contrast to the top-down methods used in epidemiology and animal models, modeling characterizes low-dose behavior by breaking biologic processes into their constituent steps. The product is an estimate of low-dose effects based on many endpoints rather than the outcome of a single rodent cancer bioassay or human epidemiologic study.

Biologically based models incorporate biomarker and physiologic information compiled from a variety of sources including *in vivo* and *in vitro* data from humans and animal models. These data may include parameters such as absorption, distribution, metabolism, excretion, and alterations in gene expression and subsequent pathogenesis. Although the output from most models represents an average population response, variability in individual parameters can be incorporated to model the dose–response variability.

Existing models of response variability estimate intermediate indicators of risk rather than the ultimate adverse health outcome. Among them are estimates of internal dose of perchloroethylene (Gearhart et al. 1993), formation of 4-aminobiphenyl adducts after exposure to aflatoxin (Bois et al. 1995), and metabolism of benzene in the liver (Seaton et al. 1995). Bois et al. (1995) used pharmacokinetic parameters such as absorption from the urine, acetylation capabilities, glucuronidation rates, and urinary pH to model interindividual variability in the concentration of DNA adducts found in the liver. The results suggest that substantial differences, up to a millionfold, in the formation of 4-aminobiphenyl adducts are possible.

The occupational exposure limits for methylene chloride were recently reduced from 500 to 25 parts per million (time-weighted average) based on a PBPK model. The model used a Bayesian analysis to generate human and rodent pharmacokinetic parameter distributions reflecting both uncertainty and variability. The results were used to estimate the risk based on the the 95th percentile of the human dose distribution (Occupational Safety and Health Administration 1997).

Biologically based models have not been used as extensively for noncancer health effects because of the complexity of processes and the variety of possible outcomes. However, the possibility of using biologically based models has been investigated and is described by Kimmel and Kimmel (1994). Models with the compound 5-fluorouracil have been reported (Shuey et al. 1994) and additional efforts are being sponsored by the EPA (Kavlock 1997).

THE FUTURE

Humans are highly diverse and, as result, variability in response to environmental toxicants should be expected. Past efforts to address response variability have tended to divide humans into those who are "susceptible" and those who are "normal" in their responses. A more accurate view of human response to contaminants recognizes a continuum extending from the most resistant to the most susceptible individuals.

Risk assessment has been the primary tool for evaluating the health impact of environmental contaminants for the past 30 years (Center for Risk Analysis 1994). Animal models and human studies that examine health outcome on the basis of exposure have made it possible to develop measures of risk that are quantitative rather than qualitative. Currently, risk assessments usually address response variability through defaults, uncertainty factors, and assumptions rather than by quantitative approaches. To assess the risk to all members of the population, risk assessments need to either include quantitative information about susceptibility and response variability or provide adequate defaults in the absence of such data.

By definition, nonquantitative approaches are used in the absence of data, making it difficult to determine whether they allow for adequate protection of the most responsive and susceptible individuals. To rectify this situation, the following chapters will examine the extent of human variability in response to several chemicals for which the database is unusually rich. From these examples, the utility of generalizations about human response variability to environmental toxicants will be addressed.

REFERENCES

Agarwal DP, Goedde HW (1992) Pharmacogenetics of alcohol metabolism and alcoholism. Pharmacogenetics 2:48–62

Allen BC, Kavlock RJ, Kimmel CA, Faustman EM (1994) Dose–response assessment for developmental toxicity. II. Comparison of generic benchmark dose estimates with no observed adverse effect levels. Fundam Appl Toxicol 23:487–495

Ames RG, Brown SK, Mengle DC, et al (1989) Cholinesterase activity depression among California agricultural pesticide applicators. Am J Ind Med 15:143–150

Amoruso MA, Ryer J, Easton D, et al (1986) Estimation of risk of glucose 6-phosphate dehydrogenase-deficient red cells to ozone and nitrogen dioxide. J Occup Med 28:473–479

Ashford NA, Spadafor CJ, Hattis DB, Caldart CC, eds (1990) Monitoring the worker for exposure and disease. Johns Hopkins University Press, Baltimore, pp 95–108

Bach JF (1994) Insulin-dependent diabetes mellitus as an autoimmune disease. Endocr Rev 15:516–542

Bailar JC III, Crouch EA, Shaikh R, Spiegelman D (1988) One-hit models of carcinogenesis: conservative or not? Risk Anal 8:485–497

Balmes JR (1993) The role of ozone exposure in the epidemiology of asthma. Environ Health Perspect 101(Suppl4):219–224

Barnes DG, Dourson M (1988) Reference dose (RfD): description and use in health risk assessments. Regul Toxicol Pharmacol 8:471–486

Bartecchi CE, MacKenzie TD, Schrier RW (1994) The human costs of tobacco use. N Engl J Med 330:907–912

Becher G, Skaare JU, Polder A, Sletten B (1995) PCDDs, PCDFs and PCBs in human milk from different parts of Norway and Lithuania. J Toxicol Environ Health 46:133–148

Bell DA, Taylor JA, Butler MA, et al (1993a) Genotype/phenotype discordance for human arylamine N-acetyltransferase (NAT2) reveals a new slow acetylator allele common in African-Americans. Carcinogenesis 14:1689–1692

Bell DA, Taylor JA, Butler MA, et al (1993a) Genotype/phenotype discordance for human arylamine N-acetyltransferase (NAT2) reveals a new slow acetylator allele common in African-Americans. Carcinogenesis 14:1689–1692

Bell DA, Taylor JA, Paulson DF, et al (1993b) Genetic risk and carcinogen exposure: a common inherited defect of the carcinogen-metabolism gene glutathione S-transferase M1 (GSTM1) that increases susceptibility to bladder cancer. J Natl Cancer Inst 85:1159–1164

Bellinger D, Leviton A, Sloman J (1990) Antecedents and correlates of improved cognitive performance in children exposed *in utero* to low levels of lead. Environ Health Perspect 89:5–11

Benet LZ, Kroetz, DL, Sheiner LB (1996) Pharmacokinetics, the dynamics of drug absorption, distribution and elimination. In Hardman JG, Limbird LE (eds),Goodman & Gilman's the pharmacological basis of therapeutics, 9th ed. McGraw-Hill, New York, pp 3–28

Black C, Pereira S, McWhirter A, et al (1986) Genetic susceptibility to scleroderma-like syndrome in symptomatic and asymptomatic workers exposed to vinyl chloride. J Rheumatol 13:1059–1062

Bogen KT, Spear RC (1987) Integrating uncertainty and interindividual variability in environmental risk assessment. Risk Anal 7:427–436

Boice JD Jr (1996) Cancer following irradiation in childhood and adolescence. Med Pediatr Oncol 1(Suppl):29–34

Bois FY, Krowech G, Zeise L (1995) Modeling human interindividual variability in metabolism and risk: the example of 4-aminobiphenyl. Risk Anal 15:205–213

Breitner JCS, Murphy EA, Woodbury MA (1991) Case-control studies of environmental influences in diseases with genetic determinants, with an application to Alzheimer's disease. Am J Epidemiol 133:246–255

Brody DJ, Pirkle JL, Kramer RA, et al (1994) Blood lead levels in the U.S. population. JAMA 272:277–283

Butler MA, Lang NP, Young JF, et al (1992) Determination of CYP1A2 and NAT2 phenotypes in human populations by analysis of caffeine urinary metabolites. Pharmacogenetics 2:116–127

Caporaso N, Hayes RB, Dosemeci M, et al (1989) Lung cancer risk, occupational exposure, and the debrisoquine metabolic phenotype. Cancer Res 49:3675–3679

Cartwright RA, Glashan RW, Rogers HJ, Ahmod RA (1982) The role of N-acetyltransferase in bladder carcinogenesis: a pharmacogenetic epidemiological approach to bladder cancer. Lancet ii:842–845

Center for Risk Analysis (1994) A historical perspective on risk assessment in the federal government. Harvard School of Public Health, Boston, MA, 58 pp

Chen YC, Yu ML, Rogan, WJ, et al (1994) A 6 year follow-up of behavior and activity disorders in the Taiwan Yu-cheng children. Am J Public Health 84:415–421

Chen YC, Guo YL, Hsu CC, Rogan WJ (1992) Cognitive development of Yu-Cheng ("oil disease") children prenatally exposed to heat-degraded PCBs. JAMA 268:3213–3218

Collins JW Jr, David RJ (1993) Race and birthweight in biracial infants. Am J Public Health 83:1125–1129

Coultas DB, Samet JM (1992) Occupational lung cancer. Clin Chest Med 13:341–354

Dean JH, Murray MJ, Ward EC (1986) Toxic responses of the immune system. In Klaassen CD, Amdur MO, Doull J (eds), Casarett and Doull's toxicology. 3rd ed. Macmillan, New York, pp 245–285

DeVito S (1994) New chemical review process for TSCA. Seminar given at the U.S. Environmental Protection Agency, Washington, D.C., August 1994

Doll R, Peto R, Wheatley K, et al (1994) Mortality in relation to smoking: 40 years' observations on male British doctors. Br Med J 309:901–911

Drechsler-Parks DM, Bedi JF, Norvath SM (1989) Pulmonary function responses of lung and older adults to mixtures of O3, NO2 and PAN. Toxicol Ind Health 5:505–517

Finkel AM (1994) Risk assessment research: only the beginning. Risk Anal 14:907–911

Flagg EW, Coates RJ, Jones DP, et al (1993) Plasma total glutathione in humans and its association with demographic and health-related factors. Br J Nutr 70:797–808

Furlong CE, Richter RJ, Seidel SL, Motulsky AG (1988) Role of genetic polymorphisms of human plasma paraoxonase/arylesterase in hydrolysis of the insecticide metabolites chlorpyrifos oxon and paraoxon. Am J Hum Genet 43:230–238

Gearhart JM, Mahle DA, Greene RJ, et al (1993) Variability of physiologically based pharmacokinetic (PBPK) model parameters and their effects on PBPK model predictions in a risk assessment for perchloroethylene (PCE). Toxicol Lett 68:131–144

Gold AR, Rotnitzky A, Damokosh AI, et al (1993) Race and gender differences in respiratory illness prevalence and their relationship to environmental exposures in children 7 to 14 years of age. Am Rev Respir Dis 148:10–18

Gorey KM, Vena JE (1994) Cancer differentials among U.S. blacks and whites: quantitative estimates of socioeconomic-related risks. J Natl Med Assoc 86:209–215

Guzelian PS, Henry CJ, Olin SS, eds (1992) Similarities and differences between children and adults. Implications for risk assessment. ILSI Press, Washington, D.C., 285 pp

Hattemer-Frey HQ, Travis CC, Land ML (1990) Benzene: environmental partitioning and human exposure. Environ Res 53:221–232

Hattis D (1996) Human interindividual variability in susceptibility to toxic effects: from annoying detail to a central determinant of risk. Toxicology 111:5–14

Hattis D (1995) Variability in susceptibility — how big, how often, for what responses to what agents? Invited paper presented at the WHO/IPCS Workshop on Variability in Susceptibility, Southampton, England, March 8–10, 1995

Higgins M, Thom T (1993) Trends in stroke risk factors in the United States. Ann Epidemiol 3:550–554

Hoel DG, Portier CJ (1994) Nonlinearity of dose–response functions for carcinogenicity. Environ Health Perspect 102(Suppl1):109–113

Horai S, Hayasaka K, Kondo R, et al (1995) Recent African origin of modern humans revealed by complete sequences of hominoid mitochondrial DNAs. Proc Natl Acad Sci U.S.A. 92:532–536

Huff J, Lucier G, Tritscher A (1994) Carcinogenicity of TCDD: experimental, mechanistic, and epidemiologic evidence. Annu Rev Pharmacol Toxicol 34:343–372

ILSI Risk Science Institute (1996) Research needs on age-related differences in susceptibility to chemical toxicants. ILSI, Washington, D.C.

IRIS database (1990) Reference dose for chronic oral exposure — nitrate. Integrated Risk Information System, Office of Health and Environmental Assessment, U.S. Environmental Protection Agency, Washington, D.C.

IRIS database (1989) Reference dose for chronic oral exposure — fluoride. Integrated Risk Information System, Office of Health and Environmental Assessment, U.S. Environmental Protection Agency, Washington, D.C.

Jackson R, Sears MR, Beaglehole R, Rea HH (1988) International trends in asthma mortality: 1970 to 1985. Chest 94:914–919

Jacobson JL, Jacobson SW (1996) Intellectual impairment in children exposed to polychlorinated biphenyls in utero. N Engl J Med 335:783–789

Jaeger MJ, Tribble D, Wittig HJ (1979) Effect of 0.5 ppm sulfur dioxide on the respiratory function of normal and asthmatic subjects. Lung 156:119–127

Kacew S (1992) General principles in pharmacology and toxicology applicable to children. In Guzelian PS, Henry CJ, Olin SS (eds), Similarities and differences between children and adults. ILSI, Washington, D.C., 285 pp

Kadlubar FF, Butler MA, Kaderlik KR, et al (1992) Polymorphisms for aromatic amine metabolism in humans: relevance for human carcinogenesis. Environ Health Perspect 98:69–74

Kauffman RE (1992) Acute acetaminophen overdose: an example of reduced toxicity related to developmental differences in drug metabolism. In Guzelian PS, Henry CJ, Olin SS (eds), Similarities and differences between children and adults. ILSI, Washington, D.C., 285 pp

Kavlock RJ (1997) Recent advances in mathematical modeling of developmental abnormalities using mechanistic information. Reprod Toxicol 11:423–434

Kimmel CA, Kimmel GL (1994) Risk assessment for developmental toxicity. In Kimmel CA, Buelke-Sam J (eds), Developmental toxicology, 2nd ed. Raven Press, New York, pp 429–453

Kociba RJ, Keyes DG, Beyer JE, et al (1978) Results of a two-year chronic toxicity study and oncogenicity study of 2,3,7,8-tetrachlorodibenzo-p-dioxin in rats. Toxicol Appl Pharmacol 46:279–303

Koenig JQ, Dumler K, Rebolledo V, et al (1993) Respiratory effects of inhaled sufuric acid on senior asthmatics and nonasthmatics. Arch Environ Health 48:171–175

Kulle TJ, Sauder LR, Hebel JR, Chatham MD (1985) Ozone response relationships in healthy nonsmokers. Am Rev Respir Dis 132:36–41

LaCroix AZ, Guralnik JM, Berkman LF, et al (1993) Maintaining mobility in late life. II. Smoking, alcohol consumption, physical activity, and body mass index. Am J Epidemiol 137:858–869

Lambert GH, Hsu CC (1992) Polyhalogenated biphenyls and the developing human. In Guzelian PS, Henry CJ, Olin SS (eds), Similarities and differences between children and adults. Implications for risk assessment. ILSI Press, Washington, D.C., 285 pp

Lewontin R (1995) Human diversity. Scientific American Library, New York, 179 pp

Lin HJ, Han CY, Lin BK, Hardy S (1993) Slow acetylator mutations in the human polymorphic N-acetyltransferase gene in 786 Asians, Blacks, Hispanics and Whites: application to metabolic epidemiology. Am J Hum Genet 52:827–834

Liu Y, Hernandez AM, Shibata D, Cortopassi GA (1994) Bcl2 translocation frequency rises with age in humans. Proc Natl Acad Sci U.S.A. 91:8910–8914

London SJ, Daly AK, Cooper J, et al (1995) Polymorphism of glutathione S-transferase M1 and lung cancer risk among African-Americans and Caucasians in Los Angeles County, California. J Natl Cancer Inst 87:1246–1253

Mannino DM, Schreiber J, Aldous K, Ashley D (1995) Human exposure to volatile organic compounds: a comparison of organic vapor monitoring badge levels with blood levels. Int Arch Occup Environ Health 67:59–64

Marabini A, Dimich-Ward H, Kwan SYL, Kennedy SM (1993) Clinical and socioecononomic features of subjects with red cedar asthma. Chest 104:821–824

Marsh DO, Myers GJ, Clarkson TW et al (1981) Dose response relationship for human fetal exposure to methylmercury. Clin Toxicol 18:1311–1318

McDonnell WF, Muller KE, Bromberg PA, Shy CM (1993) Predictors of individual differences in acute response to ozone exposure. Am Rev Respir Dis 147:818–825

Morbidity Monthly Weekly Review (1990) Asthma-United States, 1980–1987. Morbidity Monthly Weekly Review 39:493–497

Mortensen ME (1992) Mercury toxicity in children. In Guzelian PS, Henry CJ, Olin SS (eds), Similarities and differences between children and adults. Implications for risk assessment. ILSI, Washington, D.C., pp 204–213

Mushak P, Crocetti AF (1990) Methods for reducing lead exposure in young children and other risk groups: an integrated summary of a report to the U.S. Congress on childhood lead poisoning. Environ Health Perspect 89:125–135

Nakachi K, Imai K, Hayashi S, Kawajiri K (1993) Polymorphisms of the CYP1A1 and glutathione S-transferase genes associated with susceptibility to lung cancer in relation to cigarette dose in a Japanese population. Cancer Res 53:2994–2999

Nakachi K, Imai K, Hayashi S, et al (1991) Genetic susceptibility to squamous cell carcinoma of the lung in relation to cigarette smoking dose. Cancer Res 51:5177–5180

Nakamura K, Goto F, Ray WA, McAllister CB (1985) Interethnic differences in genetic polymorphism of debrisoquin and mephenytoin hydroxylation between Japanese and Caucasian populations. Clin Pharmacol Ther 38:402–408

National Research Council (1994) Science and judgment in risk assessment. National Academy Press, Washington, D.C., 651 pp

National Research Council (1993) Pesticides in the diets of infants and children. National Academy Press, Washington, D.C., 386 pp

National Research Council (1989) Biologic markers in pulmonary toxicology. National Academy Press, Washington, D.C., 179 pp

Nazar-Stewart V, Motulsky AG, Eaton DL, White E (1993) The glutathione S-transferase polymorphism as a marker for susceptibility to lung carcinoma. Cancer Res 53:2313–2318

Nebert DW (1991) Role of genetics and drug metabolism in human cancer risk. Mutat Res 247:267–281

Needleman HL, Riess JA, Tobin MJ, et al (1996) Bone lead levels and delinquent behavior. JAMA 275:363–369

Occupational Safety and Health Administration (1997) 29 CFR Parts 1910, 1915 and 1926. Occupational exposure to methylene chloride final rule. Federal Register 62:1493–1619

O'Flaherty EJ, Clark DO (1994) Pharmacokinetic/pharmacodynamic approaches for development toxicity. In Kimmel CA, Buelke-Sam, J (eds), Developmental toxicology, 2nd ed. Raven Press, New York, pp 215–244

Orban JE, Stanley JS, Schwemberger JG, Remmers JC (1994) Dioxins and dibenzofurans in adipose tissue of the general U.S. population and selected subpopulations. Am J Public Health 84:439–445

Patterson DG Jr, Todd GD, Turner WE, et al (1994) Environ Health Perspect 102(Suppl1):195–204

Perera F, Mayer J, Jaretzki A, et al (1989) Comparison of DNA adducts and sister chromatid exchange in lung cancer cases and controls. Cancer Res 49:4446–4451

Perlin SA, Setzer RW, Creason J, Sexton K (1995) Distribution of industrial air emissions by income and race in the United States: an approach using the toxic release inventory. Environ Sci Technol 29:69–80

Phillips LJ (1992) A comparison of human toxics exposure and environmental contamination by census division. Arch Environ Contam Toxicol 22:1–5

Pluim HJ, Koppe JG, Olie K, et al (1994) Clinical laboratory manifestations of exposure to background levels of dioxins in the perinatal period. Acta Paediatr 83:583–587

Polednak AP (1990) Cancer mortality in a higher-income black population in New York state. Cancer 66:1654–1660

Pope CA III, Dockery DW (1992) Acute health effects of PM10 pollution on symptomatic and asymptomatic children. Am Rev Respir Dis 145:1123–1128

Preuss HG (1993) A review of persistent, low-grade lead challenge: neurological and cardiovascular consequences. J Am Coll Nutr 3:246–254

Quinones MA, Bogden JD, Louria DB, et al (1976) Depressed cholinesterase activities among farm workers in New Jersey. Sci Total Environ 6:155–159

Reik W (1996) Genetic imprinting: the battle of the sexes rages on. Exp Physiol 81:161–172

Reintgen DS, McCarty KM, Com E, Seigler HF (1982) Malignant melanoma in Black American and White American populations: a comparative review. JAMA 248:1856–1859

Richeldi L, Sorrentino R, Saltini C (1993) HLA-DPB1 glutamate 69: a genetic marker of beryllium disease. Science 262:242–244

Richter ED, Chuwers P, Levy Y, et al (1992) Health effects from exposure to organophosphate pesticides in workers and residents in Israel. Isr J Med Sci 28:584–597

Rogan WJ, Gladen BC, Hung KL, et al (1988) Congenital poisoning by polychlorinated biphenyls and their contaminants in Taiwan. Science 241:334–336

Ross EM (1996) Pharmacodynamics: mechanisms of drug action and the relationship between drug concentration and effect. In Hardman JG, Limbird LE (eds), Goodman & Gilman's the pharmacological basis of therapeutics, 9th ed. McGraw-Hill, New York, pp 29–42

Schlaud M, Seidler A, Salje A, Behrendt W (1995) Organochlorine residues in human breast milk: analysis through a sentinel practice network. J Epidemiol Commun Health 49(Suppl1):17–21

Schwartz J, Slater D, Larson TV, et al (1993) Particulate air pollution and hospital emergency room visits for asthma in Seattle. Am Rev Respir Dis 147:826–831

Seaton MJ, Schlosser PM, Medinsky MA (1995) In vitro conjugation of benzene metabolites by human liver: potential influence of interindividual variability on benzene toxicity. Carcinogenesis 16:1519–1527

Seidergård J, Vorachek WR, Pero RW, Pearson WR (1988) Hereditary differences in the expression of the human glutathione S-transferase active on trans-stilbene oxide are due to a gene deletion. Proc Natl Acad Sci U.S.A. 85:7293–7297

Sexton K, Anderson YB (1993) Equity in environmental health: research issues and needs [foreword to special issue]. Toxicol Ind Health 9:vi

Shuey DL, Lau C, Logsdon TR, Zucker RM (1994) Biologically-based dose–response modeling in developmental toxicology: biochemical and cellular sequelae of 5-fluorouracil exposure in the developing rat. Toxicol Appl Pharmacol 126:129–144

Silbergeld EK (1991) Lead in bone: implications for toxicology during pregnancy and lactation. Environ Health Perspect 91:63–70

Silbergeld EK, Schwartz J, Mahaffey K (1988) Lead and osteoporosis: mobilization of lead from bone in postmenopausal women. Environ Res 47:79–82

Sipes IG, Gandolfi AJ (1986) Biotransformation of toxicants. In Klaassen CD, Amdur MO, Doull J (eds), Casarett and Doull's toxicology, 3rd ed. Macmillan, New York, pp 64–98

Stamler J, Dyer AR, Shekelle RBJ et al (1993) Relationship of baseline major risk factors to coronary and all-cause mortality, and to longevity: findings from long-term follow-up of Chicago cohorts. Cardiology 82:191–222

Steingrimsdottir H, Beare D, Carr AM, et al (1995) UV hypermutability of xeroderma pigmentosum cells demonstrated with a DNA-based mutation system. Oncogene 10:2057–2066

Steinmetz KA, Potter JD (1991) Vegetables, fruit, and cancer. I. Epidemiology. Cancer Causes Control 2:325–357

Sullivan NF, Willis AE (1992) Cancer predisposition in Bloom's syndrome. BioEssays 14:333–336

Swift M, Chase CL, Morrell D (1990) Cancer predisposition of ataxia-telangiectasia heterozygotes. Cancer Genet Cytogenet. 46:21–27

Tritscher AM, Seacat AM, Yager JD, et al (1996) Increased oxidative DNA damage in livers of 2,3,7,8-tetrachlorodibenzo-p-dioxin treated intact but not ovariectomized rats. Cancer Lett 98:219–225

U.S. Environmental Protection Agency (1996a) Guidelines for reproductive toxicity risk assessment. Federal Register 61:56274–56322

U.S. Environmental Protection Agency (1996b) Proposed guidelines for carcinogen risk assessment. Federal Register 61:17960–18011

U.S. Environmental Protection Agency (1995a) Draft proposed guidelines for neurotoxicity risk assessment. Risk Assessment Forum, Washington, D.C., 82 pp

U.S. Environmental Protection Agency (1995b) Policy for risk characterization. Memorandum from Carol Browner, Adminstrator, March 21, 1995, Washington, D.C.

U.S. Environmental Protection Agency (1993) The plain English guide to the Clean Air Act. EPA 400-K-93-001, Washington, D.C.

U.S. Environmental Protection Agency (1992) Guidelines for exposure assessment. Federal Register. 57:22888–22938

U.S. Environmental Protection Agency (1991) Guidelines for developmental toxicity risk assessment. Federal Register 56:63798–63826

U.S. Environmental Protection Agency (1986a) Air quality criteria for ozone and other photochemical oxidants. Office of Research and Development, Research Triangle Park, NC

U.S. Environmental Protection Agency (1986b) Guidelines for carcinogen risk assessment. Federal Register 51:33992–34003

U.S. Environmental Protection Agency (1986c) Guidelines for mutagenicity risk assessment Federal Register 51:34006–34012

Utell MJ, Morrow PE, Speers DM, et al (1983) Airway responses to sulfate and sulfuric acid aerosols in asthmatics. Am Rev Respir Dis 128:444–450

van Poppel G, Goldbohm RA (1995) Epidemiologic evidence for -carotene and cancer prevention. Am J Clin Nutr 62:1393S–1402S

Verge CF, Gianani R, Yu L, et al (1995) Late progression to diabetes and evidence for chronic beta-cell autoimmunity in identical twins of patients with type I diabetes. Diabetes 44:1176–1179

Vicellio P (1993) Handbook of medical toxicology. Little Brown, Boston, pp 36–37

Villeneuve PJ, Mao Y (1994) Lifetime probability of developing lung cancer, by smoking status. Can J Public Health 85:385–388

Walton G (1951) Survey of literature relating to infant methemoglobinemia due to nitrate-contaminated water. Am J Public Health 41:986–995

White MC, Etzel RA, Wilcox WD, Lloyd C (1994) Exacerbations of childhood asthma and ozone pollution in Atlanta. Environ Res 65:56–68

Whorton D, Milby T, Krauss R, Stubbs H (1979) Testicular function in DBCP exposed pesticide workers. J Occup Med 21:161–166

Wolff MS (1983) Occupationally derived chemicals in breast milk. Am J Ind Med 4:259–281

Wu-Williams AH, Zeise L, Thomas D (1992) Risk assessment for aflatoxin B1: a modeling approach. Risk Anal 12:559–567

Zang EA, Wynder EL (1996) Differences in lung cancer risk between men and women: examination of the evidence. J Natl Cancer Inst 88:183–192

2 Strategies for Assessing Human Variability in Susceptibility and Using Variability to Infer Human Risks

Dale Hattis

CONTENTS

0-8493-2805-5/99/$0.00+$.50
© 1999 by CRC Press LLC

INTRODUCTION TO STRATEGIES

The goal of this paper is to offer some observations that cut across the disciplinary frameworks that are the main organizing theme of the other papers in this series. Topics such as neurotoxicology, reproductive toxicity, and pulmonary diseases are plausible starting points for analysis, because they correspond to the specialties that generate, evaluate, and disseminate basic biomedical information. However, these disciplinary groupings are not necessarily natural categories to use when it comes to risk assessment modeling.

Reproductive toxicity, for example, includes a tremendous variety of toxic processes — interference with specific signaling pathways for differentiation, transplacental carcinogenesis, production of heritable genetic changes, structural and long-term functional consequences of killing key cells during development, and the stochastic effects on male fertility that occur when sperm production is inhibited. Each of these processes poses distinct types of mechanistic modeling problems and, therefore, different analytical problems for risk assessment. It is no exaggeration to say that, to risk assessors, the field of reproductive toxicology seems about as diverse as the rest of toxicology combined.

I think we can achieve some simplification in review of the estimation and uses of human variability by focusing on the following characteristics of different problems in the assessment of risks of toxic effects:

- the distinction between quantal vs. continuous outcome data;
- whether the causal models appropriate for a specific response are stochastic or deterministic on an individual basis;
- the needs/opportunities to use one-step (simple input-output) models or multistep models with intermediate parameters.

QUANTAL VS. CONTINUOUS OUTCOME DATA

By quantal data, I mean effects that are either present or absent. You either have cancer or chickenpox, or you do not. At the extreme, mortality is the ultimate quantal parameter (despite Mark Twain's quip "reports of my death are greatly exaggerated"). By contrast, some other parameters of interest are continuous — such as forced expiratory volume in the first second (FEV_1) as a measure of lung function, intelligence quotient (IQ) as a (controversial) measure of intellectual capability, hearing thresholds as an index of hearing acuity, etc. Continuous data generally require very different mathematical modeling techniques than quantal data. They are richer in potential information content, because they can take on an infinite gradation of values. Analysts should be reluctant to sacrifice some of this available information by artificially grouping continuous data into dichotomous or other aggregates.

STOCHASTIC VS. DETERMINISTIC INDIVIDUAL CAUSAL MODELS

Carcinogenesis and mutagenesis are the most well-known examples of effects that are caused in part by stochastic processes. Of course different people have different susceptibilities to genetically acting carcinogens in part because of differences in metabolic activation, inactivation, and DNA repair. However, even people who suffer the same number of DNA changes of a particular chemical type may have different disease outcomes because of chance processes. The significance of DNA changes in an individual depends crucially on the random chances of coded information having been changed at specific sites and in specific ways. When a potentially relevant piece of information has been altered in a functionally significant way in a particular person, the type of cell where those changes have occurred is important.

The DNA changes that cause carcinogenesis and mutagenesis are far from the only stochastic processes of significance to risk assessment. Whether a couple achieves conception during any reproductive cycle has important stochastic elements, for example. Conception probabilities are affected by factors such as the number and quality of sperm, but there are also chance factors that play important roles in determining whether the sperm finds the egg and whether the fertilized egg successfully implants to produce a pregnancy.

The importance of the stochastic/deterministic distinction in modeling variability is this: if the basic causal mechanism is deterministic on an individual basis, then the population dose–response relationship reveals information about the distribution of individual susceptibilities. Such an assumption is the basis of the use of classic probit analysis (Finney 1971) to analyze toxicity data for effects that are thought to be caused by the gross overwhelming of homeostatic systems. For this kind of toxic mechanism, every individual is presumed to have a specific threshold dose that, when exceeded, produces an effect. The population dose–response relationship is therefore a direct measure of the population distribution of individual thresholds. If 5%, 50%, and 95% of the exposed individuals manifest an effect at doses of 1, 2, and 4 mg/kg, we presume that (with allowance for statistical sampling uncertainties) about 5% of the members of the population have thresholds below 1 mg/kg, about 50% have thresholds below 2 mg/kg, and all but 5% have thresholds below 4 mg/kg.*

By contrast, if the basic causal mechanism includes important stochastic processes, the available dose–response data may contain no usable information about individual susceptibility. Imagine a company of soldiers advancing across an open field in the face of machine-gun fire. Some will be hit and some will not, perhaps in rough proportion to the number of machine-gun bullets. Those that are hit will suffer primarily because they were in the wrong place at the wrong time and not because they were any more "susceptible" in presenting a larger cross-sectional area to the opposing forces. Thus, it is often challenging, and sometimes impossible, to extract information about interindividual variability in susceptibility from population dose–response relationships for carcinogenesis produced by genetic mechanisms and other effects that are importantly influenced by stochastic processes.

ONE-STEP (SIMPLE INPUT-OUTPUT) MODELS VS. MULTISTEP MODELS
WITH INTERMEDIATE PARAMETERS

One-step modeling approaches — in which one simply fits some functional form to dose vs. response data — have the virtues and the disadvantages that come with simplicity. The techniques for this are generally well worked out and have well-known properties in the field of biostatistics (Rees and Hattis 1994). However, their direct use is limited to ranges of dose and duration of exposure where the effect can be directly observed. Modeling effects that occur in detectable amounts over background primarily after a prolonged exposure, or at very high doses, involve all the difficulties that result from retrospective assessment of dose (in human populations) and the nonlinearities that occur at high doses because of saturation of enzyme-mediated processes, changes in cell replication rates, etc. (Hattis 1990).

Multistep modeling approaches require the analyst to draw on multiple sources of information that quantify relationships between exposure, a series of intermediate parameters, and data for the end effect. The intermediate parameters are usually proxies for steps in the causal sequence for the end effect. Because they involve several steps, these causal models make more detailed predictions

---

* For this hypothetical example, we could characterize the distribution of individual thresholds as a lognormal distribution where the log(standard deviation) is [log(4) - log(1)]/3.29 standard deviations for a 5% to 95% range = 0.183. The corresponding probit slope is the reciprocal of this at 5.5; the corresponding geometric standard deviation is $10^{0.183} = 1.52$.

that are testable in different ways by different kinds of scientific research. The use of intermediate parameters ("biomarkers")* opens up the "black box" between exposure and effect and allows study of the individual component steps in the causal process by either experimental or epidemiologic approaches or both (Hattis and Crofton 1995, Hattis 1994, Schulte and Perera 1993b). Unfortunately "validation" of the predictive power of specific biomarkers for the adverse effect in question is often challenging (Schulte and Perera 1993a, Schulte and Mazzuckelli 1991). Quantifying the uncertainties in risk assessments based on such multistep analyses generally involves relatively innovative statistical analysis techniques, probably including Monte Carlo simulations of the effects of multiple component uncertainties in most cases.

Table 1 shows the use of these three distinctions as a framework to develop different strategies for (1) the measurement/estimation of interindividual variability in susceptibility, and (2) the use of variability information in quantitative assessments of a variety of human risks from environmental, occupational, and pharmaceutical exposures. The first six strategies briefly described in Table 1 serve as an outline for the remainder of this paper.

STRATEGY 1: QUANTAL THRESHOLD RESPONSES ESTIMATED DIRECTLY FROM THE VARIABILITY IN THOSE RESPONSES OBSERVED IN PEOPLE — e.g., PROBIT SLOPES

Classically, the underlying mechanism for many quantal responses has been conceived as a gross overwhelming of homeostatic systems. The vision is that small doses of a chemical can be accommodated without permanent harm by natural compensatory processes that tend to resist or counteract chemically induced perturbations of normal physiological processes. Individual people, however, are expected to differ both in their capabilities to resist specific chemically induced perturbations and in their functional reserve capacity for performing specific tasks in spite of some degradation of a specific function. Both of these types of differences lead to differences in individual threshold doses — the doses that are barely sufficient to cause a specific effect in particular people.

Probit analysis is simply a way of converting an assumed lognormal distribution of individual thresholds into a straight line. Lognormal distributions are expected when there are many factors that cause people to differ in their threshold doses and when those factors tend to act multiplicatively to determine the individuals' thresholds.** In this sense, a probit analysis of dose–response data can be directly interpreted in terms of a defined mechanism involving interindividual variability, although it is going too far to consider a probit analysis to represent a full biologically based model of the underlying effect in most cases.

Figure 1 illustrates the connection between an observation that a particular proportion of individuals are affected at a specific dose of toxicant, the assumed lognormal distribution of thresholds, and the Z score assigned to the response observed at that dose. The shaded portion of the normal curve simply represents the proportion of people who manifest a particular response at a specific dose. If one-half of the people show the response [i.e., at the median effective dose (ED_{50})], the shaded area in Figure 1 extends to the central line in the figure — one-half of the people have thresholds at or below that dose, and we assign a Z value of 0. If, at a higher dose, 97.5% of the people show the response, then the shaded area extends to the rightmost line in

---

* Biomarkers are broadly defined as "indicators of events in biological systems or samples" (National Research Council 1989a, 1989b). Conventionally, the "events" that are "indicated" are conceived of as either exposure, effects, or susceptibility to the effects of possible future exposure to biologically active substances, although it is recognized that there is a continuous gradation of events that occur between the uptake of a toxic substance into the body and the ultimate manifestation of impairments to health.

** For further theoretical explanation of the mechanisms producing different distributional forms, see Hattis and Burmaster (1994).

TABLE 1
Strategies for Assessing Variability in Susceptibility and Predicting Risks:
Type of Predictive Modeling

Causal process and outcome parameters	One-step modeling (simple input-output)	Multistep modeling (with causal intermediate parameters between exposure and response)
Deterministic process, quantal outcome	E.g., drug dose–response relationships Strategy 1: Classic log-probit (or logit, etc.) dose–response analysis for the end effect of concern in relationship to external dose directly yields information on the population distribution of individual thresholds.	E.g., 1. Fish ingestion → methyl mercury uptake → methyl mercury blood levels → risk of fetal/developmental impairment. 2. CO exposure → conversion of hemoglobin to carboxyhemoglobin → impaired delivery of oxygen to the myocardium → increased risk of fatality given an infarction. Strategy 2: Model overall variability in susceptibility as a function of the variability of component processes (uptake, pharmacokinetic, pharmacodynamic).
Deterministic process, continuous outcome	E.g., 1. Chronic changes in FEV_1 in response to cigarette smoking; 2. Chronic noise-induced hearing loss; 3. Dose-time response for acute pulmonary effects of ozone, acid particulates. Strategy 3: For cross-sectional data, analyze the increase in variance of the outcome parameter as a function of dose. For individual longitudinal data, directly assess individual differences in parameter change as a function of dose.	E.g., lead in house dust → lead in blood → impaired development of IQ. 2. Ingestion of ethanol → impaired reaction time → increase in risk of automobile accidents. Strategy 4: Similar to 3 but in multiple steps.
Stochastic process, quantal outcome	E.g., carcinogenesis risk assessment from conventional high-dose bioassay or epidemiologic data without special PBPK or mechanistic modeling. Strategy 5: Dose–response data cannot be interpreted in terms of variability in susceptibility. Use a generic assumption of interindividual variability, ideally derived from information about the variability of component causal processes.	E.g., 1. Changes in infant mortality as a function of toxicity-induced changes in birth weight; 2. Changes in adult mortality as a function of dust-induced changes in FEV_1; 3. Changes in male fertility as a function of toxicity-induced changes in sperm count and other sperm-quality parameters; 4. Changes in mortality as a function of cadmium-induced kidney damage; 5. Multi-step mechanistic/PBPK-based carcinogenesis modeling. Strategy 6: Combine the population "background" variability in the "functional intermediate parameter" (e.g., birth weight; urinary protein excretion as an index of kidney function) with the distribution of changes induced by the toxicant to predict changes in the incidence of the quantal outcome parameter.
Stochastic process, continuous outcome	Number and severity of automobile injuries as a function of blood alcohol levels.	Number and severity of automobile injuries as a function of defined changes in hazard recognition and response times induced by alcohol and other neuroactive substances in experimental subjects.

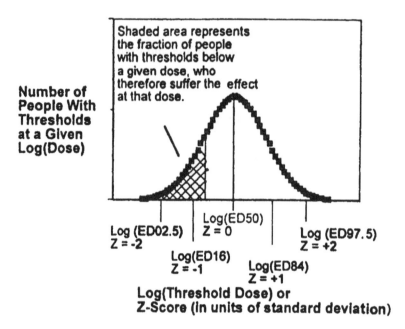

FIGURE 1   Interpretation of dose–response information for quantal effects in terms of a lognormal distribution of individual threshold doses.

Figure 1. Because this line is 2 standard deviations above the midpoint of the assumed lognormal distribution of thresholds, we therefore assign a Z value of +2. If the distribution of thresholds is in fact perfectly lognormal then, as implied in the figure, the scales of Z score and log(dose) should be directly proportional, and a plot of calculated Z scores vs. log(dose) will be a straight line. The correspondence of the points to the line in such a plot is a quick qualitative indicator of how well the lognormal assumption describes the distribution of individual thresholds indicated by the data. The slope of that line, known as the probit slope, is an estimate of the number of standard deviations of the population distribution of thresholds that is traversed by a 10-fold change in dose (1 $\log_{10}$). The probit is simply the Z score plus 5 [5 was originally added by Finney (1971), the statistician/innovator of the system, to accommodate the needs of the mechanical calculators of the time, which could not process negative numbers]. Probits therefore have units of standard deviations and the probit slope in these plots (the coefficient of the x term) is simply the reciprocal of the log[geometric standard deviation (GSD)] of the population distribution of threshold doses. Larger values of the probit slope correspond to smaller amounts of interindividual variability.

To aid in understanding the implications of various probit slopes and related measures of lognormal variability, Table 2 shows 95th percentile/5th percentile ratios (and 99th percentile/1st percentile ratios) for various amounts of lognormal variability expressed in three different ways. The GSD in the third column is the antilogarithm of the standard deviation of the distribution of $\log_{10}$ thresholds given in the second column:

$$GSD = 10^{[\log_{10}(GSD)]}$$

And, as previously mentioned, the probit slope in the first column (calculated as the coefficient of the x term in plots such as those in Figures 2 and 5 to 12 below) is the reciprocal of the standard deviation of the distribution of $\log_{10}$ values of the threshold doses, given in the second column.

The most favorable cases for direct modeling of quantal responses in relationship to dose are in clinical trials of pharmaceuticals where individuals' doses can be measured (either externally or internally), and the resulting responses can be directly observed. Figures 2, 3, and 5 show classic probit plots of two sets of observations of this type.

TABLE 2
A Scale for Understanding Lognormal Variability: Differences (-fold)
Between Particular Percentiles of Lognormal Distributions

Probit slope [1/log₁₀(GSD)]	Log₁₀ (GSD)	Geometric st. dev.	5% to 95% Range (3.3 st. dev.)	1% to 99% Range (4.6 st. dev.)
10	0.1	1.26	2.1-fold	2.9-fold
5	0.2	1.58	4.5-fold	8.5-fold
3.33	0.3	2.0	10-fold	25-fold
2.5	0.4	2.5	21-fold	73-fold
2	0.5	3.2	44-fold	210-fold
1.67	0.6	4.0	94-fold	620-fold
1.43	0.7	5.0	200-fold	1800-fold
1.25	0.8	6.3	430-fold	5300-fold
1.11	0.9	7.9	910-fold	15,000-fold
1.0	1	10.0	1900-fold	45,000-fold
0.91	1.1	12.6	4200-fold	130,000-fold
0.83	1.2	15.8	8900-fold	380,000-fold

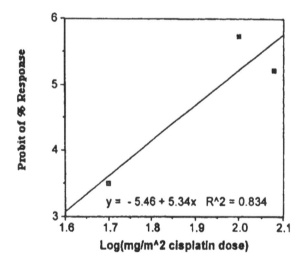

$$y = -5.46 + 5.34x \quad R^2 = 0.834$$

FIGURE 2   Log-probit dose–response plot for percentage of patients with a significant change in air conduction hearing thresholds after one course of treatment with cisplatin [data from Laurel and Jungnelius (1990)].

Figure 2 shows a traditional log-probit plot of the data of Laurel and Jungnelius (1990) on the percentage of patients with a significant change in hearing levels at any frequency after one course of treatment with cisplatin over a relatively limited range of doses. (It should be noted that classification of the individual hearing level changes as significant or not reflects just the kind of information-losing dichotomization of an otherwise continuous effect parameter that I have previously warned against.) With so few points, it is impossible to evaluate how compatible the observations are with the expectations of a lognormal distribution. Still we can observe that the indicated probit slope of about 5.3 translates into a log(GSD) of the population distribution of threshold doses of 1/5.3 = 0.19 — within the range often seen for pharmacokinetic variability alone (Hattis and Silver 1994).

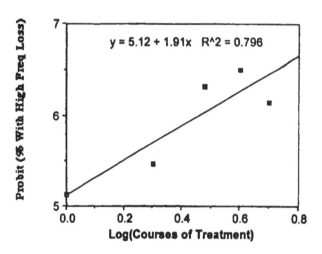

FIGURE 3 Log-probit dose–response relationship for percentage of patients with significant low-frequency hearing loss vs. number of courses of treatment with cisplatin (100 mg/m^2) [data from Blakley et al. (1994)].

The more recent data of Blakley et al. (1994) (Figure 3) allow us to examine the same system in a different light; they used a single dose of cisplatin (100 mg/m^2) but systematically collected data on the increase in the proportion of patients suffering either low- or high-frequency hearing loss separately as a function of the number of courses of treatment. With this restatement of dose, the indicated interindividual variability is considerably larger — probit slopes on the order of 1.8 indicate a log(GSD) of 0.56 — suggesting that there may be considerable interindividual variability in capabilities to resist or compensate for the damage caused by repeated treatments with cisplatin. Clearly, it is not correct in this case to simply multiply dose by the number of courses of treatment to arrive at a metric of exposure that is most directly related to risk of change in hearing loss. Different metrics of dose can lead to very different estimates of variability in susceptibility for essentially the same response.

Figure 4 is a histogram of interindividual variability in the doses of salbutamol producing maximal desired increases in FEV$_1$ among asthma sufferers as measured by Lipworth (1992). As it happens, these data are very well-described by a traditional log-probit analysis (Figure 5), and in this case the indicated interindividual variability is quite large — the log(GSD) calculated from the probit slope is 0.64. In such a distribution, there is a >100-fold range between the doses estimated to produce maximal effect in the 95th percentile person relative to the 5th percentile.

Although clinical experiments are most readily found in the pharmaceutical literature, the recent environmental health literature does contain some informative examples of studies that can help shed light on the extent of interindividual variability in susceptibility for selected undesirable responses to toxicants. Figure 6 is a log-probit plot, based on data of Nethercott et al. (1994), of the percentage of a group of 102 people who showed dermal hypersensitivity reactions to various amounts of a chromium VI compound applied to their skin. (In this and subsequent examples, I have dispensed with the addition of 5 to the Z scores to turn them into classic probits. The interpretation of the slope is the same whether the y axis is expressed in terms of the original Z scores or probits.)

The subjects in this clinical experiment were not a random sample of the general population but were drawn from people identified by their physicians as likely to have chromium hypersensitivity. Of this group of 102 patients, 54 responded at the highest concentration used. Thus, the log-probit model describes the apparent distribution of thresholds very well in this case and indicates a massive amount of interindividual variability — a log$_{10}$(GSD) of about 1. (Error bars are not given for the lowest point in this case because the lower end of even a 1 SD range would extend

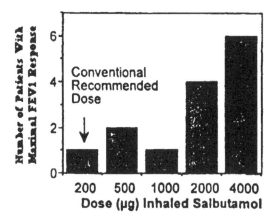

FIGURE 4   Histogram of numbers of asthmatic patients requiring various doses of salbutamol to achieve maximal FEV$_1$ response [data from Lipworth (1992)].

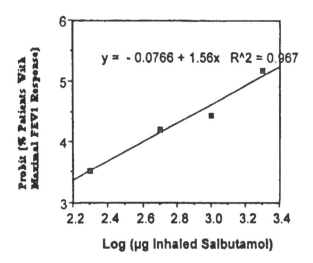

FIGURE 5   Log-probit plot of the dosages needed for maximal FEV$_1$ responses to salbutamol [based on data of Lipworth (1992) shown in Figure 4].

off the scale to very low values. The other error bars are also approximate — completely accurate error bars would not be perfectly symmetric about the observed point.)

Occupational and environmental epidemiologic data are also occasionally extensive enough to be analyzed in this way. However, in these cases, there is often some reason for concern that imperfect measurements of exposure could lead to overestimates of actual interindividual variability in susceptibility.* Figure 7 is a probit plot of the excess incidence of a specific degree of kidney

---

* Ordinary least squares linear regression analysis is well designed to handle random measurement errors in the dependent variable. However, the general assumption is that the independent variable [the log(dose) in this case] is known without error. Measurement or other errors in the independent variable tend to bias the regression in the direction of lower slopes—which are interpreted as higher estimates of the interindividual variability in susceptibility when this occurs in a probit analysis. A reasonable response to this problem is to assess the amount of uncertainty in the exposure estimates and correct for it via either analytical or simulation procedures. For a specific example where simulation approaches have been used to explore the consequences of various amounts of exposure uncertainty for probit slope observations, see Hattis and Silver (1994).

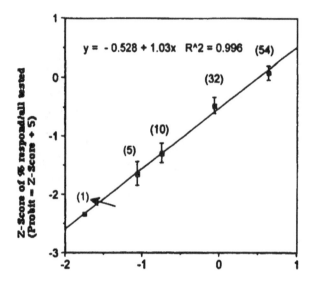

FIGURE 6   Log-probit plot of the percentage of 102 tested people who gave positive skin patch tests for chromium (VI). Error bars are approximately ±1 SD from a binomial distribution based only on counting error. The number of observed positive responses is in parentheses [data from Nethercott et al. (1994)].

tubular dysfunction* in relationship to cumulative exposures to cadmium in air in a group of >400 workers with long-term exposures to cadmium studied by Jarup et al. (1988). The indicated inter-individual probit slope in this case is <1, indicating a degree of interindividual variability that extends to the extreme values shown at the bottom of Table 2. Recall the contrasting estimates of interindividual variability in the cisplatin hearing-loss response that resulted from the measures of dose intensity vs. dose repetition in Figures 2 and 3. If the raw data for the study of Jarup et al. were available, it might be rewarding to explore whether some index other than the straightforward integral of air concentration and time were more strongly predictive of the effect incidence in this case — and therefore yields a smaller estimate of interindividual variability in susceptibility in relationship to the better dose metric. Of course it is also desirable to treat the indicator of kidney damage (β_2-microglobulin excretion) as a continuous parameter rather than to draw a single cutoff to produce a quantal representation of the underlying data. The literature on the environmental health implications of cadmium exposure is very rich and provides useful examples for other analytical strategies in the following sections.

STRATEGY 2: QUANTAL THRESHOLD RESPONSES ESTIMATED BY USING THE OBSERVED HUMAN VARIABILITY IN PUTATIVE CAUSAL COMPONENT PROCESSES

Breaking up the causal chain for a quantal noncancer response into several steps allows the analyst greater flexibility to draw on the body of information about human interindividual variability gleaned for chemicals and effects other than the one immediately under study. It also allows for the possibility of a generic analysis that can ask, "Given our understanding of how variable different causal steps have been found to be — how often, how likely is it that we will incur how much risk by using our standard 10-fold uncertainty factor to protect against the effects of human heterogeneity in susceptibility?"

---

* Kidney tubular dysfunction is indicated in this case by urinary excretion of $_2$-microglobulin polypeptide corresponding to approximately the 97.5th percentile of excretion in a population not exposed to unusual amounts of cadmium. "Excess incidence" is defined after subtracting the observed "background" incidence of high $_2$-microglobulin excretion seen in the worker group with the lowest cumulative exposure to cadmium.

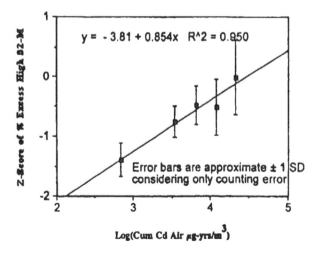

FIGURE 7   Log-probit dose–response plot of excess percentage of workers with high urinary β_2-microglobulin (β_2-M) levels vs. cumulative total air cadmium levels. Error bars are approximately ±1 SD considering only counting error [data from Jarup et al. (1988)].

In previous work (Hattis and Silver 1994), I have suggested a three-part categorization of variability components for this purpose, following the pathway from environmental exposure through production of adverse effects:

1. uptake (individual differences in the environmental concentration needed to produce a given intake of toxicant into the body, e.g., due to differences in breathing rates, dietary habits, etc.);
2. pharmacokinetic (individual differences in the amount of uptake needed to produce a particular concentration-time product of active agent in the blood or at the site of action, e.g., due to differences in metabolic activation or clearance);
3. response or pharmacodynamic (individual differences in the dose in the blood or at the active site that produces a similar risk of response).*

Data for the first two of these components (summarized in Table 3) are readily available (Hattis et al. 1987). Table 4 summarizes the more scarce information in the final category as analyzed in earlier work (Hattis and Silver 1994, Hattis 1996) and elsewhere in this paper. Both Tables 3 and 4 express available variability information in terms of the ratio of the parameter values for the 95th percentile individual in an assumed perfect lognormal population to the values for 5th percentile individuals. As previously mentioned this is a span of about 3.3 standard deviations.** For comparison, if we are willing to make the heroic assumption that some of these distributions may be truly lognormal out to the extreme tails, then 3.4 standard deviations would be expected to be the difference between 20% and a 10^{-5} incidence of effect; 3.1 standard deviations would be expected to be the difference between a 5% and a 10^{-6} incidence of effect. Data such as those in Tables 3 and 4 can be used in an analysis that combines lognormal variances from different causal steps to

---

* Each of these categories is expressed in an inverted way (e.g., the dose needed to produce a particular response, rather than the response per unit of dose) to minimize definition difficulties arising from nonlinearities in different steps of the causal process.
** The difference between the two columns in the table can be thought of either as variability among chemicals tested in the same way (or, in some cases, different tests for the same chemicals) or as the uncertainty facing an analyst in assessing the amount of interindividual variability in susceptibility to the noncarcinogenic effect of a chemical for which there are no direct measurements of human variability.

TABLE 3
Preliminary Findings of Interindividual Variability in Systemic Uptake and Pharmacokinetic Parameters

Parameter and parameter class	Width of the 90% range[a] for an average chemical or test	Width of the 90% range[a] for chemicals/tests with greater interindividual variability than 95% of all others examined
Parameters related to systemic uptake and pharmacokinetics in healthy adults		
Breathing rates	1.8-fold	2.8-fold
Skin absorption	2.5-fold	No estimate[b]
Half-life for elimination (by metabolism or excretion)	2.3-fold	5.8-fold
Maximum blood concentration (after a single dose)	2.3-fold	11-fold
Area under a concentration-time curve (AUC)	3.0-fold	8.1-fold
Blood concentration measurements combining variability in exposure concentrations, exposure-related behaviors, and pharmacokinetics		
Serum polychlorinated biphenyl concentrations	12-fold	No estimate[b]
Blood mercury and methylmercury concentrations	11-fold	17-fold
Blood lead concentrations (after gasoline lead phaseout)	13-fold	No estimate[b]

[a] Numbers are ratios of the value of each parameter for the 95th percentile individual to the value of the parameter for the 5th percentile individual based on the assumption that the values of the parameter are lognormally distributed (i.e., the logarithms of the parameter are normally distributed).
[b] There is no upper confidence limit in these cases because only a single observation of interindividual variability was available in the data set.

represent overall interindividual variability in susceptibility for a complete pathway from external exposure to end effect. If we do this, we can then make a crude first guess at the dose reduction necessary to take a typical lowest-observed-adverse-effect level or no-observable-adverse-effect level incidence close to or into the region of frequency that may be considered acceptable for some general population exposures when applied to the more serious outcome of cancer.

Table 4 cites some new analyses of data under the hypothesis of a lognormal distribution of variability in effect thresholds:

- Figure 8 shows a probit plot of the data of Liu et al. (1995) on the incidence of cataracts in a group of Chinese workers exposed to trinitrotoluene (TNT) in relationship to log(dose) assessed as hemoglobin adducts. The averaging time for production of such adducts usually corresponds to the lifespan of red blood cells in humans — about 3 months. In this case, the fit of the log-probit model to the data is not very good.
- Figure 9 shows a plot of the data of Jarup et al. (1988) of the incidence of relatively high β_2-microglobulin levels (indicative of kidney dysfunction) in relationship to the integrated blood concentration of cadmium × time. This is parallel to the analysis in Figure 7, where the dosimeter was a similar concentration × time integral of concentrations of cadmium in air. Both distributions are well-described by the lognormal hypothesis, but the apparent probit slope is much larger — indicating a much smaller amount of interindividual variability — when the x axis is defined in terms of concentrations of cadmium in blood. In part this might be because the measure of individual blood concentrations bypasses the uptake and pharmacokinetic portion of interindividual variability. Also, however, the difference may in part reflect a longer implicit averaging time for blood measurements and therefore possibly less statistical inaccuracy in the

TABLE 4
Preliminary Findings of Interindividual Variability in Parameters That Include Pharmacodynamic Variability

Parameter and parameter class	Ratio of 95th percentile to 5th percentile[a] threshold dose for an average chemical or test
Parameters combining uptake, pharmacokinetic, and pharmacodynamic variability (response in relationship to exposure to an environmental medium)	
Cadmium: kidney tubular dysfunction (high β_2-microglobulin excretion) in relationship to air concentration × time integral (Figure 7) (worker population)	7100-fold
Ozone: acute change in FEV_1 in relationship to air concentration × time (Figure 12)	13-fold
Chronic FEV_1 change in relationship to pack-years of cigarette smoking (Figure 14)	8.3-fold
Parameters combining pharmacokinetic and pharmacodynamic variability (response as a function of administered external dose)	
Cisplatin: hearing loss vs. dose (one course of treatment) (Figure 2)	4.1-fold
Cisplatin: hearing loss at one dose vs. number of courses of treatment (Figure 3)	100-fold
Salbutamol, maximum FEV_1 increase (asthma treatment) (Figure 5)	130-fold
Dermal hypersensitivity reactions to chromium (VI) (Figure 6)	1600-fold
Pure pharmacodynamic parameters (response as a function of internal dose)	
Cataracts in response to internal dose of TNT assessed as hemoglobin adducts (worker population) (Figure 8)	500-fold
Cadmium: kidney tubular dysfunction (high β_2-microglobulin excretion) in relationship to blood concentration × time integral (Figure 9) (worker population)	125-fold
Cadmium: kidney tubular dysfunction (high β_2-microglobulin excretion) in relationship to urinary cadmium excretion (Figures 10, 11) (community population)	18-fold (females) 16-fold (males)
Six neurological effects of methylmercury in adults (Iraqi data) (Hattis and Silver 1994)	12-fold
Six fetal/developmental effects of methylmercury (Iraqi data) (Hattis and Silver 1994)	approx. 460[b]
>95% maximum relief of neuropathy in diabetic patients in relationship to imipramine plasma concentration (Hattis 1996)	6.9-fold
Hemodynamic responses to serum concentrations of nitrendipine (a vasodilator) (Hattis 1996)	8.3-fold (unbound drug) 17-fold (total serum concentration)

[a] Numbers are ratios of the value of each parameter for the 95th percentile individual to the value of the parameter for the 5th percentile individual based on the assumption that the values of the parameter are lognormally distributed (i.e., the logarithms of the parameter are normally distributed).

[b] Caveat: imperfection of the dosimeter used here (maternal hair mercury) may have led to serious overestimation of the actual extent of interindividual variability [see Hattis and Silver (1994)].

estimation of true individual exposure in relationship to concentration of cadmium in blood compared with concentration of cadmium in air.

- Finally, Figures 10 and 11 show a similar probit plot of the incidence of high β_2-microglobulin concentrations in the older members of a community in Japan in relationship to cadmium dose expressed as urinary cadmium concentrations. This unusually extensive set of data reported by Nogawa et al. (1992) is also very compatible with the hypothesis of a lognormal distribution of threshold doses. In this case, however, the

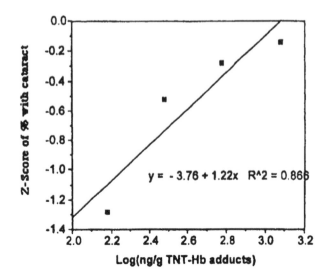

FIGURE 8 Log-probit plot of the percentage of workers exposed to TNT who developed cataracts in relationship to log(TNT-hemoglobin adduct levels) [data from Liu et al. (1995)].

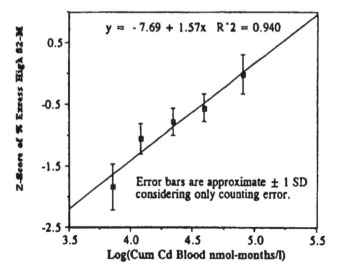

FIGURE 9 Log-probit dose–response plot of excess percentage of workers with high urinary β_2-microglobulin concentrations vs. log(cumulative total blood cadmium concentration) (without backward extrapolation of high-level exposures) [data from Jarup et al. (1988)].

apparent interindividual variability is, although still large, much smaller than observed in the worker population studied by Jarup et al. (1988) using time-integrated concentrations in blood as the dosimeter. It is possible that the urinary concentrations of cadmium are an even better reflection of the causally relevant delivered dose of cadmium to the kidney than a time integral of quantities in blood or that the difference between the blood and urinary probit slopes reflects additional pharmacokinetic variability related to the long-term storage and action of cadmium in sensitive portions of the kidney.

Overall, there are considerable data in the literature that can be interpreted under the broad framework of multistep modeling for quantal outcome data contemplated as Strategy 2.

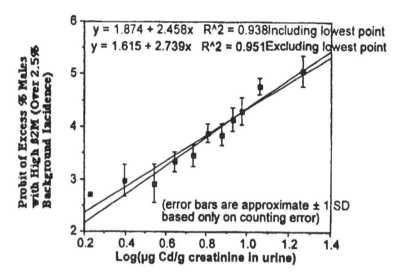

FIGURE 10 Traditional log-probit plot of excess percentage of males with high β_2-microglobulin urinary excretion [data from Nogawa et al. (1992)]. High = over the 97.5th percentile in excretion in a population with no unusual exposure. Alternative probit equations are shown including and excluding the lowest point because the lowest point is less than 1 standard deviation above the background incidence.

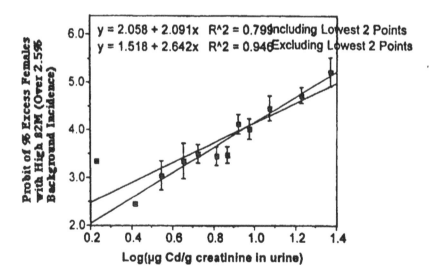

FIGURE 11 Traditional log-probit plot of excess percentage of females with high β_2-microglobulin urinary excretion [data from Nogawa et al. (1992)]. High = over the 97.5th percentile in excretion in a population with no unusual exposure. Alternative probit equations are shown including and excluding the lowest two points because the next-to-lowest point is less than 1 standard deviation above the background incidence.

STRATEGY 3: CONTINUOUS RESPONSES USED DIRECTLY AS MEASURES OF UNDESIRABLE OUTCOMES

With the development of better measurement techniques that allow more detailed observation of the operating parameters of biological systems, risk analysts will more often be able to take advantage of continuous data that quantify biological responses to toxicants. Indeed, we have already come across several examples where naturally continuous data (e.g., for β_2-microglobulin concentrations or hearing levels) had to be analyzed as quantal only because the information in the source papers was presented solely in quantal form. At least three analytical subtypes appear possible within this category:

- When full continuous data are available on individual responses observed prospectively, the analyst can directly describe the distribution of individual parameter changes in relationship to dose. For this purpose, it is helpful to first make a mathematical transformation of the dose, if necessary, so that the relationship between dose (as transformed) and the extent of individual parameter changes is linear. Second, it is helpful to have good repeated measurements of the extent of parameter change in relationship to dose in the same individuals. This allows the analyst to estimate experimental measurement error (and short-term individual variability in responsiveness). Ideally, the variance attributable to measurement errors should be subtracted from the total observed variance in parameter change per dose to arrive at a good estimate of true long-term variability in individual responsiveness. For an illustration of this variance subtraction in measurement of baseline values of an ordinary biological parameter (individual breathing rates) see Hattis and Silver (1994).
- An alternative is to define a series of effect cutoffs and use ordinary probit analysis to study the interindividual variability in the exposure or dose needed to induce those different degrees of a continuous response in different individuals. This is illustrated below with recently published data on acute ozone-induced changes in FEV_1.
- Finally, if only cross-sectional data are available, it still may be possible to model the increase in variance of a studied parameter as a function of exposure. Then one can determine what degree of interindividual variability in susceptibility to the toxicant would be consistent with induction of the observed degree of spreading in high-exposure groups relative to low-exposure groups. This is illustrated below with a summary of a previously published analysis (Hattis and Silver 1994) of apparent variability in susceptibility of long-term reduction in FEV_1 in relationship to pack-years of cigarette smoking.

PROBIT ANALYSIS OF MULTIPLE LEVELS OF ACUTE FEV_1 RESPONSE TO OZONE

Elsewhere in this volume (Bromberg, this volume) there is extensive discussion of the data of McDonnell et al. (1995) about acute FEV_1 changes in response to ozone. These data are summarized with a statistical model using a logistic regression approach.

For comparison, Figure 12 shows the results of a log-probit treatment of the frequency with which various degrees of decline in FEV_1 were observed in that set of experiments in relationship to the product of concentration × time of ozone exposure for exposure durations between 2 and 6 hours (data for the 1-hour exposures were excluded because there was no experimentally detected effect at any studied exposure level for that duration). The three sets of points and their accompanying probit regression lines represent the excesses over background (control exposures without added ozone) in the percentages of people showing different degrees of FEV_1 change — from 5% (uppermost line) to 15% (lowest line). The scatter of the points increases for the higher effect levels, almost certainly because relatively small numbers of people show the larger degrees of effect. But despite the differences in scatter, all three lines appear to be parallel, with slopes close to 2.9 — indicating approximately the same extent of interindividual variability in susceptibility at all effect levels. As shown in Table 4, this amount of variability corresponds to about a 13-fold expected spread between the doses inducing a comparable degree of response in 5th percentile vs. 95th percentile individuals.

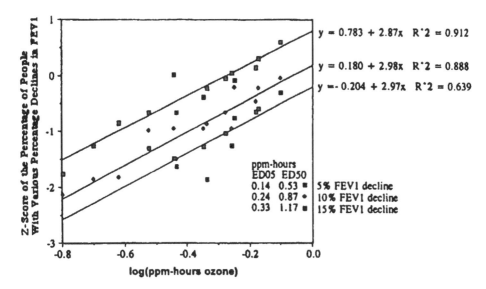

FIGURE 12   Composite log-probit plot of ozone dose-time/severity-of-response relationship [data from McDonnell et al. (1995)].

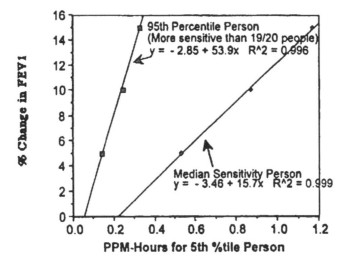

FIGURE 13   Dose-effect relationships for acute ozone FEV_1 response inferred from probit analysis of McDonnell et al. (1995).

The inset data at the bottom of Figure 12 show results of interpolations from the probit lines of the doses producing different degrees of FEV_1 change in 95th percentile sensitive vs. median individuals. These results are plotted in Figure 13. For both projected median individuals and the projected upper 95th percentile susceptible persons, the indicated dose-effect functions are consistent with a hockey-stick shape. There is an apparently linear increase in the extent of FEV_1 change with increasing parts per million (ppm)-hours of exposure above some minimal exposure designated by an intercept on the x axis. There appears to be both a smaller intercept and a larger dose-effect slope for the projected 95th percentile susceptible individual. Clearly, the more continuous treatment of dose-effect information in this example allows risk analysts to provide better insights into the likely biological consequences of different levels of exposure.

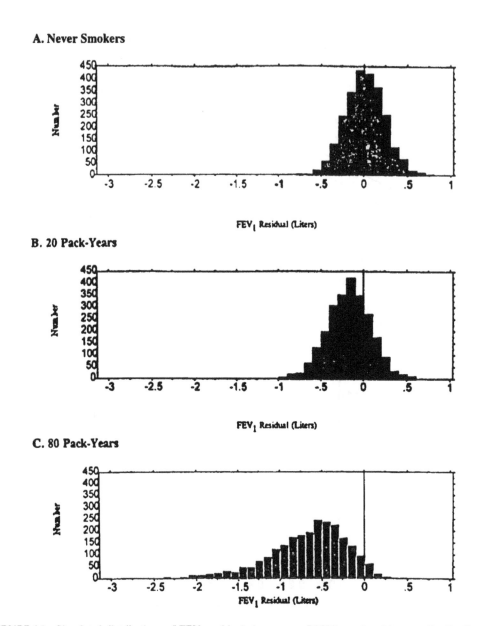

A. Never Smokers

B. 20 Pack-Years

C. 80 Pack-Years

FIGURE 14   Simulated distributions of FEV_1 residuals in groups of 2500 people with normally distributed background variability and lognormally distributed (GSD = 1.9) differences in susceptibility to FEV_1 loss from smoking.

CHRONIC FEV_1 CHANGES IN RESPONSE TO CIGARETTE SMOKE

To the extent that individuals differ in the amount of parameter change per unit exposure to an agent, cross-sectional data should indicate an increasing spread of the parameter values as dosage increases to higher concentrations. This is illustrated in Figure 14 with simulated data for the case of cigarette smoking and long-term changes in FEV_1 residuals as a function of smoking dose. (FEV_1 residuals are differences between the observed FEV_1 for an individual and the FEV_1 that would be expected on the basis of general relationships with height and age.) Real data (Dockery

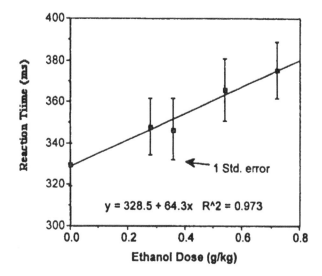

FIGURE 15 Dose-effect relationship for effect of alcohol on reaction time [data from Colrain et al. (1993)].

et al. 1988, Burrows et al. 1977) as analyzed by Hattis and Silver (1993) are similar in appearance. Using these data we asked the following questions:

- Is the degree of spreading of FEV_1 residuals with increasing cigarette dose greater than would be expected on the basis of (1) the baseline variability of FEV_1 residuals in never-smokers, and (2) the likely variability in cigarette dose within dose categories?
- What degree of interindividual variability is most compatible with the data (expressed as a GSD for an assumed lognormal distribution of individual rates of decline of FEV_1 per pack-year of cigarette smoking)?
- What are the confidence limits around our estimates of interindividual variability in susceptibility to FEV_1 decline, considering each data set separately and combined?

There are many challenges in analyses such as this. One must first do a very good job in estimating the relationships between the continuous variable under study (in this case FEV_1) and various confounding factors. Moreover, in the unexposed group, it is crucial to be able to accurately describe the distribution of departures from the basic prediction equation for different individuals. (This is the baseline variability with which the spread of values in the exposed groups are later compared.) In the case of the FEV_1 residuals for lung function, we found it necessary to use a mixture of two normal distributions instead of a single normal distribution to adequately describe our baseline variabilities for each data set. Even then, the analysis depends critically on the precise departures of the numbers of observations at the tails of the fitted distributions from those expected. In this case, our estimated GSD of 1.9 (from the more extensive of two data sets analyzed) corresponds to about an 8-fold difference between the exposures that would be expected to cause the same degree of effect in 95th percentile and 5th percentile susceptible individuals.

The expectation for an increase in variance in measured parameters with higher exposures or doses is not limited to cross-sectional data. Figure 15 shows dose-effect data for increases in reaction time in relationship to clinically administered doses of alcohol. In this case, however, there is no detectable increase in reaction time variation for individuals exposed to higher doses. I have not calculated how much interindividual variability would have been necessary to produce a statistically significant increase in the variance at higher doses of alcohol in this experiment.

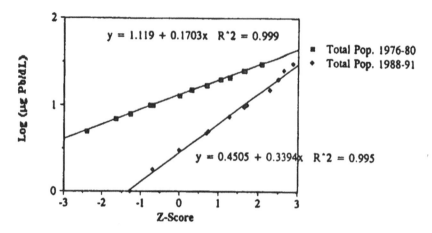

FIGURE 16 Comparison of blood lead distributions in general population of the United States in 1976–1980 vs. 1988–1991.

STRATEGY 4: CONTINUOUS RESPONSES USED DIRECTLY AS MEASURES OF UNDESIRABLE OUTCOMES — MULTISTEP MODELING

In principle, it is also possible to use continuous intermediate parameters in a multistep assessment of the implications of different kinds of individual variability for the risks of change in a continuous valued end-result parameter. For example, there could well be interindividual variability in hand-to-mouth ingestion of lead-contaminated house dust, interindividual variability in lead uptake and pharmacokinetics, and individual variability in the relationship between blood lead concentrations and changes in long-term intellectual development measured by IQ.

To this point, however, the focus of epidemiologists and risk analysts of effects of lead on children has been on the very difficult task of ascertaining the modest changes in mean IQ of the population in relationship to exposures to low levels of lead in the community with appropriate controls for numerous confounders (Schwartz 1994, Pocock et al. 1994). There has been considerable analysis of the extent of interindividual variability in overall exposure, as indexed by the lognormal distribution in blood lead levels in the population. Figure 16, based on NHANES 2 and NHANES 3 data (Pirkle et al. 1994, Brody et al. 1994, Annest et al. 1982), shows that the dramatic 5-fold reductions in median blood lead levels between the mid-1970s and late 1980s has been accompanied by an appreciable increase in variability — lessening the practical protection realized for those who are most highly exposed to the residual pathways of lead exposure. However, there has been relatively little study of whether this exposure variability is compounded by appreciable variability in pharmacodynamic susceptibility. For particularly large data sets, it is possible that investigators could explore the possibility of whether the variance in IQ (or other continuous measures of function) changes as a function of past lead exposure, similar to the analysis of long-term smoking-induced changes in FEV_1 cited above.

STRATEGY 5: QUANTAL STOCHASTIC RESPONSES (E.G., CANCER) WHOSE RISK IS AFFECTED BY OBSERVED HUMAN VARIABILITY IN PUTATIVE CAUSAL COMPONENT PROCESSES

Existing risk assessment procedures for carcinogens are intended to be conservative in the uncertainty dimension — giving estimates that are expected to be higher than true risks for typical people. However, these procedures do not consider the likely variability in susceptibility among individuals. Elsewhere (Hattis and Barlow 1996) I have published a preliminary analysis of some available data on the interindividual variability of some putative measures of metabolic activation, inactivation, and DNA repair for different types of carcinogens. Data in this section are drawn from that article.

In recent years significant amounts of new data on relevant human variability have become available. As our understanding of the molecular pathology of cancer continues to grow, it appears likely that particular population subgroups with specifically increased risk to particular classes of carcinogenic agents will be increasingly identified (Hein et al. 1992, Vineis and Ronco 1992, Kawajiri et al. 1993, Lledo 1993, Camus et al. 1993, Rautio et al. 1992). New data associating risks for specific cancers with genetic and environmentally determined levels of specific metabolic activation and inactivation [e.g., Kadlubar et al. (1992), Shields (1993)] provide an avenue for beginning to quantify the range of human susceptibilities to different categories of carcinogenic agents.

The recent publication of data about the gene responsible for ataxia-telangiectasia is a further case in point (Nowak 1995). The gene is important for detecting radiation-induced damage to DNA and for inhibiting cell replication until repair enzymes can remove the damage; impaired forms of the gene are relatively common. One copy of a mutant form of the gene is carried by about 1% of all people. Cells from heterozygotes are more sensitive to radiation-induced cell killing and to mutagenesis *in vitro*. Current preliminary epidemiology studies suggest that heterozygotes have a 3-fold increase in relative risk of cancer in general and a 5-fold increased risk of breast cancer (the breast is a relatively radiosensitive site in humans) (Savitsky et al. 1995). If these findings are confirmed, they would suggest larger relative increases in the risk as a result of specific identifiable causes, such as radiation, because the overall increases in incidence are relative to a background incidence that is related to multiple carcinogenic agents.

As discussed previously in this paper, the principal summary measure we use for expressing individual variability within data sets is the log(GSD) (see Table 2 and accompanying text). The use of this uniform summary measure of variability must not be taken to imply that all the underlying distributions are lognormal. Accurate description of the data in many cases is likely to require use of other distributional forms — particularly mixtures of two or more lognormal distributions where there is clear evidence that genetic polymorphisms are important in determining a particular type of variability.

With this simple uniform measure, Table 5 shows the amounts of variability indicated for the nonredundant data sets for the different broad types of activity. The 5th to 95th percentile ranges for observations of individual log(GSD) values for different chemicals and tests are also presented in Table 5, calculated under the assumption that the log(GSD) values are lognormally distributed. This assumption appears to be consistent with empirical observations (Figure 17). Distributions such as that shown in Figure 17 are potentially useful for assessing the uncertainty in the likely interindividual variability for a random genetically active carcinogen for which no specific human metabolic or DNA-repair variability information is available. The implications of Monte Carlo simulations of the combined effects of variability and uncertainty for cancer risk assessments for genetically acting agents are explored by Hattis and Barlow (1996).

Table 6 summarizes our results by categories that could be related to the specific enzymes involved in metabolic activation, inactivation, and DNA repair for specific classes of carcinogens. Some particularly large log(GSD) values occur where there are known polymorphisms that have been associated with increased cancer risks in some epidemiologic studies [i.e., Takamatsu and Inaba (1994), Nazar-Stewart et al. (1993)].

STRATEGY 6: USE OF CONTINUOUS FUNCTIONAL INTERMEDIATE PARAMETERS TO PREDICT RISKS OF QUANTAL STOCHASTIC NONCANCER RESPONSES

Functional intermediate parameters are continuous variables whose response to an agent is measurable in animals or people, whose background distribution in people is known, and that have

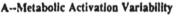

A--Metabolic Activation Variability

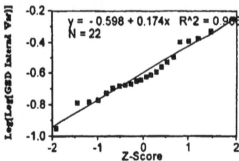

B--Detoxification Variability

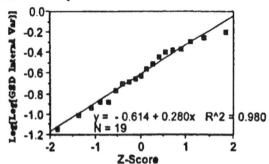

C--DNA Repair Variability

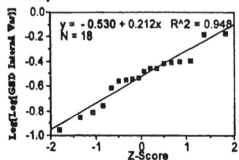

FIGURE 17 Apparent lognormal distributions of log(GSD) findings for nonredundant data sets within specific categories of activity [source: Hattis and Barlow (1996)].

known, strong relationships to specific adverse outcomes. Some examples, which are discussed below, include the following:

- birth weight as a predictor of infant mortality,
- viable sperm counts and other sperm-quality parameters as a predictor of male fertility,
- measures of impaired lung function (FEV_1) and impaired kidney function (β_2-microglobulin excretion) as predictors of mortality in adults, and
- changes in continuous measures of driving-related performance (e.g., the speed of reaction to potentially hazardous events in driving simulations) as predictors of automobile accident risks.

TABLE 5
Indicated Variability for Nonredundant Data Sets by Activity Category

Activity category	Total no. data sets	No. nonredundant data sets[a]	Geometric mean of $\log_{10}$ (GSD) (5% to 95% range)
Metabolic activation	24	22	0.253 (0.132–0.482)[b]
Detoxification	27	19	0.243 (0.087–0.681)
DNA repair	20	18	0.295 (0.134–0.650)
Complex	7	5	0.375 (0.167–0.843)
Total	78	61	

[a] In some cases we chose to analyze the data for a particular data set in two ways: (1) as a whole-population aggregate, and (2) broken down into separate subcategories by genotype, gender, cancer cases vs. controls, etc. Numbers in this column and the calculations in the final column represent distinct whole-population, nonredundant data sets.

[b] These ranges were calculated assuming that the distribution of the $\log_{10}$(GSD) values for variability for different data sets is lognormally distributed. These ranges therefore represent an estimate of the variability of the variability measurements among different specific types of activities, tissues, etc.

Adapted from Hattis et al. 1987.

In principle, use of such intermediate parameters can provide windows on the pathologic process that are earlier in the development of toxicity, more sensitive to the action of potential toxicants (when compared with attempts to observe actual cases of illness), and more accessible to direct comparative measurement in both animal models and humans.

The basic schema for using this kind of information (Institute of Medicine 1991) is as follows:

1. Elucidate the quantitative relationships between external dose of the toxicant, the internal dose and time of toxin exposure, and changes in the functional intermediate parameter.
2. Assess the preexisting background distribution of the functional intermediate parameter in the exposed human population. (This step builds in information related to the relative vulnerability of different percentiles of the exposed human population that cannot be obtained from studies in animals.)
3. Assess the relationship between the functional intermediate parameter and the incidence of the quantal health effect of concern.
4. Assess the expected change in the incidence of the quantal health effect, given the changes in the population distribution of the intermediate parameter likely to result from exposures to the toxicant.

Of course, the accuracy of the risk assessments made with this schema depend critically on the strength of the causal connection between the functional intermediate parameter and the end effect of concern. It is not essential (and, indeed, it is generally not possible) to use an intermediate parameter that is a direct link on the causal chain leading to the adverse effect. But it is important to have an intermediate parameter that has a relatively strong and consistent relationship with whatever factors really are part of the causal chain. The critical assumption is that whatever perturbs the intermediate parameter is likely also to make parallel changes in the actual causal determinants of the quantal health effect. In the future, risk assessments made by this approach will be improved if methodology can be developed to fairly assess and adjust for the uncertainties in this key assumption.

TABLE 6
Summary of Variability Results by Specific Mechanistic Categories

Activity subcategory	Number of nonredundant data sets	Geometric mean log (GSD)	Geometric standard error
Metabolic activation			
P4501A1 and aryl hydrocarbon hydroxylase	5	0.327	1.237
P4502E1 and related activities	3	0.192	1.333
P4503A4 and related activities	5	0.222	1.027
Other oxidizing activities (probable P-450)	6	0.308	1.148
Other activities potentially related to metabolic activation	3	0.180	1.080
Detoxification			
glutathione transferases			
a) nonspecific substrate (CDNB)	2	0.161	1.376
b) GSTμ (established polymorphism related to lung cancer risk)	5	0.496	1.091
c) other GSTs	4	0.167	1.503
Phase II conjugation reactions	2	0.218	1.099
Superoxide dismutase (antioxidant)	1	0.173	No est.[a]
Epoxide hydrolase	1	0.209	No est.[a]
Acetylation (established polymorphism related to bladder cancer risk)	1	0.429	No est.[a]
Decrease in mutagenic activity of various direct-acting compounds for *Salmonella*	3	0.169	1.277
DNA repair			
O6-Methylguanine methyltransferase	10	0.306	1.085
Uracil DNA glycosylase	3	0.526	1.262
Repair of benzo[a]pyrene DNA adducts	1	0.282	No est.[a]
Unrepaired or misrepaired chromosome damage or cell killing from radiation or radiomimetics	5	0.172	1.241

[a] There is no upper confidence limit in these cases because only a single observation of interindividual variability was available in the data set.

BIRTH WEIGHT AS A PREDICTOR OF INFANT MORTALITY [ADAPTED FROM REES AND HATTIS (1994) AND BALLEW AND HATTIS (1989)]

Figure 18, plotted from data of Hogue et al. (1987), shows the strong relationship that has been observed between birth weight and infant mortality (note that a logarithmic scale has been used for plotting mortality rates). Although infants with very low birth weight are at dramatically higher risk than infants in the normal weight range, even infants of about 3000 g are at higher mortality risk than somewhat heavier infants. Therefore, even these infants that are usually considered in the normal range of weight could be expected to have their risks increased somewhat by an agent that causes a marginal reduction in birth weight, if birth weights accurately reflect the status of some underlying developmental parameter that is causally related to infant mortality.

Figure 19 shows the population distribution of birth weights — both before and after a hypothetical 1% shift in the entire distribution toward lower weights that might be related to the action of a toxicant that inhibited some function that was limiting to fetal growth and development. Figures 20 and 21 show the result of multiplying the mortality rates in Figure 18 by population distribution data such as those in Figure 19. Figures 20 and 21 show the overall potential populationwide infant mortality implications of such an agent among Caucasians and African-Americans, respectively.

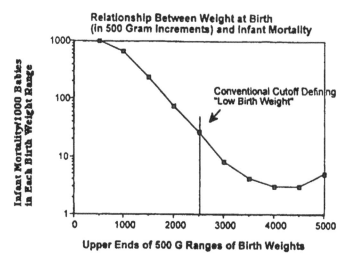

FIGURE 18 Relationship between weight at birth (in 500-g increments) and infant mortality [data from Hogue et al. (1987)].

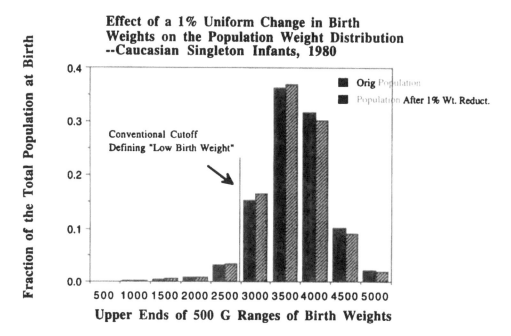

FIGURE 19 Effect of a uniform change in birth weights on population weight distributions [source: Rees and Hattis (1994)].

The weight distribution of infant deaths is bimodal — with approximately equal numbers of deaths occurring in the traditional low-birth-weight and normal-birth-weight regions among Caucasians but with somewhat larger numbers of deaths in the low-birth-weight region among African-Americans. The bimodality arises because, although the babies in the lowest birth-weight categories are at dramatically higher absolute risk of death in their first year, there are so many babies with more normal birth weights that the latter contribute a substantial proportion of total mortality.

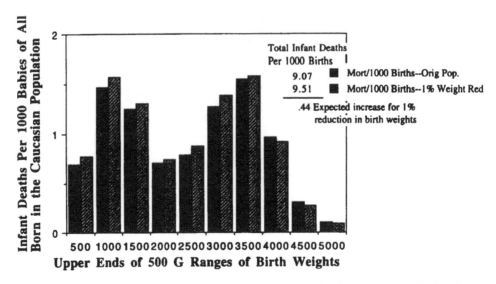

FIGURE 20 Expected effect of a 1% reduction in birth weight on distribution of overall infant deaths per 1000 babies born (Caucasian population).

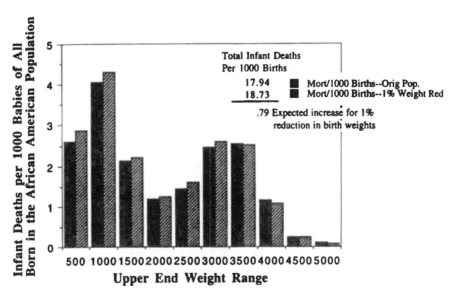

FIGURE 21 Expected effect of a 1% reduction in birth weight on distribution of overall infant deaths per 1000 babies born (African-American population).

Overall a 1% change in birth weight might be expected to be associated with nearly a 5% change over the baseline overall infant mortality — from 9.07 to 9.51 infant deaths per 1000 births for the Caucasian population in the United States and from 17.94 to 18.73 deaths per 1000 births for the African-American population in the United States. The greater expected absolute risk among African-Americans is directly related to the relatively larger numbers of infants in the low-birth-weight category. Thus, the birth-weight distribution data have led to a preliminary expectation of a moderate differential risk of response in one group vs. another.

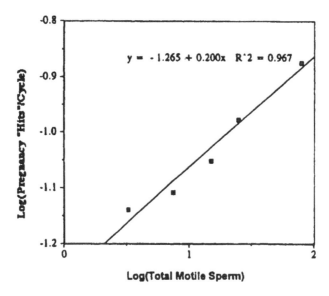

FIGURE 22   Log-log plot of relationship between total motile sperm count and pregnancy hits per menstrual cycle [data from Brasch et al. (1994)].*

SPERM COUNT AND OTHER SPERM-QUALITY PARAMETERS AS PREDICTORS OF MALE FERTILITY

Viable sperm count is one of several sperm-quality parameters that andrologists have found to be useful in diagnosing male-related subfertility and in predicting probabilities of conception (Bartoov et al. 1993). Brasch et al. (1994) have recently provided data about the frequency of conception per cycle for intrauterine insemination as a function of the total number of motile sperm that are available for use. The relationship is far less than linear — that is, increasing numbers of sperm increase the probability of conception but not proportionately (Figure 22). If 80 million sperm are available, the conception probability is a little less than double what it would be with only 3 to 8 million sperm.

Bonde (1993), in an extensive study of Danish metal workers, observed an average 28% reduction in sperm count associated with a reduced fecundity (odds ratio, 0.89; 95% cytotoxicity index, 0.83 to 0.97) in welders. It would be interesting to determine whether this observation is similar to what would be predicted based on the relationship shown in Figure 22 and similar data for other populations and circumstances of conception.

NEUROTOXICITY

Neurotoxic effects also offer a rich set of possibilities for the use of continuous intermediate parameters in risk assessments for difficult-to-study end effects of concern. These include the following:

- clinical cases of Parkinson's disease as a function of the size of the population of relevant functional neurons in the substantia nigra,
- incidence of Alzheimer's disease and amyotrophic lateral sclerosis as a function of depletion of the relevant neuronal populations for those conditions,
- noise-induced hearing loss in relationship to functional organs of Corti, and

---

* The hits-per-cycle data that form the basis of the y axis in Figure 22 are derived with the aid of a one-hit transformation of the original observations of the frequency with which couples in different groups conceived during the period of observation: observed fraction of couples who conceive = 1 — Poisson probability of 0 hits per couple = 1 - e-conception hits/couple; therefore, conception hits per couple = $-\ln(1$ — observed fraction of couples who conceive).

- for some acute neurobehavioral effects, relationships between specific alcohol-induced changes in performance and automobile accident risks (West et al. 1993, Farrimond 1990) can help analysts assess important implications of modest changes in behavioral parameters related to other exposures to solvents and drugs. Alcohol-induced neurobehavioral impairments provide a convenient scale that can be used for understanding and communicating the degree of impairment induced by a defined amount of exposure to a solvent or a drug. Expression of results in direct comparisons with the effects of various amounts of alcohol is not uncommon in the occupational medicine (Echeverria et al. 1991) and pharmaceutical (Vuurman et al. 1994) literature. This is also a good example of the stochasticity of some pathologic processes. Impaired behavior does not make an automobile accident inevitable for any individual, but it increases the risk of an accident in the presence of an array of physical and judgmental challenges presented unpredictably (from the standpoint of any individual driver) by the environment.

FEV₁ As a Predictor of General Mortality

Observations about relationships between parameters of lung function and general mortality date back to the origins of current measures of lung function. There are several large community-based studies that report associations of all-cause mortality and cardiovascular mortality in particular with FEV_1, after controlling for the effects of cigarette smoking [for both newer data and older references see Bang et al. (1993), Heederick et al. (1992), and Chau et al. (1993)]. Casiglia et al. (1993) found that many traditional cardiovascular risk factors in young adults are not as important in very old (>80 years) people. By contrast, a relatively lower FEV_1 than predicted based on age and other variables was a strong predictor of overall mortality in this group. In previous risk assessment work, I have used the association between FEV_1 and general mortality to make preliminary estimates of the likely effects of occupational exposure to coal dust on general mortality by using coal dust-induced changes in the distribution of FEV_1 as an intermediate parameter (Silver and Hattis 1991).

Cadmium-Induced Mortality with Urinary B₂-Microglobulin Used As a Functional Intermediate Parameter

Earlier (Figures 9–11) the incidence of high urinary β_2-microglobulin concentrations was used as a quantal indicator of the incidence of serious kidney tubular dysfunction in relationship to cadmium exposures in both occupational and community settings. Recently, Nakagawa et al. (1993) published the results of a large prospective study of all-cause mortality in relationship to both age and urinary β_2-microglobulin concentrations. A very helpful aspect of this paper is that in some analyses β_2-microglobulin excretion is treated as a continuous rather than a quantal parameter. The results suggest excesses in mortality down to relatively modest increases in β_2-microglobulin (that is, levels below those used to define the quantal effect data) (Figure 23). If data can be obtained (from the investigators in this study or elsewhere) showing the continuous dose-effect relationships between β_2-microglobulin excretion and cadmium exposure and excretion, then it should be possible to estimate the overall mortality effects of a broad range of rates of cadmium exposure.

STRATEGIES 7 AND 8: MODELING CONTINUOUS RESPONSES THAT OCCUR IN PART BY STOCHASTIC PROCESSES

Stochastic processes are sometimes involved in producing outcomes that are measured in graded form as continuous parameters. The example of alcohol-induced behavioral changes described previously could just as well be placed in this category if we consider the severity as well as the incidence of injuries from automobile accidents that result from chemically impaired performance.

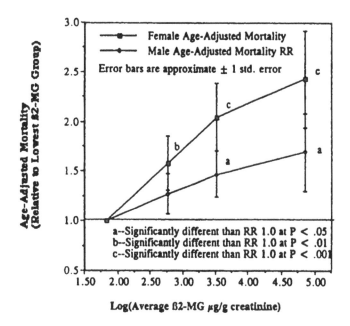

FIGURE 23 Empirical relationships between age-adjusted excess mortality relative risks (RR) and urinary β_2-microglobulin concentrations [data from Nakagawa et al. (1993)].

CONCLUSION

The examples provided above show a rich array of possibilities for improving quantitative risk assessment of both cancer and noncancer effects of environmental and occupational exposures with the aid of human data on interindividual variability. These examples are by no means exhaustive. Considerable improvements are possible if individual variability data are collected more systematically by investigators, if previously collected individual data are made more readily available for analysis, and if new targeted research is undertaken to quantify the relative vulnerability to different effects of various subpopulations (by age, pathology, gender, and other factors) within our diverse society.

REFERENCES

Annest JL, Mahaffey KR, Cox DH, et al (1982) Blood lead levels for persons 6 months–74 years of age: United States, 1976–1980. Adv Data 79:1–24

Ballew M, Hattis D (1989) Reproductive effects of glycol ethers in females — a quantitative analysis. MIT Center for Technology, Policy, and Industrial Development, Report CTPID 89-7, Cambridge, MA

Bang KM, Gergen PJ, Kramer R, et al (1993) The effect of pulmonary impairment on all-cause mortality in a national cohort. Chest 103:536–540

Bartoov G, Eltes F, Pansky M, et al (1993) Estimating fertility potential via semen analysis data. Hum Reprod 8:65–70

Blakley BW, Gupta AK, Myers SF, et al (1994) Risk factors for ototoxicity due to cisplatin. Arch Otolaryngol Head Neck Surg 120:541–546

Bonde JPE (1993) The risk of male subfecundity attributable to welding of metals. Int J Androl 16(suppl 1):1–28

Brasch JG, Rawlins R, Tarchala S, et al (1994) The relationship between total motile sperm count and the success of intrauterine insemination. Fertil Steril 62:150–154

Brody DB, Pirkle JL, Kramer RA, et al (1994) Blood lead levels in the U.S. population. Phase 1 of the Third National Health and Nutrition Examination Survey (NHANES III, 1988 to 1991). JAMA 272:277–283

Burrows B, Knudson RJ, Cline MG, et al (1977) Quantitative relationships between cigarette smoking and ventilatory function. Am Rev Resp Dis 115:195–205

Camus AM, Geneste O, Honkakoski P, et al (1993) High variability of nitrosamine metabolism among individuals: role of cytochromes P450 2A6 and 2E1 in the dealkylation of N-nitrosodimethylamine and N-nitrosodiethylamine in mice and humans. Mol Carcinogen 7:268–275

Casiglia E, Spolamore P, Ginocchio G, et al (1993) Predictors of mortality in very old subjects aged 80 years or over. Eur J Epidemiol 9:577–586

Chau N, Benamghar L, Pham QT, et al (1993) Mortality of iron miners in Lorraine (France): relations between lung function and respiratory symptoms and subsequent mortality. Br J Indust Med 50:1017–1031

Colrain IM, Taylor J, McLean S, et al (1993) Dose dependent effects of alcohol on visual evoked potentials. Psychopharmacology 112:383–388

Dockery DW, Speizer FE, Ferris BG, et al (1988) Cumulative and reversible effects of lifetime smoking and simple tests of lung function in adults. Am Rev Resp Dis 137:286–292

Echeverria D, Fine L, Langolf G, et al (1991) Acute behavioural comparisons of toluene and ethanol in human subjects. Br J Ind Med 48:750–761

Farrimond T (1990) Effect of alcohol on visual constancy values and possible relation to driving performance. Percept Motor Skills 70:291–295

Finney DJ (1971) Probit analysis, 3rd ed. Cambridge Univ Press, London

Hattis D (1996) Variability in susceptibility — how big, how often, for what responses to what agents? Environ Toxicol Pharmacol 2:135–145

Hattis D (1994) The use of well defined biomarkers (such as blood lead) in risk assessment. Environ Geochem Health 16:223–228

Hattis D (1990) Pharmacokinetic principles for dose rate extrapolation of carcinogenic risk from genetically active agents. Risk Anal 10:303–316

Hattis D, Barlow K (1996) Human interindividual variability in cancer risks — technical and management challenges. Hum Ecol Risk Assess 2:194–220

Hattis D, Burmaster DE (1994) Assessment of variability and uncertainty distributions for practical risk analyses. Risk Anal 14:713–730

Hattis D, Crofton K (1995) Use of biological markers of causal mechanisms in the quantitative assessment of neurotoxic risks. In Chang L, Slikker W (eds), Handbook of neurotoxicology, Vol. 3. Approaches and methodologies. Academic Press, New York, chap 53, pp 789–803

Hattis D, Silver K (1994) Human interindividual variability — a major source of uncertainty in assessing risks for non-cancer health effects. Risk Anal 14:421–431

Hattis D, Silver K (1993) Use of biological markers in risk assessment. In Schulte P, Perera R (eds), Molecular epidemiology: principles and practices. Academic Press, New York, chap 10, pp 251–273

Hattis D, Erdreich L, Ballew M (1987) Human variability in susceptibility to toxic chemicals — a preliminary analysis of pharmacokinetic data from normal volunteers. Risk Anal 7:415–426

Heederik D, Kromhout H, Kromhout D, et al (1992) Relations between occupation, smoking, lung function, and incidence and mortality of chronic non-specific lung disease: the Zutphen study. Br J Ind Med 49:299–308

Hein DW, Rustan TD, Doll MA, et al (1992) Acetyltransferases and susceptibility to chemicals. Toxicol Lett 64/65:123–130

Hogue CJR, Buehler JW, Strauss MA, et al (1987) Overview of the national infant mortality surveillance (NIMS) project — design, methods, results. Publ Health Rep 102:126–138

Institute of Medicine (1991) Seafood safety. National Academy Press, Washington, D.C.

Jarup L, Elinder CG, Spang G (1988) Cumulative blood-cadmium and tubular proteinuria: a dose–response relationship. Int Arch Occup Environ Health 60:223–229

Kadlubar FF, Butler MA, Kaderlik KR, et al (1992) Polymorphisms for aromatic amine metabolism in humans: relevance for human carcinogenesis. Environ Health Perspect 98:69–74

Kawajiri K, Nakachi K, Imai K, et al (1993) The CYP1A1 gene and cancer susceptibility. Crit Rev Oncol Hematol 14:77–87

Laurell G, Jungnelius U (1990) High-dose cisplatin treatment: hearing loss and plasma concentrations. Laryngoscope 100:724–734

Lipworth BJ (1992) Risks vs. benefits of inhaled β_2-agonists in the management of asthma. Drug Safety 7:54–70

Liu YY, Yao M, Fang, JL, et al (1995) Monitoring human risk and exposure to trinitrotoluene (TNT) using haemoglobin adducts as biomarkers. Toxicol Lett 77:281–287

Lledo P (1993) Variations in drug metabolism due to genetic polymorphism: review of the debriso-quine/sparteine type. Drug Invest 5:19–24

McDonnell WF, Stewart PW, Andreoni S, et al (1995) Proportion of moderately exercising individuals responding to low-level, multi-hour ozone exposure. Am J Respir Crit Care Med 152:589–596

Nakagawa H, Nishijo M, Morikawa Y, et al (1993) Urinary β_2-microglobulin concentration and mortality in a cadmium-polluted area. Arch Environ Health 48:428–435

National Research Council (1989a) Biologic markers in pulmonary toxicology. National Academy Press, Washington, D.C.

National Research Council (1989b) Biologic markers in reproductive toxicology. National Academy Press, Washington, D.C.

Nazar-Stewart V, Motulsky AG, Eaton DL, et al (1993) The glutathione S-transferase μ polymorphism as a marker for susceptibility to lung carcinoma. Cancer Res 53:2313–2318

Nethercott J, Paustenbach D, Adams R, et al (1994) A study of chromium induced allergic contact dermatitis with 54 volunteers: implications for environmental risk assessment. Occup Environ Med 51:371–380

Nogawa, K, Kido, T, Shaikh ZA (1992) Dose–response relationship for renal dysfunction in a population environmentally exposed to cadmium. IARC Sci Publ 118:311–318

Nowak R (1995) Discovery of AT gene sparks biomedical research bonanza. Science 268:1700–1701

Pirkle JL, Brody DB, Gunter EW, et al (1994) The decline in blood lead levels in the United States. The Third National Health and Nutrition Examination Survey (NHANES). JAMA 272:284–291

Pocock SJ, Smith M, Baghurst P (1994) Environmental lead and children's intelligence: a systematic review of the epidemiological evidence. Br Med J 309:1189–1197

Rautio A, Kraul H, Kojo A, et al (1992) Interindividual variability of coumarin 7-hydroxylation in healthy volunteers. Pharmacogenetics 2:227–233

Rees DC, Hattis D (1994) Developing quantitative strategies for animal to human extrapolation. In Hayes AW (ed), Principles and methods of toxicology, 3rd ed. Raven Press, New York, chap 8, pp 275–315

Savitsky K, Bar-Shira A, Gilad S, et al (1995) A single ataxia telangiectasia gene with a product similar to PI-3 kinase. Science 268:1749–1753

Schulte PA, Mazzuckelli L (1991) Validation of biological markers for quantitative risk assessment. Environ Health Perspect 90:239–246

Schulte PA, Perera FP (1993a) Validation. In Schulte P, Perera R (eds), Molecular epidemiology: principles and practices. Academic Press, New York, pp 79–107

Schulte PA, Perera FP, eds (1993b) Molecular epidemiology: principles and practices. Academic Press, New York

Schwartz J (1994) Low-level lead exposure and children's IQ: a meta-analysis and search for a threshold. Environ Res 65:42–55

Shields PG (1993) Inherited factors and environmental exposures in cancer risk. J Occup Med 35:34–41

Silver K, Hattis D (1991) Methodology for quantitative assessment of risks from chronic respiratory damage: lung function decline and associated mortality from coal dust. MIT Center for Technology, Policy, and Industrial Development, Report CTPID 90-9

Takamatsu Y, Inaba T (1994) Inter-individual variability of human hepatic glutathione S-transferase isozymes assessed by inhibitory capacity. Toxicology 88:191–200

Vineis P, Ronco G (1992) Interindividual variation in carcinogen metabolism and bladder cancer risk. Environ Health Perspect 98:95–99

Vuurman EF, Uiterwijk MM, Rosenzweig P, et al (1994) Effects of mizolastine and clemastine on actual driving and psychomotor performance in healthy volunteers. Eur J Clin Pharmacol 47:253–259

West R, Wilding J, French D, et al (1993) Effect of low and moderate doses of alcohol on driving hazard perception latency and driving speed. Addiction 88:527–532

3 Interindividual Variability in Neurotoxicity

David A. Eckerman, John R. Glowa, and W. Kent Anger

CONTENTS

INTRODUCTION

The goal of the current chapter is to provide background information on the variation from one human to the next in the effect of a neurotoxic agent. This variation is acknowledged in the current practice of risk assessment (Dourson and Stara 1983, Barnes and Dourson 1988) by dividing by 10 the largest amount of an agent found to be nontoxic [the average no-observable-adverse-effect level (NOAEL)]. The intent of this division by an uncertainty factor (UF) is to identify a reference dose (RfD) that is low enough so that, if exposure is regulated at that level, even very sensitive individuals will be protected. The practice was first advocated by Lehman and Fitzhugh (1954) in their effort to establish safe levels of chemical food additives. Although several additional or alternative UFs have been added to the RfD approach to account for various sources of potential

0-8493-2805-5/99/$0.00+$.50
© 1999 by CRC Press LLC

variability, there have been a host of criticisms [see Glowa and MacPhail (1995) for a review] — applying a fixed value independent of the distribution of effects across individuals, for example, has frequently been challenged. Even though the practice has been criticized, however, it must be acknowledged that little empirical work has directly addressed differences in human response to a chemical.

A second issue commonly raised about the current risk assessment practice is use of the NOAEL, with other starting points being recommended as a replacement. Rather than representing a certain risk, the NOAEL represents an estimate of no risk. Several alternative approaches to risk assessment have been developed, each of which incorporates individual variability into the calculation of a true risk figure. The first of these approaches was originally proposed by Dews (1986). It relies heavily on describing individual dose-effect data and assesses risks by directly determining the variability in point estimates of the dose to produce a defined level of effect. The method has been tested and expanded in several reports that provide dose-effect functions for individual subjects (e.g., Dews 1986, Glowa and MacPhail 1995). Yet, in many cases within-subject dose-effect functions are difficult to obtain (e.g., irreversible effects) and the use of this approach has not been well-developed for such cases [but see Glowa and MacPhail (1995) for a foreshadowing of such an approach]. Therefore, although the Dews–Glowa approach has much to recommend it, we focus on approaches that have been tested in situations where within-subject dose effects are not available.

An alternative method was developed by Crump (1984) and uses standard regression techniques and confidence intervals to define variability in effects at fixed doses. These are then used to create a benchmark dose to predict the likelihood of an effect at lower doses. This method is easily adapted to group data, but it does not specifically incorporate individual variability. One of the newest approaches has been described by Gaylor and Slikker (1990), an approach we will utilize below. Both of these approaches address many of the methodological (rather than empirical) concerns raised for the RfD approach based on NOAELs and lowest-observed-adverse-effect levels (LOAELs). They utilize information from throughout the range of doses, rely less on dose extrapolations, do not require assumption of a threshold or determination of a NOAEL, and allow estimation of risk for a full range of doses (Kimmel 1990; Crump 1984, 1995).

In our analysis, we rely heavily on the approach taken by Gaylor and Slikker (1990), first developed in their treatment of the effects of repeated exposure to 3,4-methylenedioxy-N-methyl-amphetamine (MDMA). In their approach, a benchmark response (BMR) was identified on the basis of variation in observations for unexposed control subjects — specifically, the response value for the poorest 1% of control subjects. This BMR defined an abnormal response, and levels of exposure to a toxic agent were judged by their increased incidence. They started, then, by first arranging the observations for control subjects along a continuum [in their initial example, they observed the concentration of 5-hydroxytryptamine (5-HT) in rat hippocampus; see Figure 1]. A frequency distribution was then created across this response dimension and a normal (Gaussian) curve was fit to the data. A point along the response continuum was selected to separate the continuum into abnormal on one side and normal and supernormal on the other. The shift in response produced by a specific exposure was then determined, and a normal curve of the same shape for control subjects was centered at the mean response for this level of exposure (see Figure 1). The increased number of subjects expected to show the BMR was then estimated from this distribution, and the increased number of subjects is taken as an estimate for increased risk. The reader should note that the distribution obtained for control subjects was suggested as the basis for this judgment, because that distribution is the one most readily established with sufficient thoroughness to provide a good estimate. That means, however, that if the agent produces a change in variability, the increased incidence of injury could be underpredicted. Still, because the low doses of specific interest in risk assessment should change the shape of the control distribution very little, the misestimation is likely to be minimal. Gaylor and Slikker's approach is presented in Figure 1, where the impact of a repeated dosing with MDMA (2.5 mg per kg of body weight) on the 5-HT concentrations in rat hippocampus is estimated from (1) the distribution of concentrations in control

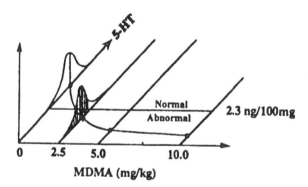

FIGURE 1   Quadratic dose–response relationship between 5-HT in ng per 100 mg of hippocampus tissue in female monkeys 1 month after two exposures to MDMA per day for 4 days. Dots represent average experimental levels; the value for 2.5 mg of MDMA per kg of body weight is estimated. A 5-HT concentration of 2.3 ng/100 mg (horizontal line) is 3 SD below the average of 4.2 ng/100 mg for control animals. All 5-HT levels below 2.3 ng/100 mg are considered abnormal and are estimated to occur in fewer than 1.3 per 1000 untreated animals. Frequency of 5-HT levels is shown for untreated animals (normal curve); the same curve is repeated, centered at the mean for the dose 2.3 ng/100 mg. Shaded area of frequency distribution represents 84% of the individuals estimated to have 5-HT concentrations in the abnormal range 1 month after a course of treatments with MDMA (2.5 mg/kg). (From Gaylor and Slikker 1990.)

subjects, (2) a separation of the 5-HT level continuum into normal and abnormal at the 1% value of the control distribution (found to be 2.3 ng/100 mg), (3) the dose-effect data determined for higher doses of MDMA, and (4) the control distribution then applied to a specific dose to estimate increased incidence of an abnormal response. In this example, a 2.5-mg/kg dose was estimated to increase the incidence of abnormal concentrations of 5-HT from 1% to approximately 68%. In our exploration of the effects of alcohol and lead, we use this Gaylor and Slikker effect-variation approach to represent the likely impact of a specific dose of the agent and thereby to provide a standard against which the UF of 10 can be compared. It is the goal of this report to provide information that will allow this fixed-value, NOAEL-based approach to be evaluated. Our specific goal is to evaluate whether the public is adequately protected by regulations that use a UF of 10 to represent interindividual differences in response to (i.e., sensitivity to) neurotoxic agents.

The primary focus here is on behavioral (functional) impairments rather than on neuropathology or psychophysiological change. Behavioral data are emphasized in part because our expertise is in this area, although the underexplored topic of human variation in effect took us away from our main background. A second reason for the focus being placed on functional impairments, however, is that determination of functional impairment provides critical information when setting exposure limits in risk assessment. In our opinion, it is difficult to base risk assessment on information about damage to the nervous system unless there is evidence showing behavioral impairment. Functional data therefore enjoy a primary role in risk assessment.

Our focus here also is on interindividual variation in response to toxic agents [to include pharmacodynamics as described by Hattis and Silver (1994) and perhaps other sources of behavioral variation as well] rather than on interindividual variation in the uptake of a toxic agent or in the pharmacokinetics of a toxic agent as they are incorporated into the body and delivered to the nervous system. We take this focus because our concern here is specifically with neurotoxicity rather than with toxicity in general; variation in uptake and pharmacokinetics would be a topic shared by all toxicities. Hattis (Hattis and Silver 1994, Hattis 1995) describes differences in response or pharmacodynamic variation, as "differences in the dose in the blood or at the active site that produces a similar risk of response" — in our case, a neurotoxic response. On the other hand, differences in uptake are defined as "differences in the environmental concentration needed to produce a given intake of toxicant" and differences in pharmacokinetics are defined as "differences in the amount of uptake needed to produce a particular concentration-time product of active agent in the blood or at the site of action." Uptake and pharmacokinetics are not our primary focus, because these differences would seemingly hold broadly across many different kinds of toxicities and would not be specific to neurotoxicity. We therefore try to see through such variation and focus, instead, on variation in response.

EXAMPLES OF OBSERVED VARIATION IN BEHAVIORAL IMPAIRMENT

We have the best knowledge for regulating exposure to an agent when we have thoroughly mapped the dose–response function over several different behavioral measures and for many individuals. We can then draw a line at an exposure level that protects a stated proportion of individuals and do so with few assumptions. There seems to be no case, however, where we have such a thorough understanding of dose–response across behaviors and across individuals. Instead, we often are forced to set a dose to protect public health on the basis of a small database derived from accidental exposures or animal experimentation. Because there are protocols judged to be safe by which alcohol can be administered by experimenters to human subjects and because public concern is (occasionally) high regarding acceptable alcohol levels for drivers, the database is perhaps the largest of any used for public health decisions. We therefore begin our review with this agent. Because public concern over the exposure of children to environmental lead was raised during the past 20 years, due in large measure to the vigorous efforts of a few investigators, a sizable database has been developed regarding this agent as well. Evaluation of the effects of lead demonstrates several difficult issues. First, the effects cannot be ethically studied experimentally in humans because no safe protocols for testing the effects of this agent are available. Second, the effects are especially difficult to characterize because the danger of this agent appears to be on behavioral development and therefore effects must be studied with young animals — a constantly changing target. Despite the difficulties, however, we also address this database.

Before starting with these examples, however, one foreshadowing comment seems appropriate. The common and accepted practices of science severely limit our knowledge about the variation in response from one individual to the next. We have a science of averages that sacrifices knowledge of individuals in its enthusiasm for average effects. Scientists focus on average data and the statistical significance of differences among groups to simplify communication of their conclusions. In most scientific reports, variability among individuals is acknowledged, if at all, by a measure such as the standard deviation around the mean (SD) (or the standard error of the mean). With such a bias for average effects, we are at a special loss when we seek to characterize what impact an agent has on the most sensitive individual rather than on the average individual. We see this limitation with special clarity in our review of the effects of alcohol, although the same bias exists in all areas we have studied.

HUMAN VARIATION IN RESPONSE TO ALCOHOL

The starting point for our review of the behavioral effects of low doses of alcohol was Moskowitz and Robinson's (1988) review of the literature for human performance "related to driving." As one might suspect, the literature is massive, with many participants and a variety of viewpoints. Focusing his review on experimental studies of the effects of alcohol on moderate drinkers, Moskowitz identified 557 citations, which were winnowed down to 177 by applying criteria regarding relevance to driving skills and quality of experimental design and measurement [e.g., blood alcohol concentrations (BACs)]. One overall observation was that studies published in the late 1970s and 1980s showed impairment at lower BACs than had been demonstrated in previous studies. A reason for this shift toward lower BACs is that "older studies [generally failed] to examine lower BACs. Indeed some older studies investigated remarkably high BACs." In addition, Moskowitz noted that experimental methods and instrumentation for measuring performance also had improved considerably in the [then] recent literature. Despite these improvements, however, the literature still had restricting limitations because most studies continued to use "only one or two dose levels other than placebo" and therefore did not give any indication of "the minimum level at which impairment might have occurred." Further, the doses used differed from country to country and across time, as investigators gravitated to doses of social rather than scientific relevance. Many studies in the United States, for example, have used 0.10% BAC as their low level of alcohol. Limiting the number of doses studied to one or two and also seeking to include a dose where an effect is expected, produces a bias toward finding a higher LOAEL. This bias occurs because the dose specified as the LOAEL must be drawn from among those doses actually tested. The published literature should be viewed, therefore, as being biased toward finding higher LOAELs than appropriate.

We should note that, in our review, we focused on studies of individuals who are normal in terms of uptake and pharmacokinetic factors. We have therefore excluded studies on groups of individuals judged to be highly sensitive to the effects of alcohol because they metabolize or eliminate alcohol in identifiably different ways. For example, considerable information is now available regarding an aldehyde dehydrogenase deficiency that increases the effect of alcohol for a large number of Asians and Native South Americans [30% to 50%; see Agarwal and Goedde (1989)]. As we consider this to be a variation in pharmacokinetics, however, we have excluded these studies.

Moskowitz divided his review into seven major areas of skilled performance: reaction time, tracking, cognitive function (including vigilance, information processing, and concentrated attention), visual functions, perception, psychomotor performance, and driver performance (including simulations and actual driving situations). These seven areas might be considered different behavioral endpoints, with the different kinds of tests providing different ways of measuring these endpoints. His conclusion is that "BACs as low as 0.03% can produce significant impairment of driver performance." The LOAEL, however, depended very much on the endpoint selected. For example, "of the 37 reviewed studies of reaction time, [only] 12 studies have reported impairment at BACs of 0.05% or less," and there are "significant numbers (16) of studies which reported finding no effects of alcohol." Those experiments that did report reaction time effects at this level utilized more complex reaction time tasks. Apparently, "simple reaction time" is not an appropriate procedure to show a low-dose alcohol effect.

Because alcohol has been administered experimentally and repeatedly to a great many human research subjects, it would appear that this literature provides an ideal opportunity to address the specific goal of this report — to assess the variation in effect from one individual to the next in the healthy population. In fact, however, inspection of the studies included in the Moskowitz review was a good exercise in demonstrating how unusual it is to find this issue addressed in articles where it could be evaluated. Whereas most reports provide means across subjects within a dose group and almost all studies report inferential statistical comparisons, few studies provide measures of variability (e.g., SDs, standard errors of the mean) much less full information on the distribution of effects. The report of significance levels from inferential statistics apparently substitutes for measures of variability. In fact, some reports provide only regression analyses and do not mention even the size of average effects. For most questions, the scientific community apparently finds a science of averages acceptable.

A discussion with Moskowitz confirmed, however, that someone who has dealt with experimental and reported data in this field over many years is comfortable with the expectation that, in a population of "normals," the distribution of effects follows a normal (Gaussian) distribution. We have therefore used this assumption in our review. We focus on three studies in our report of this review. One study presented the raw data for 10 subjects and two others presented SDs of effect.

Moskowitz and DePry (1968) reported the effect of 0.06% BAC on a subject's ability to accurately report the presence of a low-intensity tone within a burst of noise. When given as a single task, this ability was not at all affected. When the subject was asked to simultaneously keep track of spoken strings of six digits, however, as well as report on the presence of the tone (making

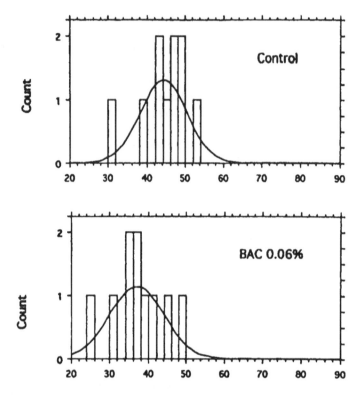

FIGURE 2 Number of correct detections of 60 tone-in-noise trials when subjects were jointly asked to recall lists of digits (divided attention task). Data are shown for each individual subject when given placebo (top) and when given alcohol (bottom) (average BAC, 0.05%). A normal curve has been fit to each distribution. (From Moskowitz and DePry 1968.)

this a divided attention task), alcohol did impair this ability. Of the 60 trials for this difficult task, in fact, the percentage of tones correctly detected dropped from an average of 74% (44.4 correct) to 61.8% (37.1 correct) as a result of the alcohol. Values for the 10 subjects are shown in the two distributions of Figure 2, with a normal curve fitted to each distribution. It might be noted that the Moskowitz and DePry study is very unusual because individual subject data are reported. It therefore provides a special opportunity in addressing our goal of characterizing individual variation in effect. With so few subjects, however, it still is difficult to determine whether a normal curve fits the data well, although there is no striking deviation from normal. In accord with Gaylor and Slikker's (1990) expectations, the calculated SDs are fairly comparable between the placebo and the low dose of 0.06% (SDs of 6.1 vs. 7.0 correct — note that any individual differences in the obtained BACs were ignored in this analysis). Did alcohol affect all subjects similarly? A correlation carried out between the placebo and alcohol scores suggested that there was a comparable effect of alcohol across subjects, because subjects generally retained their relative position in the distribution of scores between the two conditions (correlation was 0.92). A look at difference scores, however, suggests that there was some individual variation in the size of the effect of alcohol across subjects — 8 of the 10 subjects were remarkably uniform in response — 5 subjects showed a decrease in score of exactly 8, and 3 others were very close to this decrease (range, 6 to 9). Yet, there were 2 outliers. One individual showed no effect at all (the individual with the highest overall score) and another showed a decrease of 11. The message from this one small study, then, is that individuals do differ in their sensitivity to the effects of a low dose of alcohol. In this case, the most extreme outlier was apparently protected from the effect of alcohol.

Two additional studies were identified that provided SDs of effect and had somewhat larger samples (20+) and therefore allow an application of the Gaylor–Slikker analysis to demonstrate the likely increased incidence of scores judged to be abnormal (with an assumption of normality). Data were collected under placebo and low-dose alcohol (0.05% BAC in each study) to demonstrate the effect of alcohol. In one case (Linnoila et al. 1980), the subject was challenged to use a joystick to keep an unstable point of light steady at midscreen. The test continued for 2 minutes. In the other case (Landauer and Howat 1982), the subject was told to quickly move a finger from a central point to the correct one of eight buttons as soon as a digit appeared. The buttons were placed in a semicircle and digits were randomly assigned to specific buttons. The test was composed of 24 trials, with data being taken from the last 12. In each case, the test was completed both under placebo and under the low dose of alcohol (BAC = 0.05%). Figure 3 presents four distributions derived from the means and SDs across subjects for these two studies. The distributions are idealized in that they present a sample of 500 cases randomly drawn from a population having a normal-curve distribution of effects with the specific mean and SD taken from the published study. These idealized distributions are presented to demonstrate the small shift in magnitude of effects that are implied by the published values for this LOAEL. Applying the Gaylor–Slikker (1990) analysis to these distributions and setting the criterion for an abnormal value at a performance expected for the lowest 5% of the placebo distribution, the incidence of abnormal scores is expected to increase from 5% to 22% (Linnoila et al. 1980) or 11% (Landauer and Howat 1982). In the context of environmental toxicology, of course, these increases are substantial and in most cases are far above the increase that would be tolerated in risk management.

What is implied for the current RfD approach by this analysis of the effects of alcohol? First, it should be stressed that an RfD approach was not taken when the health risks for alcohol were established. But as an exercise it might be instructive to consider what would have resulted had such an approach been taken, even basing this exercise on neurotoxicity only (setting aside issues of individual differences in uptake and pharmacokinetics). Would alcohol be appropriately regulated by the current RfD procedures if we were seeking to ensure public health? In fact, the current RfD procedure would provide a remarkably conservative answer. A UF of at least 100 would be called for in this case, because the values are LOAELs (one factor of 10) that were collected with human subjects (the second factor of 10 would represent human variation; we are setting aside for the moment the fact that short-term rather than long-term regimens were used, which might call for a third factor of 10). Because all LOAELs were 0.05% BAC, a mechanical application of the RfD approach would set 0.0005% BAC as the reference concentration. Such a concentration, however, is within the range of values that would be obtained had no alcohol been consumed (background variability). Of course the regulated dose (RgD) would be different from the RfD. The RgD is influenced by the kinds of considerations brought to bear in risk management rather than in risk assessment. In the case of alcohol, the level that is commonly used to manage risk is close to the demonstrated LOAEL for a number of behavioral endpoints. The RgD is not a fraction of the LOAELs. Our point is that mechanically applying the general RfD approach would apparently overprotect for alcohol when compared with the current RgD.

HUMAN VARIATION IN RESPONSE TO LEAD

The behavioral and neurotoxic effects for environmental lead contrast with those mentioned above for alcohol in several respects. Lead is no longer self-administered in most instances (i.e., no longer used as a sweetener) but it is encountered without a person's awareness. Lead has no effect that intrudes on our awareness; instead it takes its toll insidiously. Effects of a brief moderate-dose exposure are long term, if not irreversible (although the effects of exposure to alcohol also may be long term for a fetus). Like alcohol, however, lead has long been known to be dangerous, apparently has no threshold below which it has no effect (though this is somewhat controversial), and often

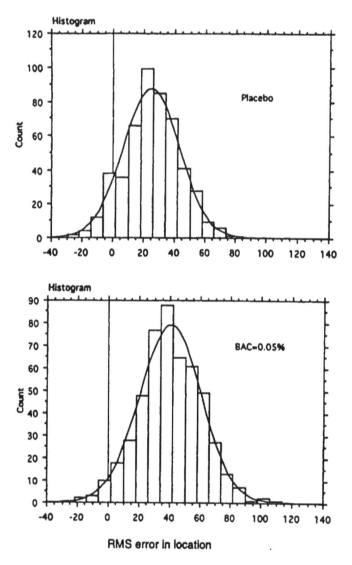

FIGURE 3A Normal curves derived from published data (produced by random generation of 500 values drawn from a normal distribution with mean and SD from the published data; Microsoft Excel 5.0). Data from Linnoila et al. (1980) are shown for root-mean-square (RMS) error in keeping the location of an unstable point of light steady in a critical tracking task requiring a high level of correction to achieve stability (their level 3) for 35- to 45-year-old subjects who were given placebo or alcohol (0.05% BAC).

has been overlooked by public health officials. Unlike alcohol, the past two decades have seen a "dramatic reduction" (Bellinger 1995) in the exposure of individuals in the United States to lead — a direct result of bold decisions that restricted the commercial use of lead during the 1970s and 1980s — decisions that were supported by a growing public awareness of the personal and societal dangers associated with lead exposure. Research in neurotoxicology, the vigorous efforts of researchers to bring their worries to the attention of the public, and an active risk assessment of the dangers of environmental lead were key elements in this remarkable public health initiative. In the course of this activity, an enormous effort was made to document the neurotoxic effects of lead, and this effort continues. Several recent reviews document the complex and subtle issues that surround interpretation

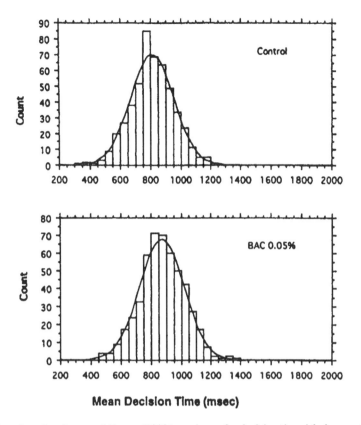

FIGURE 3B Data from Landauer and Howat (1982) are shown for decision time (choice reaction time before a central button was released) for subjects reacting quickly to eight digits, randomly assigned to different equidistant buttons. Data are in milliseconds.

of the data that have been collected [for a list, see Bellinger (1995)] and therefore no such review is necessary in this report. What is offered, however, is a brief presentation of what this literature shows regarding individual variation in neurotoxic response to lead and an evaluation of the implications offered by this variation for the use of the RfD/UF approach to risk assessment.

A single example should suffice. Bellinger (1989a) presented the full distribution of effects for four groups of infants evaluated with the Bayley scale of infant development at 12 months of age. We offer his display as Figure 4. The score used from the Bayley scale was the mental development index (the MDI residual), adjusted by regression analysis for various correlated factors that are known to affect the value obtained on the test. Zero is the expected score for an individual having a neutral position on these other controlling factors. Does lead affect this residual score? For infants that had umbilical cord lead concentrations of 14 µg/dL or less, the distributions of scores seem to be superimposed on this predicted value of zero (though this might be argued for the 10- to 14-µg/dL group). For infants with lead concentrations greater than 14 µg/dL, however, the mean of the distribution appears to have shifted to lower scores (mean shift, −7.2). The shape of the distribution, however, remained symmetrical (rather than skewed) and the overall variation around the mean remains comparable (SD approximately 12). Bellinger (1995) describes this distribution as approximately normal, and, true to this assertion when we created our own frequency distribution with values derived from Bellinger's figure (assuming all data values were at the midpoint of their MDI-residual bin), the distributions were well-described by normal fits (figure available from the authors upon request).

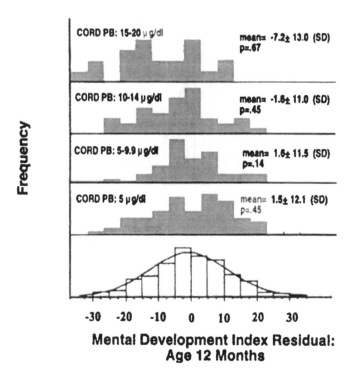

Mental Development Index Residual: Age 12 Months

FIGURE 4 Distributions of residual score at age 12 months for four groups of infants classified according to the concentration of lead in umbilical cord blood. A child's residual score was calculated by subtracting the fitted MDI score (Bayley scales of infant development) from the observed score. The fitted score was derived by using the equation describing the regression of MDI score against 12 potential confounders [maternal IQ, age, education, family social class, ethnicity, alcohol consumption during pregnancy, smoking history, home observation for measurement of the environment (HOME) score, gestational age, birth weight, birth order, and sex]. *p* values are probabilities that distributions of residual scores for exposure groups do not deviate from normality. (From Bellinger 1989.)

What are the implications of these distributions? If the distributions are normal, then the increases in incidence resulting from increased exposure would be well-predicted by the Gaylor–Slikker approach. Are they? From Bellinger's data and with this approach, we expected to find four individuals with a residual MDI below −18.5 (1.96 SDs below the mean) in the 0- to 14-µg/dL group, whereas we actually found three. We then expected to find 13 individuals with a residual MDI below −18.5 in the 15- to 20-µg/dL group; we found 10, which is very close (from a rough look at the distributions, we had expected to find low scores to be underpredicted by this approach and were surprised to find them slightly overpredicted). The closeness of these observations to their predicted values appears to confirm the appropriateness of considering the distributions to be normal.

As noted by Cory-Slechta and Pounds (1995), adults as well as children exposed to environmental lead show changes in learning and memory processes. We can then ask whether children or adults are more sensitive to the effects. Ultimately, we emphasize that, to predict its impact, it is necessary to know precisely at what stage of development exposure took place [e.g., Shaheen (1984); see Schull et al. (1990) for a similar concern regarding the effects of ionizing radiation]. The general point, however, is also made by comparing children with adults. Stollery et al. (1989) measured impairments in occupationally exposed workers and found that exposures associated with blood lead concentrations of 40 µg/dL and above resulted in "a general slowing of sensorimotor reaction time and mild impairment of attention, verbal memory, and linguistic processing. The sensorimotor slowing and incidental memory deficit persisted throughout the 8-month period of repeated evaluations following termination of exposure" [described by Cory-Slechta and Pounds (1995)].

Wyzga (1990) presented an analysis of extensor muscle weakness for workers exposed to lead. The incidence of reported weakness was 1/26 for blood lead levels of 0 to 59 and 60 to 79 μg/dL, but it increased to 6/26 for workers at about 80 μg/dL. For this survey measure then, the LOAEL was at a higher level than found by Stollery et al. (1989). It should be noted, however, that both these LOAELs for working-age adults are quite a bit higher than that for the infants studied by Bellinger. Adults, then, are relatively insensitive to the effects of lead, which means that adult data underestimate the danger of lead for children.

The question then becomes exposure at what age? Prenatal? Postnatal? Waternaux et al. (1989) presented an analysis for a series of observations for individual children to determine which exposures were most predictive of subsequent impairment. Their conclusion is that "exposure during the first six months of age, as well as prenatal exposure, are more likely to have adverse effects than later exposure. The developing brain appears to be more vulnerable prenatally than postnatally to the adverse effects of low-level lead exposure. Infants with umbilical cord lead concentrations of 10 to 25 μg/dL (levels considered acceptable until recently) displayed persistent IQ [intelligence quotient] deficits of .25 to .5 standard deviation relative to children with lead concentrations below 10 μg/dL. Children with high blood lead at 6 months experience a *decline* [emphasis added] in [mental] growth when compared to children with lower 6-month lead levels."

Confirming analyses showed these findings were not "driven by influential observations or outliers." For example, although there were two children who fell outside the range of values expected "under multivariate normality," both these children were more sensitive rather than less sensitive to the effects of lead than was predicted by the average. It is chilling to hear that "as many as 20% of newborns in urban areas may have lead levels comparable to those of the children in the high-lead group."

Again we might ask what would have occurred had a mechanically applied RfD approach with UF = 10 (for a NOAEL) or UF = 100 (for a LOAEL) been taken to set an RfD for lead effects with children. The RfD would be set at 0.1 μg/dL — a very small number. As for alcohol, this level of exposure would be undetected, because it is well within the range where measurement would be unreliable. Again, therefore, the RfD approach appears to be conservative.

HINTS OF HUMAN VARIATION FOR EFFECTS OF OTHER AGENTS: MERCURY, CISPLATIN, PARATHION, DIISOPROPYL FLUOROPHOSPHATE (DFP), PHYSOSTIGMINE, AND ACRYLAMIDE

We have found few additional examples where human variation in response has been addressed, although a more thorough search than we were able to complete might well identify more examples. In each of the cases we identified, conclusions are available only from a small set of studies and therefore are less secure than those presented above (which, the reader will recall, were not presented with much confidence). Yet, the message is, at least in three cases, less supportive of present practice than emerged from the alcohol and environmental lead review.

Mercury

Hattis and Silver (1994) provide an analysis of a key study (Marsh et al. 1987) summarizing neurotoxic effects for a group of children exposed as fetuses and for their mothers. These were children born of mothers who were poisoned by mercury after consumption of treated seed grain. Eighty-one mothers who had consumed poisoned bread during pregnancy were identified and samples of 50 to 100 strands of head hair were obtained. Single strands were analyzed to determine the highest dose and the period of pregnancy associated with the highest dose. Dose–response functions were created across deciles of exposure (range, 1 to 674 parts per million) for summary scores of problems identified in neurologic examinations. The incidence of several clinical signs of retarded development was shown to increase linearly across estimated fetal dose (in Hattis's analysis, infants were classified into low-, medium-, and high-dose groups). Regression and maximum-likelihood analyses of these data both implied a very large individual variation in effect — with thresholds for impairment spanning as much as a 10,000-fold change in dose. Such a span implies appreciable risks for sensitive individuals even when very low doses are attained. A UF = 10 approach would not identify a dose that was safe. This factor of 10 would not protect against the apparent 5-orders-of-magnitude variation in sensitivity shown in this study.

Hattis and Silver (1994) evaluated the extent to which measurement error contributed to this wide variation in sensitivity. He concluded that if a single factor were to account for this unexpectedly high degree of variation, the dose would need to be misestimated by a geometric SD of 3 — an unreasonable expectation. Hattis did suggest, however, that some increase in variance was due to grouping the children into three dose groups.

Cisplatin

Using a similar analysis, Hattis (1995) reanalyzed the effect of a course of doses with the anticancer drug cisplatin on confirmed hearing loss (hearing loss, yes/no). Two studies were identified. The study with more complete dose–response information produced a probit slope of 1.8, an outcome that again implies considerable individual variability (see Hattis, this volume). In this case 95% of the population apparently fell within 10-fold below to 10-fold above the median range — i.e., a 100-fold variation in sensitivity. Again, the 10-fold UF for variation in sensitivity is not large enough to protect the most sensitive.

Parathion, DFP, Physostigmine, and Acrylamide

The adequacy of current RfD methods for characterizing the risk of impairment for these four agents was addressed in a review article (Raffaele and Rees 1990) and therefore is presented here as a single section. The primary focus was on animal-to-human extrapolation, but a few evaluations of human dose-effect or dose–response data were included and can be addressed. Tests were classified into five endpoint categories: neurochemistry-neuropathology, physiology-consummatory, sensory-motor, electrophysiology, and learning-memory (i.e., cognitive). One overall observation was that the human NOAEL/LOAELs were consistently either the lowest or among the lowest for an agent and an endpoint. Yet, with the UFs applied, the RfDs for these substances were "comparable regardless of whether human or animal exposure data were used." Therefore, the use of animal testing was supported. Further, where available, the RfDs that emerged from this analysis were compared with previously identified dangerous levels. For parathion, the RfD was similar to the time-weighted-average standard recommended by the Occupational Safety and Health Administration. This consistency appeared to support the current RfD approach. For acrylamide, the RfD based on animal data was in the same range as the RfD established by the U.S. Environmental Protection Agency (based on neuropathology in rats), which seemed to support the use of animal data — yet the slight amount of human data available from which an RfD might be established suggested that a higher dose would be safe.

Second, although all endpoint categories were sensitive to the effects of these chemicals, there was considerable variability of NOAEL/LOAELs across different endpoints for a particular chemical. No single category stood out as consistently most sensitive. Therefore, the testing of multiple endpoints was supported.

Third, there was considerable variability in sensitivity of different tests classified in the same endpoint, with the LOAELs in some tests being lower than the NOAELs in other tests for the same species. Therefore, the use of multiple tests was encouraged.

Fourth, the fact that neurotoxicity can occur after acute doses and the possibility of tolerance to an agent building across repeated exposure suggests the need for "development of an RfD specifically for [acute] neurotoxic effects."

Fifth, and perhaps most relevant to this report, a large difference was found in the NOAEL/LOAELs of physostigmine between young and old humans and between young and old monkeys. This large effect of age suggested that the UF of 10 for interindividual differences in sensitivity was insufficient. Perhaps the section below on sources of variation will provide a more useful approach. Raffaele and Rees (1990) did conclude, however, that for the most part their exercise provided support for the current use of UFs. Their review therefore might be taken as defending the current RfD approach.

SOURCES OF VARIATION

In this section, we attempt to logically decompose the overall variation in effect from one person to another into subcomponents of variation that can be separately attributed to one source or another. What is to be gained by this exercise? What do we hope to accomplish? First, we consider this a worthwhile exercise merely to list for the reader the great variety of sources of variation that are involved in determining the outcome for an individual and thereby raise the reader's respect for the importance of variation across individuals in our understanding of the impact of toxic agents. This is a theme long championed by Weiss (1988, 1990) and Dews (1986). Recently, Hattis and Silver (1994) and Hattis (1995) and others [e.g., Gaylor and Slikker (1990), Glowa (1991)] have also emphasized the importance of this topic. We also hope that, by raising this issue, the current exercise encourages development of a model that allows the different sources of variation to be entered into our full understanding of the factors that determine toxic impairment of performance. Most urgently, however, we hope that this listing of factors will help investigators identify what should be balanced or measured when they develop a study design to reduce, model, and more accurately attribute variation in their data. Currently, little is known about the relative contributions of these sources to the overall variation in effect. We hope the framework offered below provides a structure for understanding these sources.

A VARIANCE COMPONENT MODEL FOR DATA

The following model (Equation 1) provides a useful vehicle for addressing the sources of interindividual variation in behavioral measurement.

$$Y_{ij} = E_{i.} + O_{.j} + \text{error} \tag{1}$$

In this model of the variance in scores, an individual score is Y_{ij} [where i is the specific exposure level to agent E, and j is the specific state of other variables (O) that influence the score; dots may be read as averaged over for that factor]. Thus $E_{i.}$ is the influence for a specific dose of the agent averaged over all states of the other influential variables. It might help to provide a specific example. Take Y_{ij} to be the mean value of digit span averaged over the group of individuals that are influenced by a particular level of exposure to lead (level i) and influenced comparably by their specific level of other factors (for example, the combined effects of gender, age, educational level, socioeconomic status (SES), reaction to the examiner and instructions, background in similar tasks, etc.). In Equation 1, all other influences (e.g., individual differences among persons in the group, interactions between the agent and these other influential factors, inconsistencies across successive measures for an individual) are treated as error.

$$Y_{ij.} = E_{i..} + O_{.j.} + EO_{ij.} + I_{..k} + \text{error} \tag{2}$$

The model shown in Equation 2 separates out two of the additional sources of variation that were treated as error in Equation 1. It is therefore a more complete model of variation. A specific interaction term (*EOij.*) is included to represent variance attributable to either the agent or other influential factors, but it is not possible to tell which because there is a confounding of the factors (i.e., the level of the other factors covaries with dose of the agent — for example, perhaps dose of lead varies with education level, age, etc.). This term provides a way to represent variance that cannot be uniquely attributed to one factor or the other.

In addition, a factor has been added to Equation 2 to represent the variation across individuals (*I..k* stands for individual *k* — the other terms have a dot to show that they have been averaged). In many fields of endeavor (e.g., agriculture), variation from one individual element (e.g., seed) to another is treated as mere error and it is assumed that the average effect represents the true condition. For humans, however, individuals may well differ in ways that cannot be viewed as merely part of error. Human differences are often qualitative and complex and are not well-described by a model that is quantitative and simple. Equation 2 at least allows one to expect that different influences of *E* and *O* and of their interaction *EO* might be found for different individuals.

$$Yij. = Ei.. + O.j. + EOij. + (Ii.k + I.jk + Iijk) + \text{error} \tag{3}$$

Equation 3 further divides the term for individual differences into three separate subcomponents to separately represent interindividual variation that separately influences the three prior terms in the equation — that is, there are separate terms for individual differences in the effect of the agent (*Ii.k*), in the effect of the other influences on the measure (*I.jk*), and in their interaction (*Iijk*). These terms, then, separately represent differences in individuals' reactions to the agent itself, to the other relevant variables, and to the interactive effects of the agent and these variables. It seems appropriate to consider that for the kind of variation being modeled by this formulation, individuals would differ in at least these three separate ways.

Equation 3 could be further expanded — but this level provides separate terms for the factors that seem especially important for the current topic. The sections below therefore address the separate factors noted in Equation 3. There is one small additional expansion in the last part of this section, however. For situations where several measures are taken from a single individual, it is useful to further divide the error term in Equation 3 and separately represent the sources that influence each successive measurement made for that individual. This topic of intraindividual variation is briefly addressed.

VARIATION IN AVERAGE EFFECT OF A PARTICULAR DOSE OF AN AGENT ON A SPECIFIC BEHAVIOR

Because this kind of variation was addressed in the examples above, we give it a cursory treatment in this section. Primarily we indicate the role of the agent's effect in the model presented in Equation 3 and provide a path to the more complex issues addressed in the next section — behavioral variation due to other factors. Equation 3 highlights the joint influence of the agent and of other factors. When both the agent and these other factors are varied, we may obtain one of the patterns of effect indicated in Figure 5. This figure presents idealized data showing the proportion of subjects in each dose group that were affected by the agent (i.e., the incidence of behavior *y*) — a dose–response function. The curve has the sigmoid shape commonly found when dose is not given as the logarithm. Two separate functions are drawn depending on the level of another factor (*f* or *g*) that jointly determines the incidence of the behavior. Factor *f* presents a relatively simple case, because its influence on the incidence of behavior *y* is independent of (orthogonal to) that of the agent — i.e., the dose–response function across dose *x* of the agent with and without factor *f* are parallel, with only the location being shifted. Factor *f* does not change the range of individual

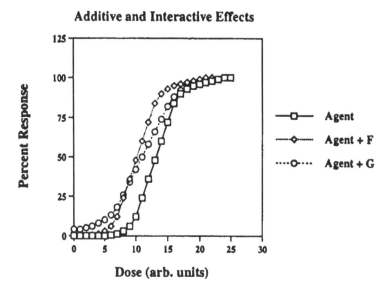

Additive and Interactive Effects

FIGURE 5   Idealized cumulative dose–response curves for percentage of a population (y) affected at different doses (x) of an agent. Curves are shown with and without factors f and g.

variation in sensitivity to the agent. Its influence is said to be additive. A more complex kind of joint influence is indicated by factor g in the figure. Factor g changes the slope of the dose–response function over x as well as its location (i.e., its effects interact with those of the agent — they are nonorthogonal) and its influence is said to be superadditive, perhaps even multiplicative. Such an interactive influence alters interindividual variation and therefore one needs to know the value of factor g to predict variation in the effect of the agent. As additional interactive factors are included (and there are several identifiable interactive factors for many neurotoxic agents), the difficulty in fully characterizing the situation regarding interindividual variation in response to the agent increases dramatically.

The situation becomes even more complex when the interaction is not linear — i.e., it is different at one point of the dose–response function than it is at another part. Such seems to be the case, for example, with the interaction between the effects of environmental lead and social class on the behavior of young children. Figure 6 [taken from Bellinger et al. (1989b)] shows an interactive effect of lead and social class (SES) in determining MDI score for infants across four successive testings (6 months apart). Scores are shown separately for three dose groups for children above and below the median SES. The pattern of effects shows a very slow (if any) improvement in MDI for children from below the median SES but a general increase in MDI score across time for children above the median SES, except for those high SES children in the high lead group. That is, whereas high social class overcame the downward influence of low-to-moderate doses of environmental lead, it did not overcome this influence for the high dose. Note that without this separation into high and low SES, the pattern of data would be very difficult to interpret and would appear to be primarily a difference in variability rather than in average MDI. These kinds of challenges are all too common in neurotoxicology.

BEHAVIORAL VARIATION: INDIVIDUAL DIFFERENCES IN A BEHAVIOR AND HOW THAT VARIATION IS INFLUENCED BY SUBJECT CHARACTERISTICS

It might be appropriate to provide some background in psychological measurement before we dive into the implications of behavioral variation. What is involved in measuring behavior?

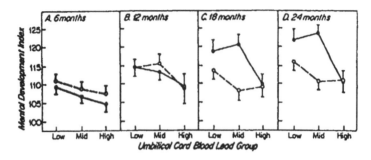

FIGURE 6 Adjusted MDI scores (Bayley scales of infant development) at ages 6, 12, 18, and 24 months for children stratified by family social class (open circles, below median; closed circles, above median) and by umbilical cord blood lead concentration (low, <3 µg/dL; mid, 6 to 7 µg/dL; high, >10 µg/dL). Error bars represent 1 standard error. (From Bellinger et al. 1989.)

Behavior Is Typically Measured As Actions Rather Than Movements

If you had a video camera available and you wanted to record behavior, you might ask individuals to behave for you so you could measure it. The video record you obtained, however, would be only the face of behavior and not in itself a measure of behavior. It would show only movements. We sometimes show an interest in movements (shaking, knee jerk, sway), but for the most part we are interested in actions. Actions accomplish something and are named by what they accomplish — they move the individual or they move an object in space, they influence someone, etc. Actions are described relative to an objective rather than to a movement. Your video recording would provide a starting point, but to offer this as a measure of behavior requires that you develop a set of categories into which you place the actions you see. You record the behavior by indicating the category and time for each action. Or, you merely count the number of occurrences for each type of action within an identified time period. This count can be expressed as a number or as a percentage of some total (e.g., percentage correct), or it can even be converted to describe a rate or speed (number completed in a given time). All these measures are approximately continuous (although simple count is an integer) and are scaled from a zero value (where zero indicates an absence of the action). The values obtained across a series of observation periods for count or rate or percentage correct for some task often display a variation that approximates a normal distribution.

Categories As Quantities

We often order these categories of action and treat the resulting continuum as though it measured quantity. It is startling how often we seem to get away with this maneuver. For example, we may order our behavioral categories along some continuum and then average the values to simplify our task of summarizing the observations we have made (e.g., average reaction time, degree of accuracy, subject's rating). These appear to be measures of quantity until a closer look indicates that they are merely combinations of behavioral categories imposed by the observer.

Place in Distribution of Scores As Behavior

Some of our most important and most commonly used behavioral measures are not really measures of behavior at all but rather are statements about where a particular behavioral outcome falls in the distribution of outcomes. For example, a modern intelligence quotient (IQ) is really a statement about how common that score is among subjects in the norming group. An IQ of 100 is assigned to the average score, an IQ of 115 is assigned to the score that was exceeded by the upper 34% of the norming group, etc. This represents a forming of the distribution of scores into a normal distribution. Instead of being a behavioral description then, an IQ is a statement regarding the location of a score in a distribution of values — in a sense a social rather than a behavioral measure.

Why Spell Out These Complexities at Such Great Length?

One reason we draw this point out and present examples below is to support the following assertion, which we would like the reader to consider. It is inappropriate to simply step around and ignore the complexities presented by behavioral measurement when commenting on the variation in the behavioral effect of an agent. Behavioral measures differ greatly and details about a behavioral measure and about measure-to-measure differences must be taken into account when interpreting what can be learned from a measure and its variation. Behavioral measures are more complex than measures in many other disciplines. To summarize this first point, we urge caution. To support our caution, we describe the variation in behavior obtained with two pairs of behavioral tests that are commonly used in neurotoxicity evaluations of adults.

There is a second reason for dwelling on this topic. We hope to provide a background for addressing the difficult issue of shared influence over behavioral outcome — that is, behavioral measures are influenced by many factors, several of which may well be confounded with degree of exposure. Interpreting the relative influence of these intertwined factors creates a minefield for most epidemiologic studies.

The Auditory Digit Span and the Visual Digit Span

Many assessments of cognitive function include an evaluation of an individual's memory span. The classic way to determine this memory capacity (Wechsler adult intelligence scale, Stanford-Binet, Wechsler memory scale) is to read a series of digits, one digit per second, to the individual and ask for immediate recall (spoken). Digits are presented in what appears to be a random order and typically are not repeated in the series. A short list (e.g., three digits) is given in the first two series. If either series is repeated correctly, the following series is then increased in length by one. This pattern continues, with the length increased by one on odd-numbered series until an error is made on two series of the same length. At that point, the test is ended and the memory span is recorded. The number recorded is typically the longest list correctly repeated, but there are variations. For example, in the frequently used neurobehavioral core test battery recommended by the World Health Organization (NCTB-WHO) (Anger et al. 1993), the measure used is the total number of series correctly repeated (a maximum of 2 times the longest list correctly repeated).

In a recent study involving 933 adults (males and females) having a variety of educational and cultural/ethnic backgrounds, Anger et al. (1996, 1997) recorded the distribution of scores that is presented in Figure 7 (top). Note that the distribution is well-described as normal and shows a mean of 9.0 ± 2.3 (SD). The same subjects also completed a different memory span test on a second testing day — the second test was very similar in format to the first, except the digits were printed in large type on a computer screen (presented one at a time). Instead of speaking, the subject pressed the appropriate digit keys on a computer keyboard (all keys except the digit keys were covered). The subject's score was recorded as the longest series of digits successfully completed. Although the distribution of scores was well-described as normal in form [see Figure 7 (bottom)], the values recorded for this visual digit span differed from those for the auditory digit span (remember the scoring difference). The mean was 6.7 ± 1.4 (SD). Thus, visual digit span (tested in this way) was different on the average from the auditory digit span (tested in its own way). If one were to merely cite the individual's digit span as being 8 then, it would be unclear which test was referred to (and hence whether the span was long or short compared with the average score).

Though scaled differently, the two memory span tests might be quite comparable. Were they? Because subjects in this evaluation took both tests (half took the auditory first, half took the visual first), we have an opportunity to determine whether their scores on the two tests placed them in the same relationship to others in the group for the two tests or in different places for the two tests. A correlation between the test scores showed that the similarity of the two tests was less than might be expected from their similarity in title ($r = 0.57$); that is, individual consistency accounted for only approximately a quarter of the variation in scores. Knowing a subject's auditory digit span does not help much in predicting the visual digit span.

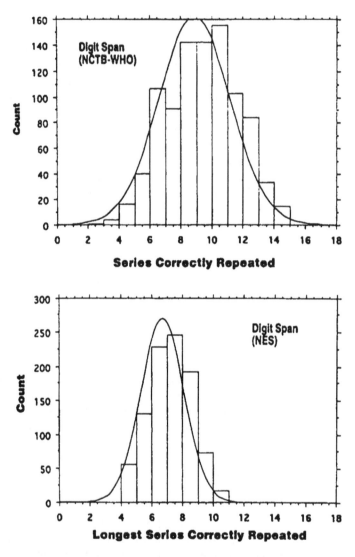

FIGURE 7 Distribution of scores for more than 900 adult participants (male and female) who completed a number of neurobehavioral tests as part of an evaluation of the role of subject characteristics in determining test scores (Anger et al. 1995a). (Top) Performance on auditory digit span test from the NCTB-WHO test battery. Measure was number of digit strings correctly repeated. (Bottom) Performance on visual digit span test from the neurobehavioral evaluation system. Measure was length of longest digit string correctly repeated.

This example has been included for several reasons: to provide an example for a kind of behavioral measurement that is common in psychological testing, to emphasize that important details of such testing are often ignored when we summarize what has been measured, and to note that two tests that appear to be quite similar may, in fact, provide measures of behavior that are noticeably different.

Digit-Symbol (D-S) and Symbol-Digit (S-D) Tests

Of all the commonly used measures for field assessments of adult neurotoxicity, the D-S test (originally from the WAIS) has perhaps the best track record for showing statistically significant toxic effects. In this test, the subject is asked to use a legend given at the top of the page that shows pairs of symbols and digits (i.e., 2 belongs with ^) to match digits with their symbols. Typically,

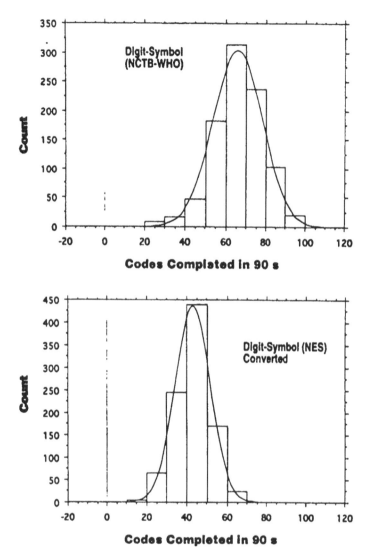

FIGURE 8 Distribution of scores for more than 900 adult participants (male and female) who completed a number of neurobehavioral tests as part of an evaluation of the role of subject characteristics in determining test scores (Anger et al. 1995a). (Top) Performance on D-S test from the NCTB-WHO test battery. Measure was number of codes correctly copied in 90 seconds. (Bottom) Performance on S-D test from the neurobehavioral evaluation system. Measure was latency to complete five trials transformed to show number of codes completed in 90 seconds.

the subject is presented with four rows of 25 digits and asked to print as quickly as possible the symbol that belongs with it below each digit. The score is the number of symbols correctly copied within 90 seconds. The symbols are one- or two-stroke line drawings a bit like — but different from — letters of our alphabet. The test therefore requires that the individual quickly write these (more or less) novel symbols, while reading back and forth between the legend and the current location in the answer set, until associations between the digits and the novel symbol are learned and therefore eye movements are limited to the answer rows.

Figure 8 (top) indicates that these counts (for the same subjects as in Figure 7) also fall into a normal distribution. The mean was 66.1 ± 12.2 (SD). Again, we now offer a comparison between this classic psychological test and a computer-based version of it — in this case, the computer-based test, however, is a S-D test. In the computerized test, the subject is asked to press the digit key (same keyboard as described above) for the digit associated with each of nine symbols (different in shape from those used in the D-S test). In this case, it should be clear, the subject does not need to demonstrate fine motor skills in writing new symbols — simply press the key. Further, although the subject does need to quickly read back and forth between the legend and the answer row (here it has nine symbols), there are only two rows shown on the screen — the legend and one answer row located directly below the legend. The test is composed of a series of five of these two-row displays, with the symbols in the coding row presented in the same order each time, but the symbols in the response row are presented in a different order in each display.

In Figure 8 (bottom), we present the distribution of scores obtained from this S-D test. The total latency across the five trials has been transformed to provide a measure that is directly comparable with that for the D-S test, showing the number of items that would have been success-fully completed within a 90-second period at the measured rate. In the computer version, fewer items were completed in 90 seconds (43 ± 8.6 SD), but again the distribution appears normal. Although the S-D test is clearly different from the D-S test and requires different subskills, it is striking that an individual's scores on these two different tests in fact correlate about as well ($r = 0.62$) as did their two scores on the digit span tests (seemingly more similar).

To complete our exploration of test similarity, however, we need to know how the two pairs of unrelated tests correlate: the tests from the NCTB-WHO correlate with $r = 0.22$ (the auditory digit span and S-D tests). The tests from the Neurobehavioral Evaluation System correlate with $r = 0.24$ (the visual digit span and D-S test). The implications of these numbers are considered in the next paragraph.

The D-S test requires considerable motor and sensory-motor coordination (learning to write the new symbols). The S-D test, however, requires only pushing the correct key. From the higher-than-expected correlation between the two tests, however, it becomes clear that the two coding tests are more similar than we thought. By the same criterion (correlation of around $r = 0.60$), the two digit span tests were found to be less similar than we at first thought. The lesson to be learned is one of caution — the names given to psychological tests do not provide a clear designation of the psychological functions they assess.

Have we met our goal in this section? The message we hoped to develop is that behavioral measures differ in many ways both among tests and among individuals. We would like the reader to believe that there are many sources for this variation — factors that determine an individual's performance on a particular test. This bountiful pool of interindividual variation is what investigators explore to understand an individual's current abilities and how these abilities may have been influenced by their exposure to toxic agents. It is our hope that we have attuned and increased the readers' respect for the challenge of such a mission rather than made them fearful.

Implications of This Analysis of Variation

We now address another thorny issue: sorting through the joint influence of intertwined factors that determine an individual's performance. A good opportunity to address this issue is provided by a project, similar to the one reported above. In this second project, data were collected from more than 700 individuals in three separate locations in the United States (Anger et al. 1997). A battery of paper-and-pencil as well as computerized tests were administered to these individuals and their scores were evaluated to determine what subject factors might influence performance. In this example, we evaluate only two tests — two of the four that were addressed above — the auditory digit span and the S-D test. The average performance for the participants in this study was in general quite similar to that in the other study. The average auditory digit span, for example, was 7.9 ± 2.33 (compared with 9.0 ± 2.3 SD above). The measure taken for the S-D test, however, cannot be directly compared because the score used was the best latency rather than the total latency. The average time

TABLE 1
Sources of Variation for Two Psychological Tests

	Unique to education	Unique to cultural/ethnic group	Shared
Digit span	10%	3%	16%
Symbol digit	9%	7%	25%

to complete the nine codes in this best-trial measure was 2.5 ± 0.84 seconds (this different measure produced a much higher rate of completion than either coding test noted above).

Participants in this study were specifically selected to be comparable to many of the nonexposed control subjects used in field assessments of neurotoxic exposures. They differed along several dimensions, including region of the United States where they were living, age, years of education, and cultural/ethnic background (African-American, Native American, Latin, or European descent). Two of these subject factors showed a particularly strong influence over the digit span and symbol digit performances — namely, level of education and culture/ethnic group. In fact, when considered separately in a regression analysis, years of education appeared to account for 26% of the variation in digit-span score and for 34% of the variation in S-D performance; culture/ethnic group appeared to account for 19% of the variation in digit span and for 32% of the variation in S-D performance.

The complexity is, of course, that level of education was unequal across culture/ethnic group for participants in this study and therefore was an intercorrelated factor complicating interpretation of the influence of culture/ethnicity. How does an investigator sort out the relative influence of the two factors when they are intertwined? The harsh answer is that one does not sort out their influence without collecting new data in another study (or several studies). One approach that can be taken to address this issue in a single study, however — and an approach we recommend — is to simply frame the problem for the reader by showing what variance is unique to each factor (accounted for by only that factor) and what variance is shared and cannot be sorted out (accounted for by both factors). These unique and shared variances can actually be measured by successive subtractions of regression factors. In the current example, for instance, the unique and shared variances are noted in Table 1.

It should be clear from the pattern of data in Table 1 that much of what might naively be attributed to the influence of culture/ethnic group is actually shared variance with the intercorrelated factor of education level. Education is not, of course, the only factor that might influence a test score that is intercorrelated with culture/ethnic group membership. Were we to seek to unconfound the influence of these other factors by merely successively removing their variation, we would in each case be removing both the unique and the shared variance for each factor. What we had left at the end of this series of subtractions would be difficult to specify and would seemingly have a distinct possibility of showing less of an effect for the agent than should be attributed to it. That would be called overcorrection, and we know of no rules to identify when the correction has been accurate and when it has not. Therefore, we advocate that shared variance not be subtracted but merely identified as shared.

DIFFERENCES FOR AN INDIVIDUAL FROM ONE OCCASION TO THE NEXT

To complete our treatment of variation in response requires one additional step — the uncertainty in placing an individual along a continuum of response to an agent. The degree to which we can accurately place an individual is limited, because an individual may produce a different behavioral score from one occasion to the next and an individual may react differently to an agent from one occasion to the next. The amount of unreliability introduced by this factor depends, of course, on many specifics — what test, what agent, what exposure regimen, what additional experiences/exposures the subject has had, etc. It should be stressed that the primary reason for including this factor in the current review is that any factor producing variability across repeated measures makes a single measurement less stable. Selection of tests, training approaches, and dosing regimens that produce data that are stable over repeated measures, therefore, can considerably improve the statistical power of an assessment. Given tests of equal sensitivity and focus, the one that shows less variability would be strongly preferred. Selection of stable but sensitive tests was an important goal for a long-term project carried out by the U.S. Navy [e.g., see Bittner et al. (1986)]. We mention this source of variation to complete our statement of the sources of differences among individuals [also see Hattis and Silver (1994) for a treatment of the benefits of reducing or modeling measurement error].

The testing of children raises a special concern regarding test-retest reliability. The use of different tests at different points in a child's development deserves special mention as a source of intraindividual variation. We often casually suggest that what is measured at different points in development is a child's intelligence. Development, however, is not a continuous change but a complex series of discontinuous leaps that require us to use wholly different sorts of tests at different ages. Intelligence for a 5-year-old, therefore, is tested by very different skills than for younger children. This theme continues for testing older children and adults. For example, for older children and adults, the IQ measure is based on tests that require a considerably expanding vocabulary. Therefore, experience that expands or limits vocabulary is critical for determining IQ in older individuals. There is, therefore, no intelligence thing that is tested across development. True to this description, tests with young children have remarkably little success in predicting subsequent intelligence. What appears as intraindividual variance across tests at different ages, then, is often better considered to be intertest variance.

EXTRAPOLATING FROM ANIMAL DATA TO ESTIMATE HUMAN RISK

Although we do not address this aspect of risk assessment here, it should be pointed out that extrapolating from one species to another when predicting risk would be placed logically at this point in our progression. Further, separate treatments might be offered for interspecies variation in uptake, pharmacokinetics, and response. It is clear there often will be many differences in uptake and pharmacokinetics. There will, of course, be differences in response to an agent as well (see above for a description of these three classes of variation). We expect, however, that there will be many cases where we find relatively little interspecies variation in response once we learn to better evaluate animal data. Perhaps these similarities result from the generally comparable evolutionary challenges that currently existing species have all met — e.g., rapid response to threat, rapid response to opportunity (i.e., a functional similarity) — rather than from strictly similar biological mechanisms (i.e., homology). To the extent that other extant species and humans are selected to operate in a common environment, we might expect comparable functionality even when mechanisms are not comparable.

HOW CAN WE IMPROVE OUR REPRESENTATION OF VARIATION IN RISK ASSESSMENT? HOW CAN WE DO THIS WITHOUT KNOWING EVERYTHING?

THE IMPORTANCE OF BIOLOGICAL MECHANISMS

The clearer we are about the biological mechanisms of action for a toxic agent, the clearer we can be about how individual differences might be involved in the effects of that agent. For example, we now understand much more about differences in alcohol sensitivity given that the role of aldehyde dehyrogenase in this variation in sensitivity is better understood. We encourage the full use of knowledge about biological mechanisms and the study of various biological markers and do not want to discourage the use of any kind of information in understanding individual variation. We agree with Silbergeld (1990) and Bellinger (1995) that knowledge about biological mechanisms plays a special role in the science of toxicology. Yet we offer two statements of caution to emphasize the importance of more molar (functional) information as well. In fact we make a case that our degree of molar information sets the problem that the molecular (biological) knowledge helps to solve. When we do not have the knowledge about interindividual variation in functional effect, then we do not know when our knowledge of biological mechanism is sufficient or to what question it might be addressed. The harsh fact that was uncovered in the review undertaken to prepare for this report is that we have little knowledge of functional variation. And where there is such knowledge (as in the developmental toxicity of lead), the information indicates that the problems are complex. For most questions in toxicology, we are not yet prepared to profit from substantial knowledge about a biological mechanism. We do not know what problems must be addressed.

We make a related point to fully express our caution. Suppose we had collected a great deal of information on interindividual variation at each currently identified stage of the biological mechanism for an agent. How would we then combine these variations to make a prediction about their functional impact? Would we add them? Multiply them? Would there be compensations at one stage (e.g., receptor upregulation) that should be taken into account in our combination rule? How would we know we had found the correct rule? We can really find these molecular measures to be useful only after we have thoroughly studied variation at the molar level. Until we have this functional knowledge, we can drastically underestimate or overestimate interindividual variation just as easily as we can accurately predict it — that is, a little knowledge (of a biological mechanism) can be dangerous.

Another related point concerns qualitative rather than quantitative differences among individuals. With the advent of positron emission tomography and magnetic resonance imaging, we know much more about how the brain "lights up" when an individual takes on a particular behavioral task. We also know that different parts of the brain light up for different individuals undertaking the same task — i.e., biological strategies differ qualitatively [e.g., males apparently exhibit very different kinds of lateralizations than females when working on a spelling task (Wood et al. 1991)]. The simpler expectation that our differences are merely quantitative cannot be defended across genders much less across individuals. Although undoubtedly using the same tools and raw materials and having a similar blueprint to work from, our highly plastic brains have carpentered themselves into different kinds of houses. Therefore, we should be prepared to find that different individuals have quite different biological mechanisms of action for toxic effects and their expression.

Thus we do not advocate that full attention be focused on the discovery of biological mechanisms; instead, we advocate that considerable attention be placed on functional evaluations of interindividual differences in toxicity. For this reason, we have organized this paper around functional themes rather than mechanisms.

Careful Use of Expert Opinion

When confronted with difficult decisions in the face of uncertainty, we often have an urge to ask friends help us decide what to do. The opinion of experts is especially sought. We turn to physicians — seemingly more often then their track record suggests we should; we turn to economists — certainly more often than their track record suggests we should. Until relevant information on human variation is available and until methods have been developed for appropriately including this information in the risk assessment process, we might try the same approach. We can expect expert opinion on what is dangerous to take into account an understanding of human variation as well as other factors.

If expert opinion were expertly used, there might be additional benefits. Experts might remind us of information we have not considered or analogies we have not drawn. The contribution from

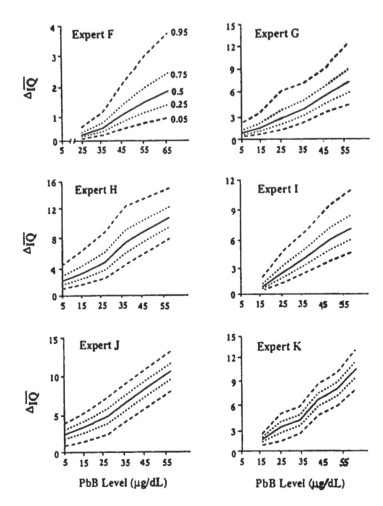

FIGURE 9 Judgments about lead-induced IQ decrements; low SES population; experts F, G, H, I, J, and K. Scales were selected to show details of each expert's judgments. Experts were asked to state their degrees of certainty that specific IQ decrements would result when the exposed group was given specific lead blood (PbB) concentrations in a hypothetical experimental exposure of children with identified attributes. The issue was caste in separate ways (judging doses, judging IQ decrements), and a succession of interviews were arranged in which displays of prior judgments were given and opportunities were provided to revise judgments until the experts were satisfied that their opinions were well expressed. Separate curves are shown for each of a number of probabilities where each value shows the expert's judgment was that the probability of decrement y at dose x would be z.

a number of experts might produce a whole greater than its parts. In addition, we might combine expert opinion to distill out an opinion that would be closer to the truth than that for any single expert. Wallsten and co-workers have been among the leaders in developing an approach to combine and distill the judgments of experts and have produced a track record that can be judged for its merits. One of these projects, for example (Wallsten and Whitfield 1986), combined the opinions of six experts on the effect of various doses of lead on IQ for low SES and high SES children. In their protocol, experts were trained to use their procedures in making estimations of probabilities and were given a hypothetical experiment involving exposure of specified children in specified situations. Every attempt was made to make the situation concrete by indicating specific tests, procedures, delays, etc. The experts were then asked to state their certainty regarding various possible outcomes involving IQ scores. These judgments were combined in a consistent manner and each expert was asked to review the implications of their judgments. There were several opportunities for back and forth discussion until all experts agreed that their opinion was well-captured by the data. These opinions were then compared and combined into an array of opinions regarding the likely impact on IQ of various exposures to lead (see Figure 9). The resulting opinions varied from expert to expert, but the modal opinion was that a blood lead concentration of between 5 and 15 µg/dL would produce a 2- to 5-point drop in IQ. A case could be made that the extensive information produced in the decade since these opinions were offered (between March 1985 and December 1986) has not shifted that modal opinion (nor, perhaps, reduced its range). If expert opinion is so ignorant, why does it remain so stable with increasing information? Perhaps others will offer theories on this point. It appears to us that the use of expert opinion should be considered as one approach to improve risk assessment. An expert's estimate of the importance of human variation will be included.

SUMMARY

Our review confirmed that the issue of human variation is rarely addressed in the literature on response to pharmacologic and toxicologic effects. In fact, the primary recommendation for research that emerges is that a major initiative be launched to change the practices of the scientific community that too often provides merely average effects and their associated statistical p values. A more complete description of the distribution of effects for individuals needed to address the topic of this paper is very rare.

Research literature on the effects of alcohol and of environmental lead on human behavioral were surveyed, and estimates of variation in response were made. For each of these agents a formulaic application of a 100-fold safety factor to the lowest average effective dose would produce an RfD value too low to be measured (and hence would be impractically conservative). Yet research about behavioral effects for other agents suggests that a 100-fold safety factor may be too small. It is clear that a formulaic application of any safety factor is inappropriate.

To fully address the issue of human variation in toxic response, an understanding is required that takes into account the many sources of human variation. A componential model is offered that describes dose–response and other biological, experiential, and measurement sources of individual differences in response. Implications of this model are described for two behavioral tests — the digit span test and the D-S test — demonstrating the subtleties involved when attributing variation in response to any specific component in the model.

Finally, a case was made that thorough understanding of behavioral as well as biological mechanisms is needed if we are to effectively address variation in response; careful use of expert opinion is urged for decisions that must be made before a thorough scientific understanding is attained.

ACKNOWLEDGMENT

The authors gratefully acknowledge the many helpful comments offered during the preparation of this review by K. Davis, C. Kovera, I. Lucki, Robert C. MacPhail, D. Rohlmann, and O. J. Sizemore. Special thanks to K. Hyer, S. Kemp, and S. Grossman for their aid in developing the section on behavioral variation. The report was written while the first author was a Visiting Scientist at the CROET, Oregon Health Sciences University, and supported by a Kenan Research Fellowship from the University of North Carolina at Chapel Hill.

REFERENCES

Agarwal DP, Goedde HW (1989) Human aldehyde dehydrogenases: their role in alcoholism. Alcohol 6:517–523

Anger WK, Sizemore OJ, Grossman SJ, et al (1997) Human neurobehavioral research methods: impact of subject variables. Environ Res 73, in press

Anger WK, Sizemore OJ, Grossmann SJ, et al (1996) Human neurobehavioral study methods: effects of subject variables on research results. American Petroleum Institute, API Publ. No. 4648

Anger WK, Cassitto MG, Liang Y-X, et al (1993) Comparison of performance from three continents on the WHO-recommended neurobehavioral core test battery. Environ Res 62:125–147

Barnes DG, Dourson M (1988) Reference dose (RfD): description and use in health risk assessments. Regul Toxicol Pharmacol 8:471–486

Bellinger DC (1995) Interpreting the literature on lead and child development: the neglected role of the "experimental system." Neurotox Teratol 17(3):201–212

Bellinger D (1989) Prenatal/early postnatal exposure to lead and risk of developmental impairment. In Paul N (ed), Research in infant assessment, Vol 25, No. 6. March of Dimes Birth Defects Foundation, White Plains, NY, p 88

Bellinger DC, Needleman HL (1992) Neurodevelopmental effects of low-level lead exposure in children. In Needleman H (ed), Human lead exposure. CRC Press, Boca Raton, FL, pp 191–208

Bellinger D, Leviton A, Waternaux C, et al (1989) Low-level lead exposure, social class, and infant development. Neurotoxicol Teratol 10:497–503

Bittner AC Jr, Carter RC, Kennedy RS, et al (1986) Performance evaluation tests for environmental research (PETER): evaluation of 114 measures. Percept Motor Skills 63:683–708

Cory-Slechta DA, Pounds JG (1995) Lead neurotoxicity. In Chang LW, Dyer RS (eds), Handbook of neurotoxicology. Marcel Dekker, New York, pp 61–90

Crump KS (1995) Calculation of benchmark doses from continuous data. Risk Anal 15(1):79–89

Crump KS (1984) A new method for determining allowable daily intakes. Fundam Appl Toxicol 4:854–871

Dews PB (1986) Some general problems of neurobehavioral toxicology. In Annau Z (ed), Neurobehavioral toxicology. Johns Hopkins Univ Press, Baltimore, MD, pp 424–434

Dourson ML, Stara JF (1983) Regulatory history and experimental support of uncertainty (safety) factors. Regul Toxicol Pharmacol 3:224–238

Gaylor DW, Slikker W Jr (1990) Risk assessment for neurotoxic effects. Neurotoxicology 11:211–218

Glowa JR (1991) Dose-effect approaches to risk assessment. Neurosci Biobehav Rev 15:153–158

Glowa JR, MacPhail RC (1995) Quantitative approaches to risk assessment in neurotoxicology. Neurotoxicology: approaches and methods, vol. 52. Academic Press, New York, pp 777–787

Hattis D (1995) Variability in susceptibility — how big, how often, for what responses to what agents? Proceedings WHO/IPCS Workshop, Southampton, England, March 8–10

Hattis D, Silver K (1994) Human interindividual variability: a major source of uncertainty in assessing risks for noncancer health effects. Risk Anal 143(4):421–431

Kimmel CA (1990) Quantitative approaches to human risk assessment for noncancer health effects. Neurotoxicology 11:189–198

Landauer AA, Howat PA (1982) Alcohol and the cognitive aspects of choice reaction time. Psychopharmacology 78:296–297

Lehman AJ, Fitzhugh OG (1954) 100-fold margin of safety. UIS Q Bull 18:33–35

Linnoila M, Erwin CW, Ramm D, Cleveland WP (1980) Effects of age and alcohol on psychomotor performance of men. J Studies Alcohol 41(5):488–495

Marsh DO, Clarkson TW, Cox C, et al (1987) Fetal methylmercury poisoning: relationship between concentration in single strands of maternal hair and child effects. Arch Neurol 44:1017–1022

McMillan DE (1987) Risk assessment for neurobehavioral toxicity. Environ Health Perspect 76:155–161

Moskowitz H, DePry D (1968) Differential effect of alcohol on auditory vigilance and divided-attention tasks. Q J Studies Alcohol 29:54–63

Moskowitz H, Robinson CD (1988) Effects of low doses of alcohol on driving-related skills: a review of the evidence. U.S. Department of Transportation, National Highway Traffic Safety Administration, National Technical Information Service, Springfield, VA, DOT HS 807 280

Raffaele KC, Rees C (1990) Neurotoxicology dose/responses assessment for several cholinesterase inhibitors: use of uncertainty factors. Neurotoxicology 11:237–256

Schuckit MA (1994) Low level of response to alcohol as a predictor of future alcoholism. Am J Psychiatry 151(2):184–189

Schull WJ, Norton S, Jensh RP (1990) Ionizing radiation and the developing brain. Neurotox Teratol 12:249–260

Shaheen S (1984) Neuromaturation and behavior development: the case of childhood lead poisoning. Dev Psychol 20:542

Silbergeld E K (1990) Toward the twenty-first century: Lessons from lead and lessons yet to learn. Environ Hlth Perspect 86(6):191-196

Slikker W Jr, Crump KS, Andersen ME, Bellinger D (1996) Biologically-based, quantitative risk assessment of neurotoxicants. Fundam Appl Toxicol 29(1):1–30

Stollery BT, Banks HA, Broadbent DE, Lee WR (1989) Cognitive functioning in lead workers. Br J Ind Med 46:698–707

Wallsten TS, Whitfield RG (1986) Assessing the risks to young children of three effects associated with elevated blood-lead levels. Report prepared for U.S. Environmental Protection Agency, Ambient Standards Branch, Strategies and Air Standards Division, Office of Air Quality Planning and Standards, Durham, NC, Argonne National Laboratory, Energy and Environmental Systems Division, Decision and System Sciences National Technical Information Service Document ANL/AA-32. U.S. Department of Commerce, Washington, D.C.

Waternaux C, Laird NM, Ware JH (1989) Methods for analysis of longitudinal data: blood-lead concentrations and cognitive development. J Am Statist Assoc 84(405):33–41

Weiss B (1990) Risk assessment: the insidious nature of neurotoxicity and the aging brain. Neurotoxicology 11:305–314

Weiss B (1988) Neurobehavioral toxicity as a basis for risk assessment. Trends Pharmacol Sci 9:59–62

Wood FB, Flowers DL, Naylor CE (1991) Cerebral laterality in functional neuroimaging. In Kitterle FL (ed), Cerebral laterality: theory and research. Erlbaum, Hillsdale, NJ, pp 103–115

Wyzga RE (1990) Towards quantitative risk assessment for neurotoxicity. Neurotoxicology 11:199–208

4 Variability in Human Response to Reproductive and Developmental Toxicity

Anthony R. Scialli and Armand Lione

CONTENTS

INTRODUCTION

WHAT DO WE MEAN BY VARIABILITY?

It has been accepted as axiomatic that humans respond differently to the same therapeutic or toxic exposures. The issue of human variability in reproductive and developmental toxicology appears to embody the simple question of the relationship of the toxic dose of an agent in the least sensitive person compared with that in the most sensitive person. Identifying such a relationship in a quantitative manner is another matter altogether. For environmental exposures, dosimetry usually is not available with the precision necessary to calculate a comparison of exposures among members of a population, some of whom are affected and some of whom are not. For pharmaceuticals, where dose estimates are more accurate, agents are given generally at similar doses. If person A is affected and person B is unaffected, it is reasonable to posit that B is less sensitive than A. Is it also reasonable to suppose that B would be affected if the exposure level were higher, and if so, how much higher?

0-8493-2805-5/99/$0.00+$.50
© 1999 by CRC Press LLC

The differences in response between A and B may represent a variability in the sensitivity of A and B as characteristics of the individuals, but the variability may have a more complex basis. If the agent is thalidomide and A and B are pregnant women, the difference in response may be due to a difference in gestational age. If A is 30 days from conception and B is seven months pregnant, and if the toxic endpoint of interest is phocomelia in the offspring, B will not respond regardless of the exposure level and the ratio between the dose affecting the least and most sensitive individuals will be infinite.

One way to address this difficulty is to consider variability among individuals who in some way share a basic susceptibility to the toxic endpoint of interest. In the thalidomide example, we could look to quantitative variability in response of 30-day pregnant women, leaving other pregnant women out of the exercise. Another strategy would be to consider all toxicity to be equivalent. Person B at seven months gestation may sleep for an excessive period of time if too much thalidomide is given. If excessive sleep is considered a toxic endpoint equivalent to that of limb defects in the offspring, a comparison of effective doses could be made. Both approaches have attractive elements and they really serve different purposes. From the point of view of studying disease mechanisms, it may be more informative to learn how people with fundamentally similar susceptibilities differ in personal dose–response curves. From the protective point of view, it makes sense to consider all adverse endpoints as equivalent if the intent is to avoid producing any kind of toxicity. The lowest exposure level needed to produce any toxic response in person A can therefore be compared with the lowest exposure level needed to produce any toxic response in person B to obtain a quantitative estimate of the variability in sensitivity between these two people.

A related concept is the idea that a line can be drawn at an exposure level that protects only one person (the least sensitive) in the population, and a second line can be drawn at an exposure level that protects everyone in the population. The ratio of these exposure levels is an estimate of response variability. Whether the difference between these lines is based on differences in fundamental susceptibility (one person is a 30-day pregnant woman and the other is a 30-year-old nonpregnant man) may not be relevant to a regulator trying to protect the public health, but it certainly is relevant to a discussion of the basis of variability in sensitivity to toxicity.

Perhaps the most satisfying model of variability in individual susceptibility to toxicity is based on the concept that a threshold amount of damage must be sustained before a toxic endpoint is manifested. In this model, each person accumulates subthreshold increments of damage without manifesting toxicity, and variability in response is due (at least in part) to differences in the point at which the individual threshold is reached. Using this approach, Hattis and Silver (1994) argued for a mathematical expression of variability with log-probit dose–response plots. An assumption of their argument is that individual thresholds are normally distributed; that is, plotting the probit of response against the exposure level (or, more probably, the logarithm of the exposure level) yields a straight line. The slope of the line can be taken as an estimate of variability, because it represents the rate at which proportions of the population are "recruited" as responders as the exposure level increases. As an example, these authors used data from Marsh et al. (1987) on neurologic deficits in children exposed antenatally to methylmercury. In this report, maternal hair mercury was used as a marker of exposure level and failure to talk by 24 months of age was used as the marker of developmental toxicity. Using the slope of the log-probit dose–response curve gave an estimate of up to five orders of magnitude difference in personal threshold exposure levels for the bulk (95%) of the population. Hattis and Silver raise the question of whether maternal hair mercury is the appropriate marker of exposure level, but as we discuss later, this measure appears to be a particularly good indicator. We revisit the methylmercury issue later in the body of this paper and offer another estimate of variability; however, the Hattis and Silver model cannot be explained away easily.

Another possibility in the modeling of toxicity response, perhaps especially relevant to toxicants with targets in the cognitive area, is a shift in the normal distribution of capabilities. Intelligence quotient (IQ) is an example of a parameter taken to be normally distributed in the population. A related parameter, developmental quotient (DQ), is used in young children as an estimate of cognitive ability. In a study on the potential effects of antenatal exposure to lead, it was suggested that umbilical blood lead concentrations of 6 to 7 or of $\geq 10\,\mu g/dL$ were associated with DQ decrements of 0.23 and 0.3 standard deviations (SD), respectively, at 2 years of age (Bellinger et al. 1987). Although this conclusion has been challenged, and although the decrement in DQ may not be a persistent reflection of developmental damage, the data raise an interesting point. If a toxicant such as lead causes a shift in the entire distribution of DQ scores, only those individuals shifted into a particularly low range will be considered impaired (i.e., developmentally delayed or retarded), but all individuals will have been affected. A child shifted from a DQ of 105 to a DQ of 100 will likely escape detection as having been damaged by the toxicant but will have been affected all the same. The issue here is not one of variability in response, inasmuch as all exposed individuals will have been affected to the same degree, but variability in underlying capabilities so that the same downward shift in DQ causes only some people to be called impaired.

It may be argued that the child with a DQ of 73 should be viewed as an individual more sensitive to a 5-point shift than a child with a DQ of 105, inasmuch as the shift to 68 will produce a diagnosis of mental retardation in this child, but this argument does not make the two children different in their response to the toxicant, only different in their baseline attributes. Of course, the study of Bellinger et al. did not prove that there was a homogeneous shift of all parts of the DQ curve, and so variability in response still may have occurred, but the possibility of such a uniform response remains intriguing.

PLAN OF ATTACK

In estimating the magnitude of variability to reproductive and developmental toxicity among humans, we take three approaches.

Baseline characteristics

First, we examine the variability in physiologic parameters and baseline occurrence of adverse outcome. With regard to pregnancy, we make the assumption that response to an agent depends to some extent on pharmacokinetic parameters. These parameters are altered by pregnancy and change over the course of a pregnancy. Although an understanding of the physiologic changes of pregnancy cannot be used to predict toxicity, differences in physiologic changes among women may give an estimate of one source of variability in response to toxicants.

In addition, it is necessary to recognize the variability in the incidence of adverse reproductive outcome in the population. Some populations have a higher incidence of certain congenital malformations, for example. It does not necessarily follow that these populations are more sensitive to agents that increase the incidence of these malformations; however, such a difference in incidence does suggest the possibility of a difference in toxicant response. Fertility-related parameters are also variable among people, and it is useful to recognize the range of these parameters in presumably normal, nontoxicant-exposed individuals.

Pharmacology and therapeutics

After consideration of the baseline characteristics, we turn to the data on variations in pharmacokinetic and pharmacodynamic measures. It can be argued that these parameters do not necessarily reflect sensitivity to toxicants; however, the field of therapeutics offers a data set that can be exploited to estimate variability in response to xenobiotics.

Toxicology

We turn finally to the data that are available on response to reproductive and developmental toxicity. Although these data may be less satisfying because precise dosimetry is not always available, some estimates of variability may still be possible.

LIMITATIONS TO OUR APPROACH

It is important to recognize at the outset that the experimental and observational data on which we rely are subject to sources of underlying variability that have nothing to do with the inherent response of the people or tissues. We refer specifically to variability introduced by experimental techniques and assays. We make the assumption, however, that the amount of variability in the results of such models gives at least a crude estimate of interindividual variability.

We also acknowledge that our approach tends to parse variability among different sources. For example, variability in fetal response to a toxicant ingested in food by the mother may be described as some combination of variability in gastric emptying time, maternal volume of distribution, placental transport, and genetic susceptibility. Although such a dissection of the process is helpful in producing manageable portions of the intoxication process, there is no clear guidance on how to add the pieces together to produce an overall estimate of variability. It is not clear whether estimates of variability for each component piece of the process should be summed, multiplied, or handled in some other fashion. We are inclined to believe that there is no simple answer to this question. If, for example, response to a toxicant depends on total drug exposure ("area under the curve" in pharmacokinetic parlance) rather than peak serum concentrations, variations in maternal gastric emptying time may not be relevant. Likewise, if the rate-limiting step for transport of a volatile testicular toxicant to its target site is diffusion into the seminiferous tubule, variations in absorption from the lung may not be important.

Finally, we recognize that our acceptance of the normality of reported distributions may introduce an error in our estimates. Means and SDs are convenient ways to describe the distribution of values in a population; however, these numbers should be applied only if the distribution is normal. When authors report means and SDs or standard errors, we assume that criteria for normality have been met, but authors do not always take the trouble to verify this assumption. In this paper, we make free use of the coefficient of variation, which is the SD divided by the mean, as an estimate of the variability of values within a study population. Our assumption of normality means that nearly the entire population falls within 2 SD of the mean; thus, a coefficient of variation of 50% suggests that the range of values in the population extends from zero to twice the mean. A coefficient of variation of 25% describes a range from 0.5 to 1.5 times the mean, which represents a half-order of magnitude between the lowest and highest individual. Large coefficients of variation may be associated with skewed distributions or with important outliers, and assumptions of normality are most tenuous in these instances.

VARIATIONS IN NORMAL

PREGNANCY PHYSIOLOGY

The alterations imposed by pregnancy have been recognized for many years as affecting therapeutics, and this topic is dealt with in more detail in the next section. In this section, we discuss the variability in the physiologic changes of pregnancy. Such variability may account for a portion of the variability among individual pregnant women or fetuses in response to toxicants.

Variability in this realm can be of two sorts. The first and most apparent variability is the change in physiologic parameters over time in the same pregnant woman. The second type of variability is reflected in differences among women at the same time in pregnancy.

Changes over time

The dramatic change in appearance of a woman over the 40 weeks of gestation are paralleled by similarly large changes in physiologic attributes. Weight, for example, increases by about 12.5 kg (Hytten 1991). Within that weight increase, the proportional distribution among maternal compartments and the conceptus change radically. At 10, 20, 30, and 40 weeks, the conceptus weighs 5, 300, 1500, and 3400 g and accounts for 0.008%, 0.5%, 2.2%, and 4.7% of the maternal mass,

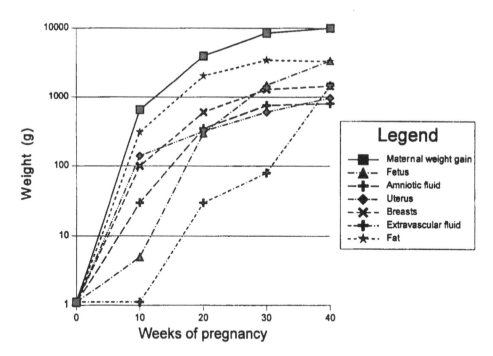

FIGURE 1   Mean components of weight gain in normal pregnancy showing variability with time. Maternal weight gain is the total of the other components. Note logarithmic scale for weight. Figure adapted from data in Hytten (1991).

respectively. Figure 1 gives the approximate relationship of the components of the increase in weight during the course of pregnancy.

As pregnancy advances, blood volume increases by about 50% (Pritchard 1965). Much of the increase is in plasma volume, but red blood cell mass also increases substantially (Figure 2). The consequent dilutional effect results in a decrease in normal hematocrit during pregnancy. There is also a dilutional decrease in serum albumin (Figure 3) and an increase in total body water (Figure 4). Changes in these parameters and other serum protein from a different source are illustrated in Table 1, along with an indication of variability among subjects in the measures. These parameters have important ramifications for drug binding and distribution, respectively, and contribute to modifications in therapeutics necessitated by pregnancy.

Other physiologic parameters that are important in drug absorption, distribution, and excretion are those involving the gastrointestinal, respiratory, dermal, and renal systems. Gastrointestinal motility is decreased during pregnancy, probably as a result of increased progesterone and decreased motilin (Christofides et al. 1982). The decrease in motility and increase in gastric emptying time are noted early in gestation, and we have been unable to locate information on changes in gastrointestinal motility or gastric emptying over the course of pregnancy. Gastric acid secretion has been found to decrease by about one-half in the middle part of pregnancy, with a return to the prepregnancy range during the third trimester (Murray et al. 1957).

Pulmonary tidal volume increases continuously during pregnancy by about 40%, from approximately 500 to 700 mL (Cugel et al. 1953, Lehmann and Fabel 1973). Because respiratory rate does not change during pregnancy, there is a continuous increase in minute ventilation from about 7.5 to 10.5 L/minute. This increase would be predicted to increase the rate of absorption of volatile agents and gases.

Dermal absorption may be increased during pregnancy as a result of increased blood flow to the skin. There appears to be substantial variation in the blood flow increase depending on the part of the body evaluated. The clearest increase in skin blood flow, of at least 6-fold, is in the hand

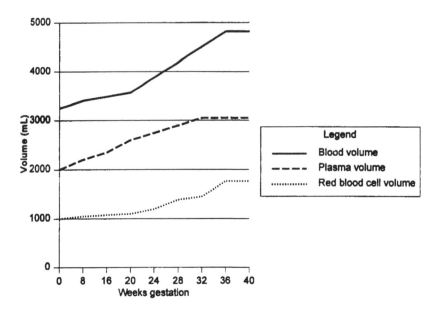

FIGURE 2   Changes in components of blood volume during pregnancy. Drawn using baseline and term values from Pritchard (1965) with slopes estimated from Scott (1972).

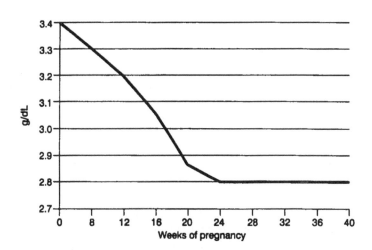

FIGURE 3   Changes in serum albumin concentration in pregnancy. Drawn from data presented by Robertson as cited in Hytten (1980). Note for comparison purposes different values obtained by other investigators and presented in Table 1.

(Abramson et al. 1943, Ginsburg and Duncan 1967). The increase in forearm skin blood flow is much less prominent (Ginsburg and Duncan 1967, Spetz and Jansson 1969).

The most striking change in the pregnant kidney is the increase in glomerular filtration rate that accompanies an increase in renal blood flow. There is a progressive increase in creatinine clearance by about 50% between the nonpregnant state and the end of the second trimester of pregnancy, with a leveling off or a slight decrease thereafter (Davison and Hytten 1974, Dunlop 1981). Renal excretion of xenobiotics that are handled primarily by glomerular filtration can be expected to occur more rapidly as pregnancy advances to the third trimester.

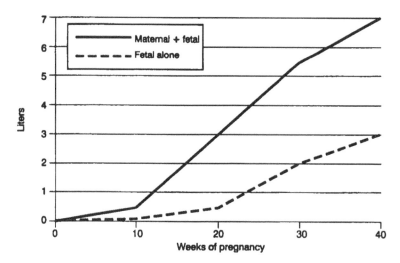

FIGURE 4 Increase in total body water during pregnancy, showing contribution of fetal compartment. Redrawn from data presented in Hytten (1991).

TABLE 1
Changes in Serum Composition in Pregnancy (mean ± SD)

Parameter	Gestation (months)

Parameter	3	4	5	6	7	8	9
Plasma volume (L)	2.635 ± 0.272	2.813 ± 0.314	3.063 ± 0.327	3.396 ± 0.460	3.645 ± 0.536	3.745 ± 0.480	3.763 ± 0.445
Total protein (g/L)	71 ± 4	67 ± 5	67 ± 5	66 ± 5	65 ± 4	64 ± 4	63 ± 4
Albumin (g/L)	44 ± 5	41 ± 3	39 ± 3	39 ± 4	36 ± 4	34 ± 4	32 ± 5
α1-Acid glycoprotein (g/L)	0.72 ± 0.21	0.63 ± 0.14	0.56 ± 0.12	0.54 ± 0.13	0.57 ± 0.17	0.60 ± 0.16	0.50 ± 0.24
Total globulins (g/L)	25.7	26.2	26.5	28.1	27.6	29.1	30.8
α1-Globulin (g/L)	2.9 ± 0.9	3.0 ± 0.8	3.2 ± 0.7	3.2 ± 0.8	3.7 ± 1.7	3.5 ± 0.9	3.6 ± 1.1
α2-Globulin (g/L)	5.6 ± 1.0	5.7 ± 0.9	6.3 ± 0.9	6.1 ± 0.8	6.3 ± 0.9	6.5 ± 1.1	6.8 ± 1.1
β-Globulin (g/L)	7.6 ± 1.1	7.7 ± 1.2	8.5 ± 1.1	8.5 ± 0.9	9.5 ± 1.4	10.7 ± 1.7	11.3 ± 1.8
γ-Globulin (g/L)	10.9 ± 2.3	9.5 ± 2.0	10.1 ± 2.1	9.3 ± 1.9	9.7 ± 2.5	9.5 ± 2.0	9.6 ± 2.5
Lipids (g/L)	6.5	6.9	7.5	7.4	9.0	9.6	10.2

From Notarianni (1990).

Liver function can be altered by pregnancy, probably as a result of increased serum estrogen concentrations. The changes include increased synthesis of many hepatic proteins, including a 2-fold increase in ceruloplasmin and a 50% increase in fibrinogen (Parisi and Creasy 1992). Biliary excretion does not appear to be altered.

Cardiac output increases over the course of pregnancy from about 5 to just over 7 L/minute (Robson et al. 1989). An increase of 0.5 L/minute is evident by the 5th week of pregnancy, with much of the balance of the increase evident by the end of the first trimester. Some studies have shown a decrease in cardiac output over the third trimester [reviewed by van Oppen et al. (1996)]; however, some or all of this decrease may be attributable to caval compression by the gravid uterus, resulting in a decrease in venous return in certain maternal positions (Ueland et al. 1969).

Variability among women

The alterations over the course of pregnancy presented above are based on means obtained in different studies, some of which were cross-sectional (studying different women at different gestational ages) and some of which were longitudinal (studying the same women over time). Substantial amounts of variability have been noted in some physiologic parameters among different women evaluated at similar points in pregnancy.

Weight gain, an easily measured parameter subject to little methodologic error, varies markedly among women. The Institute of Medicine (1990) recommended that women of normal weight should gain 25 to 35 lb during pregnancy; however, fewer than half of women were found to gain weight within this range. Observed weight gains range from almost nothing to twice the recommended gain. Maternal weight gain depends on smoking status; in one series, women who quit smoking during pregnancy gained a mean of 36.6 lb and women who continued smoking gained a mean of 28.9 lb (Mongoven et al. 1996). There was considerable variability within populations. Using the SDs to calculate the interval between 5th and 95th percentiles yielded a range of 8 to 66 lb for quitters and 7 to 53 lb for continuing smokers.

Blood volume in pregnancy shows even greater variability, ranging from 1.2 times to twice the prepregnancy volume (Pritchard 1965). To some extent, this variability may reflect measurement error; however, there are clear and large variations in total body water among pregnant women. For example, it has been demonstrated that the presence of generalized edema elevates the deuterium space increase of otherwise normal women at term from a mean of 7.5 L to just under 11 L (Hytten et al. 1966).

Respiratory parameters show less variability among pregnant women. Coefficients of variation are 13% to 29% for measured pulmonary function tests (Baldwin et al. 1977). Measurements of renal perfusion and glomerular filtration at term have shown coefficients of variation of 12% and 16%, respectively (Ezimokhai et al. 1981). In clinical practice, estimation of glomerular filtration rate by creatinine clearance may give markedly varying results in the same patient from one day to the next; however, this variability is likely to be influenced to a large extent by inadequacies in collection of 24-hour urine specimens.

In a critical review of more than 50 studies on cardiac output during pregnancy, van Oppen et al. (1996) found considerable differences in measurement technique among the studies. In spite of this obvious source of variability, coefficients of variability within studies were 2% to 23%, with most of the studies having coefficients of variation less than 15%.

Perhaps the greatest source of variability among pregnant women in physiologic parameters is disease, particularly preeclampsia. This disorder is characterized clinically by hypertension, proteinuria, and edema. Pathophysiologic features include arteriolar spasm, endothelial leakiness, and intravascular volume constriction. The decrease in blood volume is attributed to leakage of fluid into the extracellular extravascular space, and this decrease has been estimated as 20% (Pritchard et al. 1984). The resultant hemoconcentration produces an increase in hematocrit and uric acid concentration. The decrease in intravascular volume and a decrease in caliber of the renal arteriolar and capillary lumina result in decreased glomerular filtration, increasing the concentration of xenobiotics that require glomerular filtration for excretion. The measurable decrease in creatinine clearance is highly variable, with some women becoming oliguric and others not demonstrably affected by the disease.

One of the earliest abnormalities of preeclampsia is a failure of arterioles to become refractory to endogenous pressors, which is attributed to decreased prostacyclin production in the vessels and perhaps to alterations in nitric oxide synthesis. This pathophysiologic abnormality is detectable by midpregnancy (Gant et al. 1973), although clinical manifestations of the disorder are often not apparent until term. It is likely that among women destined to become preeclamptic, there are deviations from normal pregnant women in total body water, cardiovascular dynamics, and renal function that might be important in the response to xenobiotic agents in the latter part of pregnancy.

Pregnancy Outcomes

Pregnancy outcome is not uniform, and to the extent that identification of developmental toxicity relies on evaluating parameters, it is useful to evaluate the variability in these outcome parameters. In addition, if there are certain populations that have a higher incidence of adverse pregnancy outcome in the absence of exposure to a toxicant, it is possible, although at present speculative, that these groups have a greater sensitivity to toxicant-induced adverse outcome. Data primarily on birth weight and congenital anomalies are presented here. Other pregnancy outcomes (premature delivery, abortion, stillbirth, neonatal death) are to some extent reflected in these two well-studied parameters.

Birth weight

Birth weight is an important reflection of pregnancy outcome. Low birth weight, of course, may be simply a function of gestational age; however, within a given gestational age cohort, abnormally small babies are believed to have suffered impaired transplacental nutrition. Newborns with congenital anomalies [particularly aneuploidy (Droste 1992)] or infections also tend to be smaller than unaffected children at the same gestational age. Babies who are lighter than the 10th percentile are sometimes called SGA (small for gestational age). Among these babies are newborns with a syndrome of subcutaneous wasting, hypoglycemia, polycythemia, and other evidence of intrauterine malnutrition and impaired oxygenation. This syndrome is often called intrauterine growth retardation (IUGR) and is said to be the second most important cause of perinatal mortality after preterm delivery (Wolfe and Gross 1989). Mortality of the IUGR fetus/infant occurs in 1% of cases, a value 6 to 10 times higher than that in the general population (Gabbe 1991), and abnormal neurologic outcome is a risk for survivors (Allen 1984).

At the other extreme of the weight range are large-for-gestational-age babies, who may have been affected by maternal diabetes mellitus or by postmaturity and who are at greater risk for birth trauma than their normal-weight counterparts.

Birth weight is normally distributed in the population and even when low birth weight is associated with prematurity (as early as 32 weeks gestational age), there is preservation of the linearity of probability plots, indicating normality (Cogswell and Yip 1995). The SDs of the weight distributions at 32, 36, and 40 weeks are similar, and the probability plots are nearly parallel. This shift in the distribution means that small decreases in mean birth weight associated with prematurity result in a relatively larger increase in the population in the tail of the curve below a given weight limit. For example, it is customary to consider babies under 2500 g to be small (low birth weight is the usual description); this definition has little to do with the SGA designation because 2500 g is perfectly normal for a baby born at 32 weeks gestation. The use of 2500 g is based on this weight being about 2 SD below the mean of 3540 g for term newborns. At 36 weeks gestation, mean birth weight is 3010 g, 530 g lower than at term, but nearly 16% of the population is in the tail under 2500 g. Thus, for a 15% decrease in mean birth weight, there is more than a 15-fold increase in the proportion of infants classified as low birth weight (Figure 5).

Race is an important determinant of birth weight distribution in the United States. African-American babies have a weight distribution curve shifted to the left of the curve for white babies, with a mean birth weight at term of 3338 g. The effect of this shift is a tripling of the percentage of low-birth-weight babies at 40 weeks gestation from 1.1% to 3.3% (Cogswell and Yip 1995). The shift in the birth-weight distribution curve is also evident at 32 and 36 weeks. African-American women have a higher rate of preterm delivery (19% vs. about 9% for whites); thus, birth weight among African-American samples may be decreased compared with whites from the shift in the weight distribution at a given gestational age and from a decrease in gestational age. Chinese-Americans also have a leftward shift in birth-weight distribution, with a mean birth weight at term of about 3300 g (Yip et al. 1991).

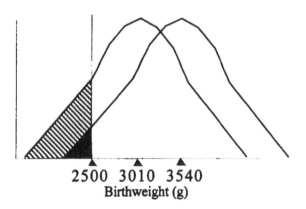

FIGURE 5   Effect of shift in weight distribution of a decrease in gestational age from 40 weeks (right curve) to 36 weeks (left curve). There is a 15-fold increase in proportion of newborns with birth weight less than 2500 g. Drawn from data presented in Cogswell and Yip (1995).

Twins and higher-order pregnancies impose additional risks of low birth weight. To a large extent, effects on weight are associated with the decreased gestational age at which these pregnancies deliver; however, there is also a downward shift in birth-weight distribution for twins and triplets independent of gestational age (Cogswell and Yip 1995). The incidence of twins is about 1% in the United States. Some investigators have identified racial variations in twinning rates; however, these variations are of low magnitude. For example, Ghai and Vidyasagar (1988) reported twinning rates in white and black Americans at 1.3%, with rates of 1.0% in Hispanic and 0.5% in Asian populations in the United States. More recently, however, the rate of twins and higher-order multifetal gestation has increased among Americans in association with increased use of ovulation induction and other assisted reproductive technologies. African-Americans have experienced a 12% increase and white Americans have experienced a 21% increase in multiple gestation as a result (Centers for Disease Control and Prevention 1994).

Infant sex has a small but potentially important effect on weight in population studies. Boys weigh 100 to 135 g more than girls among black, white, and Chinese-Americans (Cogswell and Yip 1995). This shift in distribution results in a 50% higher rate of low birth weight (<2500 g) among girls compared with boys born at full term.

Increasing maternal parity results in increasing birth weight, with close to 100 g gained between parity 1 and parity 4 or more (Cogswell and Yip 1995). This shift results in a doubling of large infants (>4500 g) among white women at high parity compared with parity 1. The effect of parity appears to be less pronounced among African-Americans. Residence at altitude produces smaller babies with a 136-g decrease in the mean birth weight when going from sea level to 1 mile (Cogswell and Yip 1995). As maternal age and education level increase, birth weight increases; teenagers have babies 200 g lighter than women over 35 and a similar shift in birth-weight distribution is associated with finishing high school. Women who are light, short, or have gained less weight than recommended also have infants with mean weights about 200 g lower than women who are tall, heavy, or gain more weight. Presentation for antenatal care in the second trimester compared with the first trimester is associated with a 140-g decrement in mean birth weight (Florey and Taylor 1994). Given that the SD values for these distributions are all 450 to 500 g, the relatively small shifts in means associated with these factors result in sizable changes in the proportion of infants shifted into the lower tail of the distribution curve, with doubling to tripling of low-birth-weight classification associated with most of these maternal characteristics.

Congenital malformations

The ascertainment of congenital malformation rates can be influenced by factors that call into question the accuracy of population differences; however, some differences are dramatic and appear

TABLE 2
Highest and Lowest Incidences of Congenital Anomalies Reported by
The International Clearinghouse for Birth Defects Monitoring System (1991)

Anomaly	Highest incidence		Lowest incidence	
	Program	Rate (per 10,000)	Program	Rate (per 10,000)
Anencephaly[a]	Mexico	18.4	Central-East France	0.8
Spina bifida[a]	N. Ireland	17.5	Tokyo	2.5
Encephalocele[a]	Mexico	3.2	Finland and Israel	0.3
Hydrocephaly	Atlanta	8.5	Finland and Spain	2.1
Microtia	Mexico	6.4	Czechoslovakia	0.1
Transposition of the great vessels[b]	Strasbourg	5.4	Japan	0 to 0.1
Hypoplastic left heart syndrome[b]	Atlanta	3.4	South America	0 to 0.1
Cleft palate	Finland	10.1	Sichuan	2.6
Cleft lip	Sichuan	16.3	Israel	5.2
Esophageal atresia	Emilia-Romagna	3.6	Sichuan	0.4
Anorectal atresia	Tokyo	5.5	Finland	1.4
Renal agenesis/dysgenesis[c]	Strasbourg	4.1	New Zealand & Tokyo	0.2
Hypospadias	Atlanta	29.7	Tokyo	2.3
Epispadias[d]	Atlanta	0.8	Czechoslovakia	0 to 0.1
Bladder exstrophy[d]	Emilia-Romagna	1	Czechoslovakia	0.1 to 0.2
Limb reduction defects	N. Ireland	8.0	Israel	3.1
Diaphragmatic hernia	Canada	4.1	Finland	0.3
Abdominal wall defects (total)	England-Wales	6.8	Central-East France & Finland	1.9
Ompalocele	Norway	3.9	Sichuan	0.8
Gastroschisis	Atlanta	1.9	Italy	0.4

[a] Rate is influenced by antenatal diagnosis and abortion of affected fetuses.
[b] Rate is strongly influenced by length of follow-up and availability of diagnostic techniques.
[c] Not including polycystic kidney.
[d] Very small numbers; rates estimated from graphic presentation in the source document.

time and again in a number of surveys. For example, the association of neural tube defects with English, Irish, or Welsh ancestry has been repeatedly identified and is discussed further below. First, however, it is of interest to note the data collected by the International Clearinghouse for Birth Defects Monitoring Systems, a nongovernmental organization of the World Health Organization. The Clearinghouse attempts to standardize the definition and reporting of anomalies by its 26 voluntary monitoring programs around the world. Data collected from 1974 to 1988 have been published (International Clearinghouse for Birth Defect Monitoring Systems 1991). The incidence of many congenital anomalies showed variability over time and reporting program. Some of this variability was likely to have been due to ascertainment problems, but some would be expected to be due to variability in the true incidence of the anomalies. Table 2 presents the highest and lowest rates for each of the 22 anomalies in the Clearinghouse report, averaged over the entire period covered by the report. Major differences among geographic locations may represent differences in exposure to an as-yet-undefined environmental cause of congenital anomalies, differences in genetic predisposition to the anomaly, or a combination of the two factors. A case for a strong difference in genetic susceptibility can be made for some anomalies, such as cleft lip and palate, where familial recurrences have been described and for which geographic differences in rate are not likely to be due to diagnostic errors.

The variation in incidence of neural tube defects within the United States has been attributed in part to patterns of migration of Americans with English, Irish, or Welsh ancestry. The highest incidence of these anomalies in the United States occurs in New England, with a diagnosis (before abortion) of neural tube defects in 16 per 10,000 pregnancies in predominantly white Bostonian women (Milunsky et al. 1989). The extensive use of antenatal testing and elective abortion of affected fetuses has resulted in a marked decrease in the incidence of these anomalies at birth; however, data from before the advent of widespread screening (Centers for Disease Control 1980) showed rates (per 10,000 births) of 10.7, 8.2, and 6.5 in the southeastern, north-central, and western parts of the country, respectively. The rate in Northern Ireland for a comparable time was 71 per 10,000 births (Nevin 1979). A decade later, the incidence at birth of spina bifida (which accounts for about one-half of all neural tube defects) had declined in 16 selected states from 5.9 to 3.2 per 10,000 births (Morbidity and Mortality Weekly Report 1992), with the reduction presumably due to antenatal diagnosis and abortion of affected fetuses. There remained, however, prominent variability among states in the incidence of spina bifida; the lowest rate was 3.0 per 10,000 in Washington State and the highest was 7.8 per 10,000 in Arkansas. Racial variation was also apparent, with an incidence (per 10,000 births) of 2.3, 3.7, 4.5, and 6.0 in Asians, blacks, whites, and Hispanics, respectively. Of course, regional and racial differences in the diagnosis at birth are likely to reflect both variability in incidence of the malformation and variability in the use of antenatal diagnostic and abortion services. Rates at birth represent a minority of affected fetuses; neural tube defects are found in about 1% of first-trimester abortuses (Byrne and Warburton 1986).

Maternal illnesses may also influence rates of congenital anomalies, apart from effects attributable to medications used to treat the illnesses. Diabetes mellitus is the best characterized example of such an increase. The offspring of women with diabetes mellitus have an incidence of major anomalies of around 9%, which is 3 to 4 times that of the general population (Simpson et al. 1983, Mills et al. 1988a). Although some of the increase in malformation rate is attributable to poor metabolic control, there appears to be an approximate doubling of major malformation risk even among diabetic women with reasonable glycemic control early in pregnancy (Mills et al. 1988a). The most commonly observed malformations are neural tube defects and cardiovascular malformations of the endocardial cushion type. The incidence of minor malformations in this population has been described as similar to that in nondiabetic populations (Hod et al. 1992). There appears to be a relationship between poor diabetic control early in pregnancy and an increased risk of miscarriage, although most diabetic women have an incidence of miscarriage similar to that of nondiabetic women (Mills et al. 1988b).

An intriguing potential interaction between maternal illness and sensitivity to xenobiotic-associated congenital malformation is illustrated by maternal epilepsy, a heterogeneous group of neurologic disorders. Although the use of anticonvulsant medication during pregnancy has been convincingly associated with an increased incidence of congenital anomalies compared with the general population, it is less clear whether there is an increase in congenital anomaly rate associated with maternal epilepsy in the absence of anticonvulsant medication [reviewed by Dansky and Finnell (1991)]. The most suggestive study in this regard (Dansky et al. 1982) found an incidence of major anomalies of 6.5% with maternal epilepsy in the absence of anticonvulsant medication with an increase to 15.9% if anticonvulsant medication had been used during pregnancy. The comparable malformation rate among children with an epileptic father and nonepileptic mother was 1.0%.

There is also a higher prevalence of malformations among the relatives of infants with anticonvulsant-associated malformations (Dansky et al. 1982), suggesting a familial, perhaps genetic, susceptibility to malformation. The association between seizure disorder and congenital malformation may be, then, a manifestation of the more generalized observation that women (and men) with a congenital anomaly are at increased risk of producing a child with a congenital anomaly. This formulation is consistent with the view that some people with epilepsy have their seizure disorder as a result of a congenital malformation of the nervous system. A number of studies [reviewed by Dansky and Finnell (1991)] have shown an incidence of congenital anomalies when the father is epileptic similar to that when the mother is epileptic, also consistent with this paradigm.

Perinatal death

The largest contributor to perinatal death is premature delivery, particularly delivery of infants weighing less than 1500 g. African-American women have a 3 times higher incidence of delivery of babies in this weight group than white American women [reviewed by Luke et al. (1993)]. Perinatal mortality is also higher among women younger than age 18 and with less than 12 years of education, but even among these women, the risks for pregnancies among black women is 24% to 36% higher than for white women (Kleinman et al. 1988). In a study conducted among Swedish women, where race is homogeneous, maternal age of 35 years or more is associated with a 50% increase in perinatal mortality (Cnattingius et al. 1993).

Age and parity are also associated with an increase in stillbirth as well as other pregnancy complications. Pregnant women over 35 years old have about a 50% increase in the incidence of hypertensive disorders and placental complications, IUGR, and stillbirth compared with women of age 34 or younger. Nulliparous women have about a 20% increase in these complications compared with parous women (Raymond et al. 1994). The risk for stillbirth increases significantly with age beginning in the 20s: risk estimates of 1.1, 1.4, 1.7, and 2.0 are found for the age groups 28 to 31, 32 to 36, 37 to 41, and 42 to 45, respectively, using age 20 to 24 as the reference group (Raymond et al. 1994).

FERTILITY-ASSOCIATED PARAMETERS

It has been difficult to identify measures that express the fertility of a given population. Inasmuch as the only proof of fertility is identification of a pregnancy, the first difficulty is the necessity to include only noncontracepting couples in surveys using this endpoint. Among populations of such couples, apparent fertility changes over time, as the most fertile couples conceive and drop out of the cohort under observation. In addition, infertility is a heterogeneous condition. It is easier and arguably more useful to examine the variability in the individual parameters that underlie fertility.

It should be recognized that fertility-associated parameters rarely give evidence of infertility, the absolute inability to produce a pregnancy. At best, these parameters are associated with sub-fertility, or a decrease in the likelihood that a pregnancy will be produced in a given observation period.

Semen parameters

The dependence of semen analysis on abstinence period and technical factors involved in the examination of semen is well-known (Amann 1981). Even with controlled periods of abstinence, there is a large degree of variability in semen parameters among and within men. In one survey, there was a 5-fold difference between the upper and lower bounds of the 95% confidence interval for sperm concentration when multiple ejaculates were collected from 36 normal men (Schwartz et al. 1979). In another study, semen samples were collected each month for 9 months from 45 presumably normal men (Schrader et al. 1988). The coefficient of variation for the total sample was 79%, with a 44% coefficient of variation within subjects. Johnson (1982) was able to stabilize daily sperm output in men by using daily ejaculation; when the last three of five such ejaculates were evaluated, the coefficient of variation for sperm concentration within subjects fell to 12%. This finding suggests that a large portion of the variability noted within men is due to variation in the conditions under which measurements are made rather than to variations in the sperm-producing capacity of the testis; however, the limitation imposed by the high degree of measurement variability is a practical consideration of importance in evaluating reproductive studies.

Sperm morphology is even more problematic because human sperm shape shows a continuum of variation. There are, to be sure, extreme forms that can be recognized as abnormal, but subtle variations in shape and size, which are detectable with computerized methods, may not be so neatly categorized. Different morphometry methods have been proposed (Schmassmann et al. 1979, Katz et al. 1981, Schrader et al. 1984, Jagoe et al. 1986, DeStefano et al. 1987, Turner et al. 1988, Moruzi et al. 1988). In the absence of formal computer-assisted morphometry, i.e., using percentage normal forms as the measured variable, the coefficient of variation among men has been reported at 19%; the within-subject coefficient of variation has been reported at 14% (Schrader et al. 1988).

Motility, reported as percentage motile, has a 45% among-subject coefficient of variation and a within-subject coefficient of variation of 26% (Schrader et al. 1988). Computer-assisted sperm analysis has permitted measurement of various kinds of sperm movement; however, these measurements are highly dependent on machine settings (Vantman et al. 1988, Knuth et al. 1987). In a 9-month longitudinal study of 45 normal men, using computer-assisted sperm analysis of motility parameters, the mean percentage motile was 60%, with a coefficient of variation among men of 45% and within subjects of 26% (Schrader et al. 1991). Other computer-derived motion characteristics (e.g., curvilinear velocity, straight-line velocity, amplitude of lateral head displacement) showed less variability among subjects but similar coefficients of variation within subjects. Most of the variability in these motion parameters was attributed to different swimming patterns and speeds of individual cells within the sample, which is a normal characteristic of sperm in the ejaculate. Although percentage motility, the traditional parameter, was judged to be too variable to permit a single measure to represent an individual in a study, the more detailed motion parameters were believed by these authors to be more promising in this regard.

Recently, attention has been paid to the question of whether semen parameters, particularly sperm concentration, have shown deterioration in the population over time. In part, this issue involves disagreement about modeling the behavior of semen parameter results from different studies, and at least one editorialist has commented on the difficulty of interpreting widely scattered data and incorporating the potential confounders of length of abstinence and subject age (Sherins 1995). The study that prompted the Sherins editorial evaluated semen collected in one unit in Paris by similar evaluation techniques and taking into consideration (in a regression model) subject age and length of abstinence (Auger et al. 1995). The authors concluded that sperm concentration decreased 2.1% per year and motility and normal morphology decreased 0.6% and 0.5% per year, respectively. Sperm concentration among study subjects, however, averaged 98.8×10^6 per mL with a SD of 73.5×10^6 per mL, a 74% coefficient of variation. The 95% confidence interval for such a distribution would extend from well under 0 to well over 200×10^6 per mL, biologically implausible limits to say the least. As in the Schrader et al. (1988) study, variability for percentage motile and percentage normally formed sperm was less extreme: the means ± SD for these parameters were 66% ± 12% and 61% ± 13%, respectively, giving coefficients of variation of 18% and 21%, respectively.

Although a decrease in semen quality in the population over time is debated, semen quality decreases with age for a given subject. Testis weight does not change and, interestingly enough, shows little variability from one man to the next, with coefficients of variation of 5% or less (Johnson et al. 1984a). The preservation of testis weight with age may be due to an increase in weight of the tunica albuginea at the expense of testis parenchyma. The volume of the seminiferous epithelium is reduced in older men because of a reduction in tubule length rather than a change in tubule diameter (Johnson et al. 1984a). Daily sperm production decreases with age (Amann 1981, Neaves et al. 1984, Johnson et al. 1984a) and is associated with a decrease in Sertoli cell number (Johnson et al. 1984b). It has been known for more than half a century that older men are more likely to be azoospermic than younger men (Blum 1936). Men in their 90s have fathered children, indicating that fertility can be preserved into old age; however, the shift in the distribution of sperm production toward lower levels with age makes it necessary to consider this variable in studies using semen parameters as endpoints.

Menstrual cycle parameters

In female humans and other primates, the cyclic recruitment, maturation, and ovulation of a dominant follicle is associated with a uterine discharge in nonconceptive cycles. A number of cycle-associated

TABLE 3
Cycle-Associated Parameters Used or Proposed for Evaluation of Reproductive Normalcy in Women

Parameter	Rationale	Limitations
Cycle length	Very short cycles may be associated with luteal phase inadequacy; very long cycles may be associated with anovulation	Requires prospective recording for accuracy.
Premenstrual symptoms	Symptoms are believed to require luteal hormones; therefore, regular molimina suggests that the woman is ovulatory.	Adequacy of luteal function is not evaluated
Basal temperature recording	Early morning temperature increases in response to progesterone after ovulation.	The phenomenon does not occur in all cycles in all women; the adequacy of the luteal phase is not evaluated; the ovulatory temperature increase is easily obscured by minor infections or by a restless night's sleep.
Salivary or vaginal electrical resistance change	Changes in electrolyte composition of body fluids accompany periovulatory events.	Instrumentation is required. The adequacy of the luteal phase is not assessed.
Assessment of the character of cervical mucus or vaginal secretions	Changes in mucus composition during the preovulatory estrogen increase produce an abundant, slippery secretion.	Luteal phase adequacy is not assessed.
Plasma, salivary, or urinary hormone concentrations	A profile of the cycle can be constructed using gonadotropin levels and estrogen and progesterone parent compounds or metabolites.	Frequent sampling is necessary. In population studies, storage and transport of samples can be problematic.
Timed endometrial biopsy	Endometrial histology in the luteal phase changes sufficiently to date the endometrium with respect to ovulation.	An invasive procedure is required.
Ovarian imaging using ultrasound	Follicle growth and collapse are associated with ovulation.	Requires instrumentation and daily visits of the subject to the imaging center.

phenomena have been considered for clinical or research use in evaluation of reproductive normalcy (Table 3). For some of these parameters, there are data on variability within and among women.

Cycle-length data in presumably normal women were collected in a large prospective study conducted at the University of Minnesota (Treloar et al. 1967). Cycle length varied between 22 and 35 days in 90% of the subjects, with excursions outside this range in the remainder. Mean cycle length was 29.1 days with a SD of 7.46, giving a coefficient of variation of 26% (Chiazze et al. 1968). Most women had cycles of varying length; only 13% of women followed for 10 study cycles had less than 6 days variation in cycle length.

Cycle length increases at the extremes of reproductive life: mean cycle length at age 12 is 35.1 days, from which it decreases steadily to 27.1 days at age 43 and increases to 51.9 days at age 55 (Vollman 1977). It should be recognized that prospective recording of menstrual cycle data is important for accuracy; women asked to recall previously recorded menstrual cycle information showed poor accuracy in reporting cycle lengths (Bean et al. 1979).

Many of the other parameters in Table 3 are used to suggest that ovulation has occurred and, to some extent, to retrospectively time when ovulation occurred. In the latter regard, inaccuracies of the methods may limit the ability to determine variability among women. For example, basal temperature recording produces a shift in the temperature level that presumably occurs just after ovulation; however, ultrasound evaluation of the ovary suggests that physical collapse of the follicle may occur several days away from the basal temperature shift (Leader et al. 1985).

Dating the endometrium by histologic criteria relies on a paper published in 1950 (Noyes et al. 1950). Accuracy of this method was originally established by using basal temperature criteria (Noyes and Haman 1953). Two observers disagreed by more than 1 day in 38% of specimens, and a similar number of specimens disagreed with the basal temperature ovulation estimation by more than 1 day. Using the day of the luteinizing hormone surge to establish ovulation improved the apparent accuracy of endometrial dating (Koninckx et al. 1977); however, the chief limitation of this test appears to be the innate variability within subjects. It is common for endometrial dating to give different results in two different cycles in the same woman.

Within women, cycle changes due to lifestyle and other factors are commonly encountered. Weight gain or loss, travel, physical conditioning, and psychological stress are known suppressors or modifiers of ovulation [reviewed by Sacks (1993)]. Many of these alterations are mediated through corticotropin-releasing hormone activity in the central nervous system. Extremes of lighting conditions, such as prolonged arctic darkness, can also disrupt cyclicity, an effect presumably mediated by melatonin.

Menarche and menopause, at the extremes of reproductive life, occur at widely varying ages. The onset of menstrual cycles occurs normally between ages 10 and 16 with a mean of 12.5 years. The last menstrual period occurs at a mean age of 50, with the normal range extending from age 40 to 60.

Infertility, spontaneous pregnancy loss, and age

A special example of variation in reproductive success is that associated with age, particularly of women (Menken et al. 1986). Women older than 40 years of age have a decreased likelihood of conceiving, attributed to deficient follicular recruitment (Batista et al. 1995). In addition, pregnancies in older women are more likely to spontaneously abort. Even after the detection of embryonic cardiac activity by ultrasound, which is ordinarily a good predictor of a normal pregnancy, older women may abort. Rates for spontaneous pregnancy loss under these conditions increased from zero at age 30 or less to about 4%, 15%, and 20% for women aged 31 to 35, 36 to 39, and ≥40, respectively (Smith and Buyalos 1996). It is generally believed that the propensity of older women to produce aneuploid conceptuses is responsible for the increase in abortion rate. Even with oocyte donation from young women, however, women 40 and over have a 10-fold increase in miscarriage rate (from 4.7% to 48%), suggesting a uterine contributor to the reproductive failure (Cano et al. 1995).

PHARMACOLOGY

The history of pharmacology has often focused on the monitored administration of foreign compounds to individuals. However, few reports outside of those addressing the distribution of a pharmacogenetic trait have provided sufficient primary data to permit a reliable estimate of the scope of interindividual variation. Commentators on pharmacokinetic variables, such as the elimination half-life of a drug and total body clearance, have suggested that the coefficient of variation of population averages outside of pregnancy may be as large as 50% with a 5-fold range of variation (Mammen 1990).

The best-defined examples of variation in the pharmacodynamics of pregnancy have been derived from clinical and analytic studies involving women who need to continually use effective doses of various medications during gestation. Prominent examples include women with asthma or epilepsy. The need for effective doses of antibiotics and psychotherapeutic agents during gestation also may be important for maternal and fetal health. Data on the variations in blood concentrations of some of these drugs can be used to explore the magnitude of variation of pharmacokinetic characteristics of women when they are pregnant.

A unique aspect of the reproductive period is the maternal production of breast milk. Milk is a readily sampled material, and concern about the possible transfer of drugs and potential toxicants through breast milk has also provided us with a source of data for considering interindividual variation.

VARIATION IN PHARMACOKINETICS AND THERAPEUTIC DRUG LEVELS DURING PREGNANCY

Many of the physiologic changes of pregnancy, such as an increased volume of distribution, increased cardiac and renal blood flow, or reductions in plasma proteins (detailed in the previous section), can alter each of the pharmacokinetic variables that influence drug levels in the body. Most of the physiologic modifications in pregnancy are progressive and have been shown to vary consistently as gestation advances (Table 1; Notarianni 1990).

Data on the protein binding of drugs in pregnancy have shown that total plasma proteins decrease by about 11% at term (see Table 1). Approximately one-half of total plasma proteins consists of albumin; thus, the decrease in total plasma proteins largely involves a reduction of albumin. In contrast with the changes in albumin and α1-acid glycoprotein, however, certain globulins, such as α1-, α2-, and β-globulin, increase throughout pregnancy, by 24%, 21%, and 49%, respectively (Notarianni 1990). In addition to changes in the plasma content of binding proteins in pregnancy, there are also indications that the drug-binding capacity of serum proteins may be reduced by the appearance of endogenous inhibitors of drug binding, such as free fatty acids (Perucca and Crema 1982). For example, the protein-bound fraction of diazepam has been shown to decrease during pregnancy because of both decreased levels of albumin and increased free fatty acids (Kuhnz and Nau 1983).

The pharmacokinetic changes of pregnancy may be most critical for those women who depend on the presence of effective levels of antiepileptic drugs (e.g., phenytoin and valproic acid) for seizure control during gestation. Available evidence indicates that uncontrolled seizures can pose a significant risk to both mother and fetus (Pauerstein 1987). Asthmatic women are another population of patients who depend on adequate control of their symptoms throughout pregnancy. Because of the interest in monitoring the serum of these patients during gestation, we have a database for analysis of concentrations of various drugs among women using comparable doses of certain antiepileptic drugs as well as theophylline, which was at one time a mainstay in the treatment of asthma. In addition, Table 4 summarizes available reports on the pharmacokinetic variables for other agents that are used during pregnancy as antimicrobial and psychotherapeutic agents.

Specific factors altering pharmacokinetic properties for individual drugs in pregnancy may vary widely. For example, increases in renal and hepatic blood flow may lead to the increased renal clearance and metabolism of an agent. Respiratory changes in pregnancy may alter maternal acid-base equilibrium, changing the protein binding, distribution, and metabolism of certain drugs. To simplify our efforts at delineating the range of interindividual variation in this area, we first focus on values for the total plasma clearance (CLp) of certain drugs. Changes in this pharmacokinetic parameter are generally reflective of the sum of the various alterations in drug handling in both pregnant and nonpregnant individuals. After considering CLp data, we discuss the specific changes in pharmacokinetic variables that apply to a few specific agents.

Of the drugs listed in the table, significant changes in CLp during pregnancy were shown for ampicillin, cefuroxime, clorazepate, and oxazepam. The coefficient of variations for the clearance data vary widely. Only a small number of coefficients of variation are <25% and several are >40%. A comparison of the nonpregnant subjects shows that two drugs have coefficients of variation below 10% (clorazepate and oxazepam), one antibiotic (cefuroxime) has a coefficient of variation of 14%, and the remaining drugs have values between 25% and 42%. Thus, typical data for clearance values have a relatively high degree of variation. When the coefficients of variation for nonpregnant subjects are compared with the remaining groups of pregnant women exposed to the same drugs, comparable or higher levels of variation are seen for most agents in the table. This analysis shows that the level

TABLE 4

Pregnancy-Associated Pharmacokinetic Parameters for Some Pharmaceutical Agents

Drug	Dose	Gestation (weeks)	Elimination half-life (hours)	Volume of distribution (L) ± SD	CL_p (mL/min) ± SD	Coefficient of variation in CL_p (%)
Ampicillin	10 mg im	30 ± 4	1.17	0.99 ± 0.31	10.1 ± 3.8	38
		not pregnant	1.45	0.68 ± 0.20	5.5 ± 1.6	29
	500 mg iv	9–33	0.65	35.3 ± 18.4	613 ± 302	49
		not pregnant	0.74	25.5 ± 13.0	394 ± 165	42
Cefuroxime	750 mg iv	11–35	0.73	17.8 ± 1.9	282 ± 34	12
		at delivery	0.87	19.3 ± 3.1	259 ± 35	14
		not pregnant	0.97	16.3 ± 2.1	198 ± 27	14
Metronidazole	250 po	8–14	5.70	not reported	118 ± 52	44
		not pregnant	5.92		183 ± 80	44
	1000 po	8–14	7.73		75 ± 6	8
		not pregnant	7.80		62 ± 17	27
Diazepam	10 iv	38 ± 1	65	149 ± 65	28 ± 10	36
		not pregnant	46	78 ± 9	20 ± 5	25
Clorazepate	20 im	37–42	1.3	0.43 ± 0.17	3.33 ± 0.53	16
		not pregnant	2.0	0.33 ± 0.17	1.67 ± 0.09	5
Oxazepam	25 po	40	6.5	1.22 ± 0.46	2.13 ± 0.73	34
	30 po	not pregnant	9.7	± 0.06	0.82 ± 0.07	9
	15 po	not pregnant	9.3	1.20 ± 0.54	1.58 ± 0.06	4
Phenobarbitone	not specified	16–24	not reported	not reported	5.3 ± 1.6	30
		25–32			± 2.1	36
		33–40			5.7 ± 1.7	30
Phenytoin	not specified	16–24	not specified	not specified	28 ± 11	39
		25–32			22 ± 7	32
		33–40			27 ± 10	37
		at delivery			37 ± 14	38
		not pregnant			19 ± 5	26

Note: im = intramuscularly; iv = intravenously; po = orally.

From Cummings 1983.

of variation reported for these drugs increased during pregnancy for two agents, both benzodiazepine derivatives — clorazepate and oxazepam. The respective increases were 3- and 9-fold. It should be noted that the data for oxazepam, which has the largest variation, were derived from three separate reports. Thus, methodologic variations, rather than interindividual variation, may play a predominant role in that example.

In a report on pharmacokinetic data from epileptic women, the increase in rates of CLp for phenytoin and valproic acid ranged between 30% and 50%, and the maximum effect occurred 4 to 6 weeks before delivery (Chiba et al. 1982). These increased clearance rates were not clearly attributable to alterations in the hepatic metabolism of the drugs.

Progressive decreases in the plasma binding of phenytoin during pregnancy have been reported by several groups of investigators (Chen et al. 1982, Perucca et al. 1981a, Ruprah et al. 1980). A major effect of this decreased binding is believed to be renal excretion of the free drug, explaining the increased rate of clearance. For phenytoin, a roughly 3-fold increase in dosage from 450 to 1250 mg/day may be necessary for some pregnant women to maintain plasma levels in the therapeutic range (Dalessio 1985, Levy and Yerby 1985).

Valproic acid is normally highly bound by albumin and other acid-binding plasma proteins. During pregnancy, binding may change from 90% to 60% (Perucca et al. 1981b, Froescher et al. 1984). These changes appear to be higher than expected to result from reductions in plasma binding proteins and have been associated with displacement by free fatty acids, which increase during pregnancy (Albani et al. 1984). Although impaired protein binding increases the available free fraction of the drug, an elevation in CLp may offset these changes, keeping the free valproic acid concentrations relatively constant during pregnancy (Johannessen 1992). However, a reduction of dose after delivery may be necessary to avoid toxicity (Johannessen 1992).

By contrast, alterations in the pharmacokinetics of theophylline in pregnant asthmatic women may require a reduction in dose to avoid plasma concentrations that are toxic to mother and fetus. In addition to a reduction in plasma binding, which increases free theophylline, the distribution of this agent in the increased body water that accompanies pregnancy somewhat offsets the elevation of free theophylline. However, an overall reduction in CLp of this agent may cause an undue increase in the steady-state levels of the drug (Gardner et al. 1987). In one analysis, a 40% increase in free theophylline concentrations was predicted if maintenance doses were unchanged in pregnancy (Frederiksen et al. 1986). Case reports have indicated that a 30% reduction in the administered dose was needed to maintain serum concentration in the therapeutic range (Carter et al. 1983).

Overall, changes in administered doses to compensate for the altered pharmacokinetics of pregnancy are unlikely to exceed dose reductions by a factor of 30% to 50%, as discussed for theophylline, or increases larger than 3-fold, as discussed for phenytoin.

BREAST MILK

As a drug excretion system the mammary glands have several unique properties. The structure of the holding sacks, the alveoli, allows milk to separate into a fatty layer on top of an aqueous layer, which holds most of the protein. In such a system, more fatty components (lipid-soluble drugs and many persistent environmental contaminants) are taken in by the suckling infant after the storage chambers have been partially emptied. Hind-milk, produced toward the end of a feeding, has been shown to contain about 2 to 3 times as much fat as fore-milk (Hartmann et al. 1985, Neville et al. 1988). In addition to variations in fat and protein content, the pH of milk varies with the content of certain components, such as citric acid (Morriss et al. 1986).

Any analysis of the content of milk samples is complicated by these factors. Because breast milk is not of uniform composition, small samples are likely to be unrepresentative of the whole (Ferris and Jensen 1984). Although the fat content of milk increases during feeding, the fatty acid composition of the lipids of breast milk does not vary between fore- and hind-milk (Gibson and Kneebone 1980). Obtaining full samples of human breast milk by repeatedly emptying the breasts by pumping is exceptional in drug studies (Hartmann et al. 1985). Thus, available data on milk composition and drug content are often based on estimates taken from samples of fore- and hind-milk or by extrapolating from single samples.

During pregnancy, the high concentrations of progesterone released by the placenta inhibit lactalbumin synthesis in the breasts. Once the placenta is delivered and progesterone levels fall, the mammary glands rapidly become productive. The onset of lactation during the first 5 days after delivery is highly variable, with milk yields reported between 98 and 755 mL/day (Hartmann et al. 1985). Milk yield also varies with demand. A woman with twins has increased milk production sufficient to feed both. The frequency of breast-feeding and the completeness of emptying of the breasts at each feeding have been shown to be key factors in increasing yield (Hartmann et al. 1985).

In the breast milk of seven women who collected fore-milk and hind-milk samples for 22 days, the mean coefficients of variation were 12.6% for protein and 6.6% for lactose (Hartmann et al. 1985). Large variability is expected in drug disposition in milk both within the same woman and among individuals, because variations in milk protein, lipid content, and pH are important determinants of drug disposition (Table 5).

TABLE 5
Breast Milk Parameters

Patient	Postpartum interval at sampling	Milk pH	Volume percent fat	Total protein (g/dL)
1	3 days	7.54	8.86	2.55
2	7 days	6.97	12.45	2.38
3	10 days	7.14	8.13	1.58
4	11 days	7.21	7.99	2.03
5	1 month	7.49	3.49	0.93
6	3 months	7.46	8.65	0.99
7	4 months	7.13	6.53	1.01
8	4 months	7.62	8.34	0.92

Adapted from Anderson (1991) and Fleishaker et al. (1987).

One of the methods used to express the amount of a xenobiotic in milk is the milk/plasma (M/P) ratio, which compares the concentration in milk with the concentration in maternal plasma. A sample of M/P ratios is presented in Table 6. A comparison of M/P ratios can give an idea of the magnitude of variability in drug excretion in milk; however, it should be remembered that these ratios are constructed in different ways by different investigators. In many instances, the numbers put into the ratio are concentrations obtained nearly simultaneously, regardless of where the drug is in its distribution or excretion phase. Often peak plasma concentration is compared with a simultaneous milk sample. It is more representative of xenobiotic distribution in milk to use a ratio of the area under the time-concentration curve for the agent in both milk and plasma, but this technique is not often used.

When M/P ratios are expressed as a range (Table 6), the data were often collected from more than one woman and represent an estimate of interindividual variability; however, a range sometimes indicates the difference between fore- and hind-milk in the same woman. Single numbers reported in Table 6 are sometimes mean concentrations of samples from more than one woman and are sometimes single measurements in one subject.

It is not unusual for M/P values to cover a 3- to 7-fold range. Six drugs indicated in the table — aspirin (salicylate), cephalexin, chloroquine, diazepam, N-desmethyldoxepin (DDP), and zuclopenthixol — had M/P ratios with a range covering an order of magnitude or more. Each of these values was evaluated to determine what factors explained the wide range of variation reported for the six agents.

Aspirin (acetylsalicylic acid) and other salicylates are very common components of analgesic products and some gastrointestinal aids (e.g., Pepto-Bismol®) that might be used during pregnancy and lactation. The range of M/P ratios in Table 6 is 0.03 to 0.34 [taken from Findlay et al. (1981)]. Levels of salicylate in milk derived from drug ingestion by the mother reach a maximum value within 2 hours and remains relatively constant for the next 12 hours. Because salicylates are eliminated more slowly from milk than from plasma, the M/P ratio increases in spite of a decline, albeit slow, in milk concentration. The large and relatively rapid fluctuation in plasma salicylate concentrations produces ratios ranging from 0.03 to 0.08 at 3 hours, which increase to a ratio of 0.34 at 12 hours. The range of variation in M/P ratios in this case originated because of variation over time based on the pharmacokinetics of the agent and not because of interindividual variation.

Cephalexin is a moderately lipid-soluble cephalosporin antibiotic. The range of M/P ratios listed for this agent is 0.008 to 0.14. These values were derived from one study in which six mothers were given a single 1-g oral dose of cephalexin on their third postpartum day (Kafetzis et al. 1981). Peak serum concentrations of the drug occurred at 1 hour and declined 10-fold in 3 hours. In contrast, concentrations in milk increased slowly to a peak value 4 hours after the dose. The wide range of M/P ratios also originated from the different elimination rates in serum and milk. However, the usefulness of these data is limited because they were obtained after a single dose of the agent. In clinical practice, the cephalosporin, like other antibiotics, would be used repeatedly over several days. No data were found about the distribution of this drug after repeated doses to produce a maintenance (steady state) concentration in blood.

Chloroquine, an antimalarial agent used throughout the world, was listed as having M/P ratios between 0.268 and 4.26. This range was obtained from two reports. In one study, six nursing mothers 17 days postpartum were given a 5 mg/kg intramuscular injection of chloroquine (Akintonwa et al. 1988). Two hours later, mean milk and serum concentrations were 0.227 µg/mL (range, 0.163 to 0.319) and 0.648 µg/mL (range, 0.46 to 0.95), respectively. The mean M/P ratio was 0.358 (range, 0.268 to 0.462). In an earlier study, three women had been given a single dose of 600 mg of chloroquine 2 to 5 days postpartum (Edstein et al. 1986). Serum and milk samples were collected for the following 9.5 days, and an overall comparison of drug excretion (using the total drug in both compartments) yielded M/P ratios that ranged from 1.96 to 4.26. The values shown in Table 6 reflect the lowest M/P ratio in the first study and the highest value reported in the second study. The variation in this case is due at least in part to observations based on two different phases of drug excretion. However, in the first study, because doses were comparable on a mg/kg basis, the data allow an estimate of a 2-fold interindividual variation in M/P ratios at the single time point evaluated.

DDP is a major active metabolite of the tricyclic antidepressant doxepin. The range of M/P ratios for this compound is 0.12 to 2.35. These data were derived from two reports (Matheson et al. 1985, Kemp et al. 1985). One report was based on the hospitalization of an 8-week-old infant because of respiratory depression and sedation. Her mother had been taking doxepin since 2 weeks postpartum. Four days before the hospitalization, the dose had been increased from 10 to 75 mg/day. Subsequent analysis of maternal serum and milk found M/P ratios of only 0.12 and 0.17 for DDP, but the infant's serum concentration of DDP was comparable to that of the mother (62 µg/L). Doxepin levels were barely measurable in the infant. DDP is known to have a long elimination half-life in adults that depends on hydroxylation and conjugation with glucuronic acid, metabolic steps that are limited in both the fetus and the newborn (Juchau et al. 1980).

The second case (Kemp et al. 1985) included an infant whose mother was treated with larger daily doses of doxepin from 1 to 7 months postpartum. Milk and serum samples were collected between 7 days and 3 months after drug therapy was begun. Milk samples were collected before and after a feeding. More doxepin and DDP were found in postfeed samples, suggesting that these lipid-soluble agents were distributed in the fat-rich hind-milk samples (Hartmann et al. 1985, Neville et al. 1988). Review of the original data suggests that an M/P ratio of 1.28, from an average of the collected pre- and postfeed values reported by Kemp et al. (1985), is probably the best estimate of the M/P ratio for this compound. At 43 days of age, the second infant had serum assayed for doxepin and DDP; doxepin was below the limits of detection of the assay and the DDP concentration was only 15 µg/L, approximately 4 times lower than that found in the first case report, despite a 2-fold higher daily dose (150 mg) being ingested by the mother. This infant did not display any signs of drug intoxication.

The variability between M/P ratios found in these two studies is difficult to assess given the important difference in the timing of milk and plasma collection and the differences in infant age, which may influence milk composition. Still, it is clear from this example that a single M/P ratio or even a narrow range would fail to capture marked differences in infant response to the exposure.

Diazepam is a benzodiazepine that is very widely used as an anxiolytic. The range of M/P ratios is 0.14 to 2.7. These values were calculated from a single short report on nine lactating mothers (days postpartum not noted) (Cole and Hailey 1975). Regarding the variation in these values, the author stated that the data "do not permit confident prediction and may be related to

TABLE 6
M/P Ratios of Selected Drugs

Acebutolol	2.3–9.2	Doxepin	0.51–2.39
N-acetylacebutolol	1.5–13.5	N-Desmethyldoxepin	0.12–2.35
Acetaminophen	0.91–1.42	Doxorubicin	4.43
Acetazolamide	0.25	Doxycycline	0.3–0.4
Acyclovir	0.6–4.1	Encainide	1
Amikacin	trace	Erythromycin	0.4–1.6
p-Aminosalicylic acid	0.016	Ethambutol	1
Amiodarone	2.3–9.1	Ethanol	1
Amitriptyline	1.0	Ethosuximide	0.78–1.0
Amoxapine		Fenoprofen	0.017
as 8-hydroxyamoxapine	0.45–0.86	Flunitrazepam	0.61–0.9
Amoxicillin	0.013–0.043	Flupenthixol	0.5–1.62
Amphetamine	2.8–7.5	Gold	0.014–0.104
Ampicillin	0.2	Haloperidol	0.6–0.7
Aspirin (salicylate)	0.03–0.34	Heparin	0
Atenolol	2.9–3.6	Hydralazine	1.4
Butorphanol	0.7–1.9	Hydrochlorothiazide	0.25
Caffeine	0.5–0.76	Ibuprofen	trace
Captopril	0.012	Imipramine	1
Carbamazepine	0.24–0.69	Indomethacin	1
Cefadroxil	0.009–0.019	Insulin	0
Cefamandole	0.02	Iodides	4–23
Cefazolin	0.02	Isoniazid	1–2
Cefotaxime	0.027–0.16	Kanamycin	0.05–0.40
Cephalexin	0.008–0.14	Lincomycin	0.9
Cephalothin	0.073–0.53	Liothyronine	0.36
Cephapirin	0.068–0.480	Lithium	0.4
Cephradine	0.2	Magnesium	1.9–2.1
Chloral hydrate		Maprotiline	1.3–1.5
as trichloroethanol	0.6–0.8	Meperidine	>1
Chloramphenicol	0.51–0.61	Mepindolol	0.35–0.61
Chloroquine	0.268–4.26	Meprobamate	2–4
Chlorthalidone	0.05	Methadone	0.83
Chlorpromazine	<0.5	Methabarbital	
Chlorprothixene	1.2–2.6	as barbital	trace
as the sulfoxide	0.5–0.8	Methimazole	0.3–1.16
Chlortetracycline	0.4	Methotrexate	0.08
Cimetidine	4.6–7.44	Metoclopramide	1.8–1.9
Cisplatin	1.1	Metoprolol	2.0–3.7
Clemastine	0.25–0.5	Metronidazole	1
Clonazepam	0.33	Minoxidil	0.67–1.0
Clonidine	1.5	Morphine	trace
Codeine	trace	Nadolol	4.6
Colistimethate	0.17–0.18	Nalidixic acid	0.08–0.13
Cyclosporine	0.17–0.40	Naproxen	0.01
Desethylamiodarone	0.8–3.8	Nitrofurantoin	trace
Desipramine	0.4–0.9	Nortriptyline	0.7–3.71
Diazepam	0.14–2.7	Novobiocin	0.1–0.25
Digoxin	0.6–0.9	Oxprenolol	0.14–0.45
Disopyramide	0.4–0.9	Dothiepin	0.33
N-Monodesalkyl	5.6	Penicillin	0.02–0.13

TABLE 6 (continued)
M/P Ratios of Selected Drugs

Pentoxyfylline	0.54–1.13	Spironolactone	
Phenacetin	0.16–0.90	as Canrenone	0.51–0.72
Phencyclidine	10 (in mice)	Streptomycin	0.5–1.0
Phenobarbital	0.4–0.6	Sulfanilamide	0.5–0.6
Phenytoin	0.18–0.54	Sulfapyridine	0.4–0.6
Prednisolone	trace	Sulfasalazine	0.3
Prednisone	trace	Sulfisoxazole	0.06–0.22
Primidone	0.8	Terbutaline	1.4–2.9
Procainamide	1.0–7.3	Tetracycline	0.25–1.5
N-acetylprocainamide	1.0–6.2	Theophylline	0.7
Propoxyphene	0.5	Ticarcillin	trace
Propranolol	0.2–1.5	Timolol	0.80–0.83
Propylthiouracil	0.55	Tolbutamide	0.09–0.40
Pseudoephedrine	2.6–3.9	Tolmetin	0.005–0.007
Pyrazinamide	0.036	Trimethoprim	1.25
Pyridostigmine	0.36–1.13	Triprolidine	0.5–1.2
Pyrimethamine	0.46–0.66	Valproic acid	0.15
Quinidine	0.71	Vancomycin	1
Ranitidine	1.9–6.7	Verapamil	0.23
Rifampin	0.20	Zuclopenthixol	0.12–2.20
Oxycodone	3.4		

From Scialli (1992).

the lack of equilibration between maternal milk and blood in this short study." Neither the size of the dose nor the frequency was reported. Other data, presented in a more thoughtful and thorough publication (Bennet 1988), indicate that a range for M/P ratios between 0.13 and 0.20 may be a more reasonable estimate for diazepam.

Zuclopenthixol, a thioxanthene tranquilizer with properties similar to chlorpromazine, is marketed in some northern European countries. The range of M/P ratios is 0.12 to 2.20. The low value was derived from a single sample from a patient who had delivered 4 days previously, when milk composition and corresponding drug excretion were still highly variable (Aaes-Jorgensen et al. 1986). No information was given on what fraction of the milk was sampled. The high value comes from a second case report (Matheson and Skjaeraasen 1988), in which widely varying milk concentrations were obtained from seven samples from the same subject. We wonder whether poor experimental design and analytic error accounted for the wide range of the reported values; however, if the analyses are accurate, this case represents large intraindividual variability in milk concentration of this agent.

Repeated samples of plasma and breast milk are necessary to establish a credible value for an M/P ratio. Early in lactation milk is susceptible to wide variations in protein and fat content, and, even in established lactation, fluctuations in fat content of milk from early and late samples in one feeding may cause a single M/P ratio to be unrepresentative of total exposure. In spite of the technical difficulties in obtaining meaningful estimates of disposition of an agent in milk, case reports suggest that there is an important degree of variability in the milk content of xenobiotic agents. It is useful to remember, however, that manifestations of toxicity in the infant depend on parameters such as absorption of the agent from milk, distribution within the infant, biotransformation, excretion, and individual infant susceptibility. M/P ratios give only an estimate of administered dose.

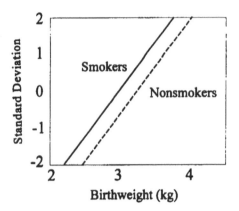

FIGURE 6   Probability curve of birth weight according to smoking status. From Cogswell and Yip (1995).

RESPONSE TO TOXICANTS

To assess variation in developmental or reproductive response to toxic exposures, we would like to have reliable information on exposure levels as well as response. We have considered a few agents for which dose and response information is sufficient to estimate the variability in human response.

CIGARETTE SMOKING

Cigarette smoking during pregnancy has been associated with increased risk of miscarriage, bleeding during pregnancy, placental implantation abnormalities, and low birth weight. In addition, there may be a genetically susceptible population at risk for an increase in facial clefts in response to maternal smoking.

Birth weight

Children born to women who smoke are 150 to 250 g lighter than the newborns of nonsmoking women delivering at the same gestational age (Kramer 1987, Cogswell and Yip 1995, Aronson et al. 1993). Data from the U.S. Natality Survey show the probability plots for birth weight to be linear and parallel when smokers are compared with nonsmokers (Cogswell and Yip 1995; Figure 6). The birth-weight distributions for infants of smokers and nonsmokers in the state of Wisconsin show the same parallel relationship (Aronson et al. 1993). The finding that the distributions are parallel throughout the range of birth weights suggests that smoking shifts the entire weight distribution by a similar amount in all pregnancies. Such an effect is consistent with similar susceptibility to smoking in the entire population, even though a diagnosis of low birth weight is made only at the tail of the distribution below 2500 g, resulting in an approximate doubling of this complication. These data do not contain information on smoking dose, however, and it is possible that dose and differential individual susceptibility interact to produce a distribution that happens to be parallel to the nonsmoker distribution; however, this possibility appears far-fetched to us. Even if we assume that the parallel shift in birth-weight curves is due to similar sensitivity across the population, a few individuals who are particularly sensitive (or particularly resistant) to the birth-weight effects of smoking will not be detected in this type of population study.

A greater reduction in birth weight and other somatic measurements has been observed when the mother smoked ≥20 cigarettes per day than when she smoked 1 to 19 cigarettes per day (Cliver et al. 1995); however, more than 70% of the birth-weight effect of smoking ≥20 cigarettes per day was found in women smoking 1 to 19 cigarettes per day. Effects of smoking on birth weight were greater in white women than in black women in this study, but it is not clear whether cigarette

---

TABLE 7
**Decrements in Birth Weight[a] by Race
and Number of Cigarettes Smoked**

Cigarettes smoked per day (self-report)[b]

Race	Black	White
1 to 9	7	4
10 to 19	7	5
≥20	9	6

Cigarettes smoked per day (self-report)[c]

Race	Black	White
1 to 10	5	4
11 to 20	5	6
>20	6	8

[a] Expressed as percentage decrease compared with nonsmokers.
[b] From English et al. (1994).
[c] From Abel (1980).

---

consumption differences between black and white women could explain the difference. A more detailed evaluation of the effects by race of cigarette dose showed a larger proportional birth-weight decrement among the children of black than among the children of white smokers within the same self-reported cigarette groups (1 to 9, 10 to 19, and ≥20 cigarettes per day) (English et al. 1994). Black women also had higher serum cotinine concentrations at any given self-reported dose of cigarettes. When expressed as a function of cotinine, birth-weight reduction did not show a difference by race. One explanation for this observation is that black women may absorb more nicotine from cigarettes or metabolize it differently, resulting in a greater effect from the same number of cigarettes.

If there is a racial difference in sensitivity of birth weight to maternal cigarette smoking, we can estimate the magnitude of this difference from the data presented by English et al. (1994). Table 7 compares the decrement in birth weight in each smoking group by cigarette smoking and race. An initial observation is that smoking a few cigarettes per day gives about 70% of the effect of smoking ≥20 cigarettes per day, as noted in the Cliver et al. study. This phenomenon is consistent with the early attainment of a threshold toxic effect (see below). With regard to the comparison by race, black women are about twice as sensitive as white women to the birth-weight-reducing effects of cigarettes; that is, in a black woman who smokes 10 cigarettes per day the effect on birth weight is about the same as in a white woman who smokes 20 cigarettes per day. This estimate is crude at best, given the imprecision of the dosimetry and the relatively large SD (500 to 600 g) around the mean birth weights. In addition, an older study by Abel (1980) does not support this estimate. Figure 7 displays the birth-weight effect of smoking by race from the Abel study and Table 7 presents the percentage decrement in birth weight of smoking by self-reported dose. It is not clear from these data that there is a greater effect on black women than on white women with respect to birth-weight decrement from similar amounts of cigarette consumption.

Maternal age also has been found to modify the impact of smoking on birth weight (Fox et al. 1994). Between the teenage years and age 40 or more, smoking-associated birth-weight decrements increased 3-fold, from 117 to 376 g. Dose of smoking was not ascertained.

The idea that there is a relatively low threshold dose at which smoking produces its major effects on birth weight is supported by a study conducted with urban black women and their infants in which effects of smoking were evaluated, controlling for maternal age and concurrent use of

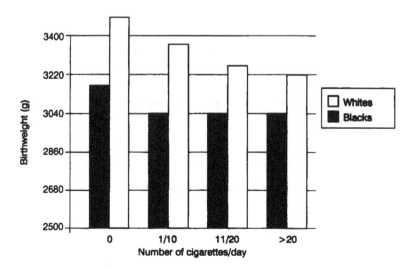

FIGURE 7 Effect of self-reported smoking dose on birth weight by race. Drawn from data presented in Abel (1980).

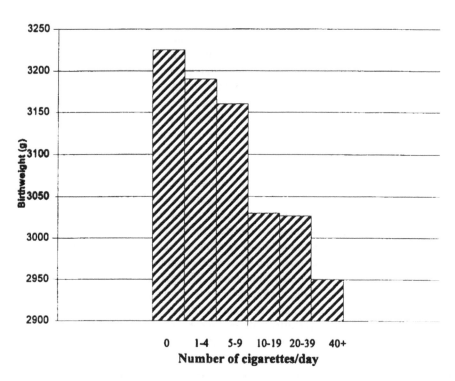

FIGURE 8 Relationship between number of cigarettes smoked per day during pregnancy and birth weight in black urban women, controlled for use of ethanol, cocaine, and opioids. From JL Jacobson et al. (1994).

ethanol, opioids, and cocaine (JL Jacobson et al. 1994). Figure 8, drawn from that study, shows an apparent effect of cigarette smoking on birth weight at 1 to 4 cigarettes per day, with a major effect at about 10 cigarettes per day. Heavier smoking is not associated with a proportional additional effect.

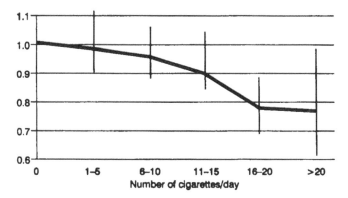

FIGURE 9   Relative rate of conception among smokers. Rate among nonsmokers is taken as 1.0. Bars represent 95% confidence intervals. Drawn from data in Howe et al. (1985).

If there is a low threshold dose at which smoking effects on birth weight occur, a small degree of variability in response would be anticipated. Let us postulate that there is as much as a 2- or 3-fold difference in personal thresholds; if those thresholds are 1, 2, or 3 cigarettes per day, most smokers would far exceed their personal thresholds and would have similar responses.

Support for the hypothesis of a low threshold and a similar response among pregnant smokers can be found in a study of the effects of a single cigarette on fetal vascular response. Experienced smokers had Doppler flow velocity measurements of the umbilical artery before and after smoking 1 cigarette (Morrow et al. 1988). Fetal heart rate increased from a mean of 138 to 146 beats per minute. The ratio between systolic and diastolic flow velocity, which is proportional to downstream resistance, increased 26%. Coefficients of variation for fetal heart rate and systolic/diastolic ratio were 8% and 21%, respectively. Although the presumed change in hemodynamic parameters may not represent all of the smoking effect on birth weight, or even an important part of the effect, the large alteration in subjects who had been habituated to smoking exposure suggests a powerful effect of what is considered to be a small dose.

Perinatal mortality

The smoking-associated risk of perinatal mortality varies within the population as a result of interactions with race and age. Among women smoking less than 1 pack of cigarettes per day, black women have a risk estimate for perinatal death 25% higher than white women; at ≥1 pack per day, the risk estimate is 56% higher among blacks than among whites (Kleinman et al. 1988). Among Swedish women, nulliparity and age interacted with smoking to increase the risk of stillbirth, with a 3.5-fold increase in the risk estimate for nulliparous 35-year-old women who smoked 10 or more cigarettes per day compared with multiparous women in their early twenties who did not smoke (Cnattingius et al. 1993). Among nulliparous 35-year-old women who did not smoke, the risk estimate was 2.4 times that in the reference group; thus, about one-third of the excess risk in this age and parity group can be attributed to smoking. Above age 40, the difference between smokers and nonsmokers disappears (Raymond et al. 1994), suggesting that the stillbirth risk attributable to this age group is high enough to make inapparent the attributable risk of smoking.

Subfertility

A number of studies have associated smoking among women or men with a decrease in the likelihood of conceiving a pregnancy [reviewed by Stillman et al. (1986)]. Among those studies showing a statistically significant effect, risk estimates for measures of subfertility among smokers range from 1.3 to 3.3. A graphic example from one such study is presented in Figure 9, demonstrating a dose-related decline in the rate of conception among smokers. The 95% confidence intervals suggest that smokers are heterogeneous in their fertility rates. Part of the heterogeneity

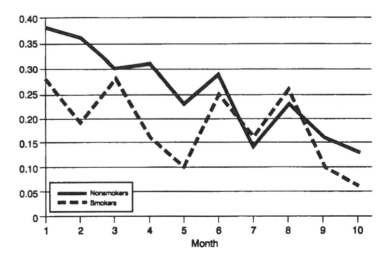

FIGURE 10   Probability of conception per cycle in smokers and nonsmokers. Drawn from data in Baird and Wilcox (1985). Only the first 10 months are shown; this period represents about half the period under study in this paper.

is likely because of the heterogeneity in rates of fertility in the general population. Figure 10 illustrates the behavior of the monthly probability of conception from the data of Baird and Wilcox (1985). As fertile couples conceive and drop out of the sample, the fecundity of the population declines. At most of the time points in this display, the probability of conception among smokers appears lower than among nonsmokers.

The adverse effects of smoking on female fertility appear to be due to toxicity to the ovary and specifically to the follicular apparatus. Experience with stimulated ovulation shows that smokers develop fewer follicles that yield fewer oocytes in response to the same degree of stimulation (Van Voorhis et al. 1992). The difference in the number of follicles (18.7 vs. 14.3 in nonsmokers and smokers, respectively) was significant but the coefficients of variation were large (45% and 62%, respectively), suggesting considerable heterogeneity in response to ovulation-inducing agents in both smokers and nonsmokers.

Toxicity to the ovarian follicle is also suggested by an earlier age at menopause among women who smoke. Figure 11 shows the relationship between smoking dose and the number of women menopausal at any given age in two populations evaluated by the Boston Collaborative Drug Surveillance Project (Jick and Porter 1977). In both samples, smoking one pack of cigarettes per day appears to advance menopause by about 2 years. The curves in these figures are arguably parallel, suggesting that the effect of smoking on the population may be uniform. Of course, finding a uniform reduction in age at menopause in a population study such as this does not preclude the existence of a small number of women who have a particularly high or low sensitivity to the ovarian toxicity of cigarette smoke.

Transplacental genotoxicity and teratogenicity

A large number of studies have evaluated the genotoxicity of cigarette smoke and its components. We address only the evidence for variability in genotoxicity measures associated with maternal smoking and not the question of the extent to which placental or transplacental genotoxicity reflects susceptibility to malignant or other disease in the offspring. Examination of DNA adduction in placental tissue from smokers and nonsmokers shows certain adducts to be present only or almost only in association with maternal smoking (Everson et al. 1988). The relative amount of DNA adduct formation was variable among smokers and not related to the number of cigarettes smoked (Figure 12). There was an association, however, between adduction and biochemical indices of smoke exposure (cotinine, thiocyanate, and carboxyhemoglobin), suggesting a difference among

A. Boston Series

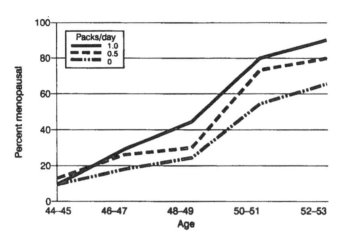

B. International Series

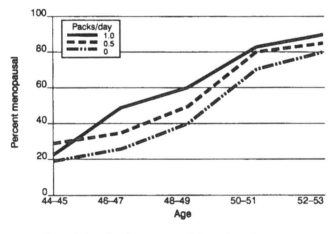

FIGURE 11   Percentage of population that is menopausal in a given 2-year age range by smoking status. (A) Women from the Boston area. (B) Women from a multinational sample. Drawn from data presented by Jick and Porter (1977).

women in method of smoking, smoke constituent toxicokinetics, or self-reporting of cigarette consumption.

An evaluation of mutations in the hypoxanthine phosphoribosyltransferase (*hprt*) locus in lymphocytes of newborns showed an increase with maternal smoking status (Ammenheuser et al. 1994). The variant frequency among newborns was 2.17×10^{-6} if the mother smoked and 0.77×10^{-6} if she did not. The coefficient of variation for this frequency was 35% for smokers and 50% for nonsmokers, suggesting a large range in the distribution of *hprt* variants in nonexposed and exposed individuals. Much of the variability among infants exposed to cigarettes may have been attributable to differences in number of cigarettes consumed by the mother; however, there were not enough individuals in each dose group to estimate the magnitude of the dose effect.

In a study of chorionic villus micronuclei, paternal but not maternal smoking was found to be associated with genotoxicity (Cui et al. 1990). The magnitude of the effect appeared to be small; however, only an English abstract was available to us and we could not estimate the extent or

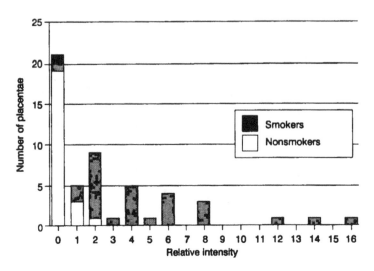

FIGURE 12   Variability in relative intensity of adduct 1 in placentae from nonsmoking and smoking women. Drawn from data of Everson et al. (1988).

variability of this effect from the available information. Still, the conclusions of these authors raise the possibility that passive smoke exposure or paternal genetic effects are a source of variability in the placental response to cigarette smoking.

Of considerable interest is the evidence that an interaction between exposure to cigarette smoke and genetic characteristics is responsible for variability in the risk of congenital malformations, specifically oral clefting. Attention has been focused on the *Taq*I site in the transforming growth factor α (TGFα) locus, for which the so-called *C2* allele, an uncommon variant, appears to confer a risk of clefting. The relationship between *Taq*I polymorphisms and cleft lip and palate was suggested by Ardinger et al. (1989) but was not observed consistently by other investigators. When children born with oral clefts were compared with children born with noncleft congenital malformations, an association between the *C2* allele and clefting was identified among pregnancies in which the mother smoked (Hwang et al. 1995). The association in this population, born in Maryland, was strongest for isolated cleft palate. The risk estimate for the association between this abnormality, smoking, and the *C2* allele was 5.6, with nonsmoking mothers who had children with only the common *C1* allele used as a reference group. When cigarette dose was considered, the risk estimate for ≤10 cigarettes per day was 6.16 and for >10 cigarettes per day the risk estimate was 8.69.

A California sample with normal births used as a reference group found a similar risk estimate for isolated cleft lip with or without cleft palate associated with maternal smoking of ≥20 cigarettes per day and the *C2* allele (Shaw et al. 1996). In this study, maternal smoking also was said to be associated with an increased risk of clefting among children with only the *C1* allele, although the 95% confidence interval included unity. An association between maternal smoking and palatal clefting has been identified in some but not all studies on the subject [reviewed by Shaw et al. (1996)]. Additional evaluation of the findings from this study are presented in Table 8, suggesting a strong association between the *C2* allelic variation and susceptibility to smoking-associated oral clefting.

Although these results suggest an important source of variability in susceptibility to smoking-related developmental toxicity, it is not possible to estimate the magnitude of the variability in terms of cigarette dose. The uncommon *Taq*I allele is present in 12% to 20% of the population, representing a considerable number of individuals potentially at risk. It is not known what dose of cigarettes (if any) would make the clefting incidence in individuals with the common allele equivalent to that of non-smoking-exposed, or minimally exposed, individuals with the uncommon allele.

TABLE 8

Risk Estimate for Oral Clefts Associated with Maternal
Smoking and the Uncommon (*C2*) *Taql* Polymorphism
of TGF-α: Odds Ratio with 95% Confidence Interval

TGF-α allele	Smoking ≥ 20 cigarettes/day	Nonsmoker
Isolated cleft lip with or without cleft palate		
Uncommon	5.7 (1.4–18.7)	0.88 (0.53–1.5)
Common	1.8 (1.0–3.2)	reference
Isolated cleft palate		
Uncommon	10.2 (2.4–37.3)	1.1 (0.58–2.2)
Common	1.8 (0.85–4.0)	reference

From Shaw and Lammer (1997).

ETHANOL

The association between ethanol abuse during pregnancy and developmental toxicity is well-known. In spite of the voluminous literature on this subject, there are important difficulties in evaluating variability in response because of unreliable reporting of ethanol doses. Still, it appears inescapable that variability exists because reports on even very heavy drinkers show rates of alcohol-associated developmental toxicity of <100%. Without attempting to review the entire literature on ethanol reproductive and developmental toxicity, we will comment on estimates of the magnitude of variability in response.

Developmental toxicity — somatic

Fetal alcohol syndrome (FAS) refers to a group of anatomic abnormalities, growth impairment, and cognitive/behavioral deficits in the offspring of women with heavy ethanol consumption during pregnancy. Within this group, there is considerable heterogeneity of FAS manifestations. In one detailed evaluation (Hanson et al. 1976), short palpebral fissures were noted in almost all of 41 affected children, but no other abnormality was present in more than two-thirds of the sample. Part of the problem in evaluating the literature is related to different definitions that have been used for fetal effects produced by alcohol and for maternal alcoholism. This area, then, is encumbered by inconsistencies and inaccuracies in ascertaining exposures and in defining outcomes, so we are not optimistic that very useful estimates on variability of response will be forthcoming. We will review, however, some of the data in an effort to delineate sources of variability that may be involved in ethanol response.

A number of early reports on pregnancy outcome in ethanol-using women did not consider possible confounding effects of smoking and other maternal factors [reviewed by Roman et al. (1988)]. These reports were largely descriptive. For example, Hanson et al. (1978) presented 163 children selected from a 1529-pregnancy survey on ethanol and caffeine exposure. In this group, 30 children were estimated to have been exposed during pregnancy to maternal ethanol consumption of ≥1 ounce of absolute alcohol per day, and 3 (10%) of these children showed physical abnormalities suggesting an ethanol effect. By contrast, 4 of 79 children (5%) exposed to an estimated maternal ethanol dose of 0.1 ounce or less had such physical features.

In a more detailed evaluation of anatomic abnormalities in 359 children, Ernhart et al. (1987b) examined the association between the presence and number of congenital defects and maternal factors including reported dose of ethanol, maternal age, race, parity, nutrition, cigarette smoking, and use of illicit drugs during pregnancy. Ethanol dose was identified as an important predictor of anatomic defects, particularly with regard to craniofacial defects. The other potential covariants were not found to be significant. It can be estimated from the graphic depiction (not shown) of the relationship between anomaly count and ethanol dose in this paper that the coefficient of variation for each dose group was about 100%, suggesting large variations in the number of anomalies noted, even though there was a relationship between this parameter and reported ethanol intake.

One of the most constant features of ethanol developmental toxicity is a reduction in newborn somatic parameters — specifically, weight, length, and head circumference. Even controlling for smoking, maternal age, maternal height, parity, gestational age, and sex of the child, consumption of 1 ounce of ethanol per day early and late in pregnancy has been associated, respectively, with a birth-weight reduction of 91 and 160 g (Little 1977). Using a regression analysis, this author later suggested that 10 g of ethanol (somewhat less than 1 drink) per day during the week before pregnancy recognition was associated with a birth-weight reduction of 225 g (Little et al. 1986). These data are reminiscent of the effects of smoking, giving some hope that an examination of variability in birth-weight effects might be similarly informative.

There appears, however, to be substantial variation in birth weight that is not easily explained by known contributors to low birth weight. Smith et al. (1986) reported data from pregnancies in which women did not drink ethanol, stopped ethanol use during the pregnancy, or continued to drink. Birth weight varied with drinking status: data for the three groups were 3282, 3091, and 3038 g, respectively, in the offspring of nondrinkers, former drinkers, and continuing drinkers. The coefficients of variation in the groups suggested, however, an increase in variability in women who continued to use ethanol, increasing in the three groups from 15% in the nondrinkers to 18% in former drinkers and 19% in continuing drinkers. Although this study was not designed to examine variability among subjects, the results are consistent with the drinking group consisting of a more heterogeneous population or a population more heterogeneous in its response to ethanol. In an evaluation of children born to alcoholic women in Sweden, where the population is more homogeneous, mean birth weight was 2912 g with a SD of 607 g, giving a coefficient of variation of 21% (Kyllerman et al. 1985). Compared with controls and a population standard, these children had birth weights, lengths, and head circumferences 1.2, 1.4, and 1.4 SD below the mean, respectively, with 95% confidence intervals of about 1 SD for each parameter. Thus, nearly the entire population of children of alcoholic women in this racially homogeneous sample fall between the mean and 2 SD below the mean for these somatic parameters.

The missing information from this study, however, is ethanol dose. A similar exercise (Barr et al. 1984) identified a dose-related decrease in weight, length, and head circumference at birth with a persistence of the weight and height deficits at 8 months of age. The birth data were largely attributable to differences in cigarette consumption among the dose groups, but the 8-month data were independent of the effects of smoking and caffeine. Examination of the weight data from this study (Table 9) shows a birth-weight distribution consistent with smoking, with relatively uniform SDs consistent with a parallel shift in the distribution curve. The 8-month data, however, show a near doubling of the SD and coefficient of variation in the high-dose group compared with lower dose groups, suggesting an important source of variation in this portion of the sample. Our concern is that difficult-to-measure differences in lifestyles and other exposures among women who drink heavily will obscure efforts to identify effects specifically resulting from ethanol exposure in these children.

The variability in long-term somatic growth in children with ethanol-associated dysmorphic features has been described by Streissguth et al. (1978). These children remain small, both in height and weight, but the severity of growth impairment varies and appears only loosely related to the severity of the dysmorphic features (Figure 13). The limitations of this presentation are the inclusion of children at different ages and the omission of estimates of maternal ethanol consumption during pregnancy; however, it appears reasonable to conclude that variable degrees of dysmorphism with variable degrees of growth impairment are the rule.

TABLE 9
Weight at Birth and 8 Months of Age of Children Exposed to Ethanol During Gestation

	Birth weight			Weight at 8 months of age		
Ethanol intake	Mean (g)	SD (g)	CV (%)	Mean (kg)	SD (kg)	CV (%)
<0.1	3550	442	12	8.8	0.6	7
0.1 to 0.9	3500	361	10	8.6	0.8	9
1 to 2	3350	410	12	8.6	0.6	7
≥2	3380	490	14	8.5	1.1	13

Note: Ethanol intake expressed in g of absolute alcohol per day. CV = coefficient of variation.

Data estimated from figures presented in Barr et al. (1984).

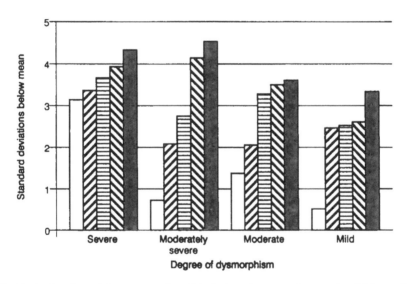

FIGURE 13   Reduction in weight, expressed as SDs below the mean for age, in children with different degrees of ethanol-associated dysmorphism. Each bar represents one child. Drawn from data presented by Streissguth et al. (1978).

Developmental toxicity — behavioral and neurologic

Infants born to women who consume large doses of ethanol show abnormalities on the Brazelton Neonatal Behavioral Assessment Scale (Smith et al. 1986). In some cases, poor performance may have been due to infant withdrawal from ethanol on day 3 of life. Associations were found among length and amount of maternal drinking, birth weight, and poor autonomic regulation and orientation on testing. These authors present the mean scores in each of the six components of the Brazelton scale according to drinking status (nondrinker, stopped drinking during pregnancy, continued to drink). It should be recognized that variability in the scores was common, even among the offspring of nondrinking women, with coefficients of variation as high as 50% for some indices. In the autonomic regulation and orientation spheres, variability in scores increased with drinking status. The coefficients of variation in nondrinkers, former drinkers, and continued drinkers were 28%, 35%, and 44%, respectively, for autonomic regulation and 21%, 26%, and 33%, respectively, for orientation. The data for birth weight are consistent with heterogeneity in the population that continued to drink. Use of the Brazelton scale by a different group (Streissguth et al. 1983) also showed abnormalities among the offspring of ethanol-using women, although these abnormalities were in the different areas of arousal and habituation.

Most studies of neurobehavioral development after ethanol exposure have used the Bayley scales of infant development, which include a psychomotor and a mental component designed to evaluate the attainment of age-appropriate skills. The mean score on the Bayley scale is 100, with a SD of 16. Poor performance on the Bayley scale has been noted in the offspring of heavy drinkers (Coles et al. 1987). Moderate maternal drinking (about 0.25 ounce of absolute alcohol per day) has also been associated with a decrease in Bayley scores (Streissguth et al. 1980, Fried and Watkinson 1988). It is interesting that the scores presented by Streissguth et al. (1980) do not show the increase in variability at high ethanol intakes about which we speculated above; except for one ethanol dose group, the SD values at each level of intake are close to 16, and there is no apparent trend in the magnitude of the SD with dose.

There is evidence that the offspring of women over 30 years old are at particular risk of scoring poorly on developmental tests after antenatal exposure to ethanol (O'Connor et al. 1986, JL Jacobson et al. 1993). In a rigorous consideration of factors contributing to performance on the Bayley scale at 13 to 18 months of life, JL Jacobson et al. (1993) found maternal ethanol exposure during pregnancy to account for only 1% to 2% of the variance in scores. For women over 30 years old, ethanol accounted for up to 7% of the variance. These authors point out that the Bayley scale may be too global in its scope and that a more detailed test of specific ethanol-disrupted functions may be more informative. An evaluation of the processing ability of infants (SW Jacobson et al. 1993) identified an impairment associated with prenatal ethanol exposure, but this exposure did not contribute much to the observed variance when other factors were considered. Infant reaction time (SW Jacobson et al. 1994) yielded more promising results, with 4% to 13% of the observed variance in scores attributable to maternal ethanol use during pregnancy.

Mental retardation has been noted consistently in children and adults with FAS (Coles et al. 1991), and these individuals are reported to have abnormalities in behavior and attention, noted to a greater extent by some investigators (Streissguth et al. 1989) than others (Brown et al. 1991). The cognitive deficits may parallel the degree of dysmorphism, but, as one investigator wrote, "the relationship is far from linear" (Coles 1993); Figure 14 shows the distribution of IQ in 20 individuals of different ages, grouped according to degree of dysmorphism. The relationship between IQ and physical abnormalities appears stronger than that between size and physical abnormalities (Figure 13), but considerable variability is still suggested among and within dysmorphic groups.

Interpretation of the variability in ethanol-associated developmental toxicity

The variability of the manifestations of prenatal ethanol exposure and the inconsistent ability to attribute effects to ethanol dose has led some reviewers to wonder whether ethanol is the causative factor in FAS or whether other attributes of the pregnant woman's lifestyle, nutrition, medical condition, or genetic endowment may be involved to a greater extent than the drug exposure (Roman et al. 1988). One reviewer (Neugut 1981) stated the reservations in this way: "The practice of heavy consumption of alcohol is *assuredly* ill-advised. Nonetheless, actually disentangling the causal associations between *in utero* alcohol exposure and untoward pregnancy outcomes from observed associations where the alcohol exposure is merely a 'passenger variable' getting a free ride on other causal associations…largely remains to be accomplished." We know, for example, that women who drink ethanol during pregnancy tend to be older, less often married, of higher parity, and more likely to have a history of medical, psychiatric, and obstetric disorders than women who do not use ethanol during pregnancy (Sokol et al. 1980).

Although we believe the characteristic facial and neurobehavioral features plus the support of experimental animal studies make it inescapable that ethanol is causally involved in human developmental toxicity, we agree that it has been difficult to evaluate sources of variability in the very

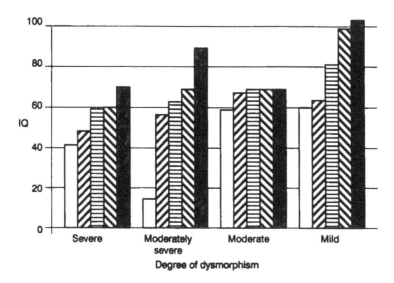

FIGURE 14 IQ in children with different degrees of dysmorphism. Each bar represents one child. Drawn from data presented by Streissguth et al. (1978).

heterogeneous outcomes that have been reported. We are mindful, however, that the technical difficulty of quantitating exposure and of standardizing outcome reporting may be responsible for much of the variability and for our inability to satisfactorily explain it. Patterns of ethanol use are likely to influence manifestations of toxicity; after all, 14 beers every 2 weeks is 1 drink per day, but if all 14 are taken on a single Saturday night, the consequences for the conceptus may be different than if they are taken on a daily basis. The importance of peak ethanol concentration vs. total exposure (area under the curve) for human pregnancy outcome has been evaluated in some studies, but this relationship is not well-understood and has not been characterized successfully in a quantitative manner [reviewed by Gladstone et al. (1996)].

There also may be important variations in genetic susceptibility to ethanol toxicity. It is not known to what extent such toxicity is due to ethanol itself or to its biotransformation product acetaldehyde. Biotransformation of ethanol to acetaldehyde is catalyzed by alcohol dehydrogenase (ADH) and acetaldehyde is biotransformed by aldehyde dehydrogenase (ALDH) to acetic acid. Both enzymes are known to have isoforms of different activity. If acetaldehyde is one of the agents responsible for ethanol toxicity, it would be anticipated that the more active ADH and the less active ALDH isoforms would be associated with the greatest risk. There is an ADH variant, called atypical ADH or $ADH2_2$ that has about 100 times the activity of the typical enzyme (Yoshida et al. 1981). The atypical enzyme is found in up to 85% of Japanese individuals (Harada et al. 1980, Agarwal 1981) and may be responsible for the flushing reaction to ethanol found among some people of Asian origin. By contrast, Native North Americans have not been found to have a high prevalence of this isoform (Rex et al. 1985, Bosron et al. 1988).

Variants of ALDH, however, are believed to be more important in determining susceptibility to acetaldehyde-related flushing after ingestion of ethanol. The ALDH2 isoform has a variant ($ALDH2_2$) that is inactive and is found in about one-half of Orientals (Agarwal et al. 1984; Goedde et al. 1980, 1983a, 1983b, 1984). The typical isoform, $ALDH2_1$, is found in most North Americans, including Native Americans (Rex et al. 1985).

To our knowledge, no association between ADH or ALDH isoforms and manifestations of fetal ethanol toxicity has been published, although we understand that identification of such an association has been attempted (EM Faustman, personal communication, 1996). If enzyme differences or other genetic factors confer an enhanced (or reduced) sensitivity to ethanol developmental toxicity, racial differences in the incidence of such toxicity would be expected. In fact, racial differences have been reported. The incidence of FAS is 0.029% among whites and 0.048% among blacks (Abel and Sokol 1991) in spite of the observation that women who continue to drink during pregnancy are more likely to be white (Prager et al. 1984). Among Native North Americans living on reservations, the prevalence of FAS and other manifestations of ethanol fetopathy among children is reported to be close to 20% (Robinson et al. 1987). This very high prevalence may reflect a particularly high rate of heavy ethanol use during pregnancy in the population, increased sensitivity to ethanol toxicity, or a combination of the two. In addition, we are mindful of the tendency to find abnormalities in proportion to the effort invested in looking for them. Inasmuch as many of the features associated with ethanol fetopathy are nonspecific, it is tempting to wonder whether overdiagnosis among Native American populations has been fueled by the belief that these people are at particular risk.

Finally, ethanol is a toxicant that produces tolerance when used chronically. Tolerance is often described as a phenomenon wherein increasing doses are used to gain the same euphoric effect, with increasingly higher plasma ethanol concentrations producing less impairment of the sensorium. It is possible that tolerance also includes a decrease in the sensitivity of conceptal tissues to ethanol toxicity. Thus, two pregnant women consuming the same amount of ethanol may not expose their fetuses to the same risk of toxicity if one is tolerant and the other is not. We have not located data, however, that permit estimation of how large such an effect might be if it occurs at all.

LEAD

Lead is a ubiquitous environmental contaminant associated with industrial activities. With the removal of lead from gasoline in developed countries and better attention to limiting exposure from paints and other lead-containing materials, lower levels of this agent are encountered by reproducing individuals. Possible adverse developmental effects at these low levels have raised concern and have prompted considerable investigation. In our consideration of susceptibility to lead toxicity, we use tissue concentration of lead, and particularly blood lead concentration, as a measure of exposure. We are aware that there is disagreement about the extent to which such measures represent the amount of lead at targets such as the fetal brain. Alternative proposals, such as measuring bone lead by x-ray fluorescence (Todd and Chettle 1994) show promise, but there is little experience with use of these measures in dose–response analyses of developmental toxicity.

Placental transport

At term, cord blood concentrations of lead have been found to be similar to maternal concentrations; however, coefficients of variation for nonexposed populations are in the 35% to 50% range (Korpela et al. 1986, Gershanik et al. 1974, Buchet et al. 1978, Zarembski et al. 1983). At increased blood lead concentrations, a correlation between maternal and cord blood and the large coefficients of variation continues to be found (Clark 1977). A significant predictor of maternal blood lead concentration is urban residence. Maternal age, birth weight, infant sex, and race have not been found to be associated with maternal or neonatal blood lead concentrations (Gershanik et al. 1974, Rabinowitz et al. 1987). There is an association between maternal blood lead concentration and use of tobacco, ethanol, and caffeine (Rabinowitz et al. 1987, Ernhart et al. 1985). Within the fetus, the amount of lead deposited increases with gestational age, probably as a function of fetal weight, with particular avidity of lead for bone and liver (Barltrop 1969). First-trimester conceptal tissues obtained after voluntary abortion show lead concentrations unrelated to maternal blood concentration (Borella et al. 1986). Conceptal lead concentration is not normally distributed and the range is wide, covering nearly 2 orders of magnitude. Placental lead concentration also is not normally distributed and the range of values in normal women spans 1.5 orders of magnitude (Roels et al. 1978). Placental lead is weakly correlated to maternal blood lead concentration but does not appear to be correlated with smoking or urban residence.

Embryo/fetal survival

A relationship between high maternal lead exposure or increased maternal blood lead concentration and abortion or stillbirth has been described, sometimes apocryphally, in a number of sources; however, we have not located information that permits an estimate of variability in response with regard to these outcomes. There is a report (Wibberly et al. 1977) describing an increase in placental lead concentration in pregnancies ending in stillbirth that shows the usual large variation in concentrations in this population [1.49 µg/g ± 0.69 (SD)].

Neurologic/cognitive development

The relationship between exposure to lead during pregnancy and neurologic outcome in the off-spring has created a controversy about the clinical significance and robustness of decrements described with low levels of lead. We do not propose to address this controversy, but we review the distribution of test results in an attempt to identify the degree of variability among those children with similar prenatal lead exposures as estimated by cord blood lead concentrations.

Many tests of cognitive development are influenced by factors such as parental intelligence and home environment, and the putative effects of lead also have been shown to be influenced by these factors (Wigg et al. 1988; McMichael et al. 1988). In addition, tests developed and validated in one racial or socioeconomic group may not perform as reliably in other groups. There are, then, factors in the measurement instruments that introduce imprecision and call into question the veracity of small observed effects. For the sake of this exercise, however, we assume that reported scores are reliable and not overly influenced by extraneous factors. We bear in mind that as older children are tested, variability is introduced by postnatal exposures to lead, a factor considered by investigators working in this area.

Table 10 summarizes the findings in Port Pirie, Australia, and in Boston, where cognitive testing was evaluated in three groups of children categorized by cord blood lead concentrations. The cognitive test used was the Bayley scale, with its population mean of 100 and SD of 16. The distribution of scores suggests reduced variability, if anything, at younger ages, with an approximation of the normal variance at 18 and 24 months. A number of other studies have evaluated the relationship between cord blood lead concentration, or other surrogates of prenatal exposure, and cognitive testing with regression models to isolate the contribution of exposure to lead (Bergomi et al. 1989; Ernhart et al. 1987a, 1989; Cooney et al. 1989; Rothenberg et al. 1989; Fergusson et al. 1988; Dietrich et al. 1986, 1987). Although not all studies have agreed that there is a significant effect of prenatal exposure to lead on cognitive function measurements in the child, a large number of studies from various parts of the world have identified distributions in test scores that have variances similar to those encountered in reference populations.

The effect of prenatal lead exposure on cognitive development, then, appears not to be subject to large-scale variability when cord blood lead concentration or a similar biomarker is used as an estimator of exposure. We recognize the possibility, however, that the lack of evidence of substantial variability is due to imprecision in measurements or to the greater relative importance of other determinants of cognitive performance.

METHYLMERCURY AND CONGENITAL NEUROLOGIC TOXICITY

An outbreak of neurologic disease in the Minimata Bay area of Japan in 1956 was caused by contamination of local seafood with methylmercury from a nearby factory (Harada 1995). This episode recurred in the mid-1960s in Japan and was repeated in the early 1970s in Iraq after consumption of bread made from grain treated with methylmercury (Greenwood 1985). These episodes were associated with adult neurologic illness as well as with a permanent cerebral palsy-like congenital disorder in the offspring of poisoned women. Organic mercurials appear to be a common contaminant of seafood, and other episodes of exposure have been reported. For example, Cree Indian boys have been found to have minor neurologic abnormalities associated with prenatal

TABLE 10
Bayley Scales of Infant Development and Relationship to Lead Concentration in Cord Blood

Reference (site)	Age at testing (months)	Blood lead (μg/dL), mean ± SD	Score Mean	Score SD
Baghurst et al. (1987) (Port Pirie)	24	high	102	11
		intermediate	109	13
		low	112	12
Bellinger et al. (1987) (Boston; summarizes several time points presented in earlier papers)	6	± 0.6 (low)	109.2	12.9
		± 0.3 (medium)	108.6	12.0
		14.6 ± 3.6 (high)	106.1	11.1
	12	same groups as above	113.1	12.5
			115.4	12.9
			108.7	12.8
	18	same groups as above	113.4	15.5
			116.6	16.7
			109.5	17.5
	24	same groups as above	115.9	17.2
			119.9	14.4
			110.6	16.5

exposure to methylmercury (McKeown-Eyssen et al. 1983). The level of exposure of this population appears to have been substantially lower than the levels in the Japanese and Iraqi poisonings.

Methylmercury crosses the placenta readily and maternal and cord blood concentrations are similar [reviewed by Clarkson (1987)]. In addition, hair mercury concentration closely parallels blood mercury, with entry of mercury into the hair above the scalp requiring about 20 days from ingestion of contaminated fish. The approximate relationship between hair and blood is 200:1; that is, 1 part per million (ppm) of mercury in hair corresponds to about 5 parts per billion (ppb) mercury in blood (Clarkson 1987). Because hair grows at about 1 cm/month, a strand of maternal hair cut at the scalp at birth can give a profile of maternal mercury blood concentrations throughout pregnancy.

By this method of dosimetry, the Iraqi poisoning has been evaluated in detail and is used here to evaluate sensitivity to the effects of methylmercury. It should be noted that such methods were not used in Japan because of the late discovery of the etiologic relationship between methylmercury intoxication and the congenital neurologic disorder (Harada 1995). About a decade after the Iraqi episode, Marsh et al. (1981) published their evaluation of the relationship between maternal peak hair mercury and the number and severity of neurologic signs in offspring. Mean concentrations in hair were 3 to 293 ppm, with a range from 0.4 to 640 ppm. A subsequent paper used a precise analytic technique for the hair measurements and constructed dose–response curves that used frequency of retarded walking as one outcome parameter and frequency of central nervous system signs as another (Cox et al. 1989). These curves were used to estimate probable thresholds for toxicity but can also be used to estimate differences in sensitivity within the population. One method of making this estimate is to use the slope of a probit or log-probit plot, as discussed in the introduction to this paper; however, we prefer the more visually appealing estimation by considering the most sensitive individual to be at the upper bound of the 95% confidence interval near the threshold and the least sensitive individual to be at the lower bound of the 95% confidence interval where the dose–response curve plateaus (Figure 15a).

In this instance, however, there is no plateau and so the least sensitive individual can be taken as being at the lower bound of the 95% confidence interval only at the upper end of the observed range (Figure 15b). With our proposed method, there is a difference of about 2 orders of magnitude

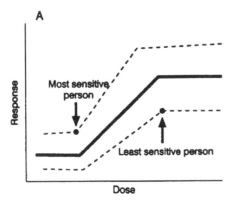

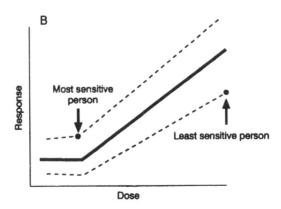

FIGURE 15   Sample dose–response curves showing a method of estimating the most and least sensitive individuals in a population. Mean is designated by a solid line with the 95% confidence interval represented by the dotted lines. (A) Response plateaus. (B) No plateau within the range of observed exposures.

between the most and the least sensitive individual in the Iraqi population. This estimate may be low, because the absence of a plateau suggests that children were still being recruited into the ranks of responders at the top end of the observed range. Thus, the 5-order-of-magnitude estimate that Hattis and Silver (1994) based on the slope of the log-probit curve is not ruled out by our estimate.

The use of these dose–response data involves some potential problems. The most important is the relative imprecision of the identification of late walking [or late talking, evaluated by Marsh et al. (1987) and used by Hattis and Silver (1994)], which was by parental report. It is possible, even likely, that knowledge of the outcome of concern affected the responses of women who perceived themselves as having been poisoned. The second problem is the possibility that the poisoning episode in Iraq was different in an important way from methylmercury exposure under other conditions. Although estimates of the methylmercury dose in the Japanese poisonings are less trustworthy because they were made some time after birth of the affected infants, it has been noted that the effect levels in Japan appeared to have been lower by nearly an order of magnitude (Grandjean et al. 1994).

The lower bound of the confidence interval on the low point in the effect range (the exposure level at which the most sensitive person shows toxicity) has gained support from two other exposed populations. Analysis of a sample from New Zealand calculated a benchmark dose under a variety of specifications but did not come out with a concentration lower than 10 ppm (Gearhart et al. 1995). This concentration is about the same as that found for very mildly affected Cree Indian boys in Canada (McKeown-Eyssen et al. 1983) and for the sample used in the model of Cox et al. (1989).

VARIATIONS IN ENZYME ACTIVITY

There are two examples of human congenital malformations for which susceptibility has been associated with altered enzyme activities. Although the data do not permit a quantitative estimate of the difference in sensitivity, we mention these examples to illustrate a source of variability in response that may prove to be important.

The anticonvulsant phenytoin is administered to patients, including pregnant women, at fairly uniform doses, often guided by serum concentrations. Thus, embryos and fetuses are exposed to approximately the same concentrations of this agent. Peak drug concentrations and total exposures would be expected to be similar, yet there is considerable variation in the manifestations of phenytoin developmental toxicity. Some exposed infants have major anatomic malformations, microcephaly, and cognitive delay and others are apparently normal.

One proposed explanation is that the active embryo-toxicant is an arene oxide intermediate that results from monooxygenase biotransformation of the parent drug. This intermediate is detoxified by the enzyme epoxide hydrolase, encoded by a gene for which there are two alleles. One of the alleles codes for an active form and the other codes for a less active form. Individuals homozygous for the inactive gene are believed to have an increased sensitivity to phenytoin toxicity. This hypothesis was supported by an evaluation of 24 children with antenatal phenytoin exposure. Ten of the children had evidence of adequate amounts of epoxide hydrolase, evaluated indirectly by the ability of peripheral lymphocytes to resist phenytoin cytotoxicity, and 14 children had inadequate resistance. The incidence of major malformations was lower in the first group (20%) than in the second group (86%) (Strickler et al. 1985). In another study, amniocytes (fetal skin cells) were harvested in the midtrimester of 19 pregnancies in which the mother used phenytoin. In four cases, the fetal cells were estimated to be deficient in epoxide hydrolase activity; in these children, phenytoin-associated abnormalities were evident at postnatal evaluation. The remaining 15 children were apparently unaffected (Buehler et al. 1990).

Although these reports indicate considerable variability in response based on epoxide hydrolase phenotype, estimation of the magnitude of variability is limited by lack of information on the plasma levels of phenytoin in women with affected fetuses and by lack of information about whether fetuses with ample epoxide hydrolase activity would be affected if the phenytoin dose were raised high enough. We can, however, make some assumptions to get an idea of how much variability there may be. We assume that the women with affected children had a plasma phenytoin concentration at the low end of the therapeutic range (10 to 20 µg/dL). We also assume that all fetuses would be affected by some manifestation of developmental toxicity at a dose that caused severe maternal toxicity, although this developmental toxicity might not be a malformation syndrome. According to the manufacturer's label, adult toxicity starts to appear at plasma concentrations of 20 µg/dL and consists of nystagmus on lateral gaze. The label implies that it is rare not to have toxicity at a plasma concentration of 50 µg/dL. Thus, if we assume that the phenytoin developmental effects start to appear at 10 µg/dL and that all adults, and hence fetuses, show toxicity at a concentration of 50 µg/dL, a 5-fold difference in sensitivity is suggested. Of course, this estimate does not take into consideration the possibility that conceptus toxicity might occur at plasma concentrations below those that are therapeutic in adults.

The second example is a recently described mutation in the gene for methylenetetrahydrofolate reductase (van der Put et al. 1995). Individuals with two copies of the mutated gene are at increased risk for spina bifida. It is presumed that the mechanism for spina bifida susceptibility conferred by this mutation is similar to that associated with low intake of folic acid. It is not known if individuals with a mutation are at greater risk of toxicant-associated neural tube defects and, if so, what the magnitude of the increased risk is. Five percent of control subjects were found to be homozygous for the mutated gene compared with 13% of individuals with spina bifida. The odds ratio was 2.9, with a 95% confidence interval of 1.0 to 7.9.

GONADAL SENSITIVITY

The most useful information on variability in human reproductive endpoints is that involving radiation or cytotoxic chemical effects on the testis or ovary. In these instances, dosimetry may be precise enough to make reasonable estimates on variability in response.

Testicular radiation

The most accurate dosimetry information in this area comes from experiments performed in the 1960s and 1970s under the sponsorship of the Atomic Energy Commission. Prison inmates who volunteered to participate had testicular irradiation one or more times followed by serial testicular biopsies, hormone determinations, and semen analyses. The reports of these experiments are summarized by Meistrich and van Beek (1990) with additional details appearing in Clifton and Bremner (1983) and Rowley et al. (1974). Experiments were performed at the Washington and Oregon State Penitentiaries, with some variations in the protocols between the two institutions. At both sites, the number of inmates at any given dose group was small. The focus of the studies was on the time course of testicular damage and recovery as well as on identification of sensitive sites within the testis. The difference in response among men within a given dose group was not an endpoint of interest, but some information in this regard can be extracted from the available reports.

The lowest radiation groups at the two sites were 7.5 and 8 radiation absorbed dose (rad; 1 rad = 0.01 Gy), respectively, and these doses apparently were considered inconsequential given that hormone measurements and biopsies were not routinely done in the men in these groups. The next highest dose group (10 rad) included three men, all of whom showed gonadotropin increases consistent with damage to the seminiferous epithelium. At 15 rad, eight men had decrements in semen parameters about 20 days after irradiation. Thus, we can assume that all men responded, at least transiently, in the range of 10 to 20 rad. In one study, an exposure to 8 rad was described as not producing an important change in semen parameters. In the other study, 7.5 rad appeared to produce a transient decrease in sperm concentration in some but not all men, at least according to our interpretation of the scattergram of results presented by Clifton and Bremner (1983). If all men responded at 10 to 15 rad and only some men responded at 7.5 rad, there is at least a 1.3- to 1.5-fold variation in sensitivity to the toxic effects of radiation.

Another way to use the data from these experiments and from accidental radiation and therapeutic radiation follows a model introduced by Meistrich and van Beek (1990). These reviewers graphed the percentage of men becoming azoospermic after testicular radiation doses from the low level of the Oregon and Washington experiments to the high end of these experiments (600 rad). Invoking a modification of the method recommended by Hattis and Silver (1994), we have used this graph to construct a probit curve, assuming that testicular sensitivity to radiation is normally distributed (Figure 16). The assumption of normality appears reasonable; the correlation coefficient for the line drawn in Figure 16 is 0.87. From the slope of the curve, we can see that one probit unit is gained with every 50 rad. If we assume that the most sensitive and the least sensitive man are 4 SD apart, the magnitude of the dose difference in sensitivity is 200 rad. If the no-effect level is near 7.5 rad, this range represents a variability of 1.4 orders of magnitude.

It can be argued that a construction such as that in Figure 16 exceeds what can be permitted given the quality of the data. Dosimetry for some of the underlying dose points was at times calculated and imprecise. In addition, the endpoint of azoospermia may be insensitive.

Cytotoxic chemotherapy in men

Men treated for cancer with cytotoxic agents may develop azoospermia with various capacities for recovery. The impact on the seminiferous epithelium appears to be the greatest for alkylating agents,

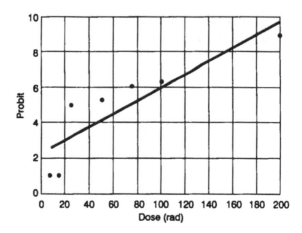

FIGURE 16   Probit of azoospermic response to testicular radiation constructed from data graphed by Meistrich and van Beek (1990).

particularly if radiation therapy is also used (Pryzant et al. 1993). Cyclophosphamide has received considerable attention in this regard, in part because it is commonly used in the treatment of boys and young men with lymphoma.

Although most men demonstrate oligospermia after cytotoxic chemotherapy, not all do, and rates of recovery vary depending on the agents used, the dose, and the individual. In one series of 35 men treated with cyclophosphamide, doxorubicin (Adriamycin), and dacarbazine, some of whom received radiation and some of whom received additional drugs, sperm concentration months after treatment varied from zero to well into the normal range (Meistrich et al. 1992). Cyclophosphamide doses of >7.5 g/m² were associated with a decreased likelihood of recovery; however, there was a man who received a 12.5-g/m² dose of this agent whose sperm concentration recovered to 63 million per mL 7 years after therapy. Another man who received cyclophosphamide at 4.1 g/m² was azoospermic thereafter (although his follow-up was not as long). There are other instances in series of this sort where men receiving very similar treatment regimens have different responses, with some becoming azoospermic and others recovering normal sperm concentrations (Meistrich et al. 1989). Age may be a factor, but there are enough departures from an age-related generalization in these reports to consider that other factors must also play a role.

Radiation and chemotherapy in women

There is a definite relationship between age and susceptibility to permanent reproductive impairment in women [reviewed by Damewood and Grochow (1986)]. The ovary is less sensitive to radiation damage than is the testis. At doses of 60 rad, there is no apparent adverse effect. At 150 rad, women over 40 years old may show gonadal failure. At more than 800 rad, virtually all women experience ovarian failure. Thus, there is about a 5-fold difference between the most sensitive and the most resistant ovary as a function of age.

The reports on female reproductive function after chemotherapy are similar to those in men. The use of alkylating agents and radiation therapy in combination appears to impose the greatest likelihood of premature ovarian failure (Byrne et al. 1992). Age appears to be the major determinant of likelihood of ovarian failure after chemotherapy (Chapman et al. 1979, Schilsky et al. 1981). Differences in dose regimens and medical condition complicate efforts to calculate the degree of variation in sensitivity by age.

CONCLUSION

A number of sources of variation in response exist in reproductive and developmental toxicology. Among the most evident are the changes in physiologic and pharmacokinetic parameters associated with pregnancy. The need to alter therapeutics for pregnant women confirms the clinical relevance of these alterations in physiology and pharmacology measures.

Variability in response to toxicants has been evaluated in detail, or can be inferred from existing data, for a few agents. It is interesting that the magnitude of this variability is variable, depending on the agent and perhaps on the populations at risk for exposure to the agent. We find, for example, that the effects of smoking on birth weight appear to be fairly uniform and we postulate that this uniformity is due to small amounts of smoking being sufficient to bring all women over their personal thresholds for toxicity. The response to ethanol, however, is extremely variable, and we suspect that there are a number of determinants of ethanol-associated toxicity, many of which are features of the population at risk. In spite of a large number of studies, some of which have been very carefully performed, it is not possible to arrive at an estimate of the variability in developmentally toxic response to ethanol independent of these features of the population.

The best quantitative estimates of variability, in our view, are obtained from work on the developmental toxicity of methylmercury and the testicular toxicity of x-rays. Although it is possible to argue about a number of the assumptions made in the estimates, it appears that variability in human response to these agents spans 1.5 to 2 orders of magnitude and may, in fact, be even greater with respect to methylmercury. Whether these two agents are representative of most agents producing reproductive or developmental toxicity is unknown.

In our review, we have been impressed with the importance of age and smoking status on a number of outcome parameters. Given the tremendous variability of these two parameters in the population, we suspect that a substantial amount of variability in human response to toxicants can be attributed to these factors. Not all studies on the adverse effects of other toxicants (e.g., ethanol) have included a careful analysis of age and smoking status. Any effort to estimate the magnitude of human response to xenobiotics, it seems, must at a minimum consider these two potential sources of variability.

REFERENCES

Aaes-Jorgensen T, Bjorndal F, Bartels U (1986) Zuclopenthixol levels in serum and breast milk. Psychopharmacology 90:417–418

Abel EL (1980) Smoking during pregnancy: a review of effects on growth and development of offspring. Hum Biol 52:593–625

Abel EL, Sokol RJ (1990) A revised conservative estimate of the incidence of FAS and its economic impact. Alcohol Clin Exp Res 15:514–524

Abramson DI, Flachs K, Fierst SM (1943) Peripheral blood flow during gestation. Am J Obstet Gynecol 45:666–671

Agarwal DP (1981) Racial differences in biological sensitivity to ethanol: the role of alcohol dehydrogenase and aldehyde dehydrogenase isozymes. Alcohol Clin Exp Res 5:12–15

Agarwal DP, Eckey R, Harada S, Goedde HW (1984) Basis of aldehyde dehydrogenase deficiency in Orientals: immunochemical studies. Alcohol 1:111–118

Akintonwa A, Gbajumo SA, Biola Mabadeje AF (1988) Placental and milk transfer of chloroquine in humans. Ther Drug Monit 10:147–149

Albani F, Riva R, Contin M, et al (1984) Differential transplacental binding of valproic acid: influence of free fatty acids. Br J Clin Pharmacol 17:759–762

Allen MC (1984) Developmental outcome and followup of the small for gestational age infant. Semin Perinatol 8:102–156

Amann RP (1981) A critical review of methods for evaluation of spermatogenesis from seminal characteristics. J Androl 2:37–58

Ammenheuser MM, Berenson AB, Stiglich NJ, et al (1994) Elevated frequencies of *hprt* mutant lymphocytes in cigarette-smoking mothers and their newborns. Mutat Res 304:285–294

Anderson PO (1991) Drug use during breast-feeding. Clin Pharmacol 10:594–624

Ardinger HH, Buetow KH, Bell GI et al (1989) Association of genetic variation of the transforming-growth factor-alpha gene with cleft lip and palate. Am J Hum Genet 45:348–353

Aronson RA, Uttrech S, Soref M (1991) The effect of maternal cigarette smoking on low birth weight and preterm birth in Wisconsin, 1991. Wis Med J 92:613–617

Auger J, Kunstmann JM, Czyglik F, Jouannet P (1995) Decline in semen quality among fertile men in Paris during the past 20 years. N Engl J Med 332:281–335

Baghurst PA, Robertson EF, McMichael AJ, et al (1987) The Port Pirie Cohort Study: lead effects on pregnancy outcome and early childhood development. Neurotoxicology 8:395–402

Baird DD, Wilcox AJ (1985) Cigarette smoking associated with delayed conception. JAMA 253:2979–2983

Baldwin GR, Moorthi DS, Whelton JA, MacDonnell KF (1977) New lung functions and pregnancy. Am J Obstet Gynecol 127:235–239

Barltrop D (1969) Transfer of lead to the human foetus. In Barltrop D, Burland WL (eds), Mineral metabolism in pediatrics. Blackwell Scientific, Oxford, pp 135–151

Barr HM, Streissguth AP, Martin DC, Herman CS (1984) Infant size at 8 months of age: relationship to maternal use of alcohol, nicotine, and caffeine during pregnancy. Pediatrics 74:336–341

Batista MC, Cartledge TP, Zellmer AW, et al (1995) Effects of aging on menstrual cycle hormones and endometrial maturation. Fertil Steril 64:492–499

Bean JA, Leeper JD, Wallace RB, et al (1979) Variations in reporting of menstrual histories. Am J Epidemiol 109:181–185

Bellinger D, Leviton A, Waternaux C, et al (1987) Longitudinal analysis of prenatal and postnatal lead exposure and early cognitive development. N Engl J Med 316:1037–1043

Bennet PN, ed (1988) Drugs and human lactation. Elsevier, New York, pp 358–360

Bergomi M, Borella P, Fantuzzi G, et al (1989) Relationship between lead exposure indicators and neuropsychological performance in children. Dev Med Child Neurol 31:181–190

Blum V (1936) Das problem des mannlichen klimakteriums. Wien Klin Wochenschr 2:1133–1140

Borella P, Picco P, Masellis G (1986) Lead content in abortion material from urban women in early pregnancy. Int Arch Occup Environ Health 57:93–99

Bosron WF, Rex DK, Harden BA, Li TK (1988) Alcohol and aldehyde dehydrogenase isoenzymes in Sioux North American Indians. Alcohol Clin Exp Res 12:454–455

Brown RT, Coles CD, Smith IE, et al (1991) Effects of prenatal alcohol exposure at school age. II. Attention and behavior. Neurotoxicol Teratol 13:369–376

Buchet JP, Roels H, Hubermont, Lauwerys R (1978) Placental transfer of lead, mercury, cadmium, and carbon monoxide in women. Environ Res 15:494–503

Buehler BA, Delimont D, van Waes M, Finnell RH (1990) Prenatal prediction of risk of the fetal hydantoin syndrome. N Engl J Med 322:1567–1572

Byrne J, Fears TR, Gail MH, et al (1992) Early menopause in long-term survivors of cancer during adolescence. Am J Obstet Gynecol 166:788–793

Byrne J, Warburton D (1986) Neural tube defects in spontaneous abortion. Am J Med Genet 25:327–333

Cano F, Simón C, Remohí J, Pellicer A (1995) Effect of aging on the female reproductive system: evidence for a role of uterine senescence in the decline in female fecundity. Fertil Steril 64:584–589

Carter BL, Driscoll CE, Smith GD (1983) Delayed elimination of theophylline in pregnancy. Can Med Assoc J 128:1142

Centers for Disease Control and Prevention (1994) Increasing incidence of low birthweight — United States 1981–1991. Morbid Mortal Weekly Rep 43:335–339

Centers for Disease Control (1980) Congenital malformations surveillance report, January–December 1980. Centers for Disease Control, Atlanta, GA

Chapman RM, Sutcliffe SB, Malpas JS (1979) Cytotoxic-induced ovarian failure in women with Hodgkin's disease. I. Hormone function. JAMA 242:1877–1881

Chen SS, Perucca E, Lee JN, Richens A (1982) Serum protein binding and free concentration of phenytoin and phenobarbitone in pregnancy. Br J Clin Pharmacol 13:547–554

Chiazze L Jr, Brayer FT, Macisco JJ Jr, et al (1968) The length and variability of the human menstrual cycle. JAMA 203:89–92

Chiba K, Ishizaki T, Tabuchi T et al (1982) Antipyrine disposition in relation to lowered anticonvulsant plasma level during pregnancy. Obstet Gynecol 60:620–626

Christofides ND, Ghatei MA, Bloom SI, et al (1982) Decreased plasma motilin concentrations in pregnancy. Br Med J 285:1453–1454

Clark ARL (1977) Placental transfer of lead and its effects on newborns. Postgrad Med J 53:674–678

Clarkson TW (1987) The role of biomarkers in reproductive and developmental toxicology. Environ Health Perspect 74:103–107

Clifton DK, Bremner WJ (1983) The effect of testicular x-irradiation on spermatogenesis in man. A comparison with the mouse. J Androl 4:387–392

Cliver SP, Goldenberg RL, Cutter GR, et al (1995) The effect of cigarette smoking on neonatal anthropometric measurements. Obstet Gynecol 85:625–630

Cnattingius S, Forman MR, Berendes HW, et al (1993) Effect of age, parity, and smoking on pregnancy outcome: a population-based study. Am J Obstet Gynecol 168:16–21

Cogswell ME, Yip R (1995) The influence of fetal and maternal factors on the distribution of birthweight. Semin Perinatol 19:222–240

Cole AP, Hailey DM (1975) Diazepam and active metabolites in breast milk and their transfer to the neonate. Arch Dis Child 50:741–742

Coles CD (1993) Impact of prenatal alcohol exposure on the newborn and the child. Clin Obstet Gynecol 36:255–266

Coles CD, Brown RT, Smith IE, (1991) Effects of prenatal alcohol exposure at school age. I. Physical and cognitive development. Neurotoxicol Teratol 13:357–367

Coles CD, Smith IE, Falek A (1987) A neonatal marker for cognitive vulnerability to alcohol's teratogenic effects. Alcohol Clin Exp Res 11:197

Cooney GH, Bell A, McBride W, Carter C (1989) Neurobehavioural consequences of prenatal low level exposures to lead. Neurotoxicol Teratol 11:95–104

Cox C, Clarkson TW, Marsh DO, et al (1989) Dose–response analysis of infants prenatally exposed to methylmercury: an application of a single compartment model to single-strand hair analysis. Environ Res 49:318–332

Cugell DW, Frank NR, Gaensler EA, Badger TL (1953) Pulmonary function in pregnancy. I. Serial observations in normal women. Am Rev Tuberculosis Pulm Dis 67:568–597

Cui YQ, Dong ZW, Liu SB, et al (1990) [Assessment of the mutagenic effect of maternal factors on human chorionic villi by micronucleus test.] I Chuan Hsueh Pao 17:238–242

Cummings AJ (1983) A survey of pharmacokinetic data from pregnant women. Clin Pharmacokinet 8:344–354

Dalessio DJ (1985) Current concepts: seizure disorders and pregnancy. N Engl J Med 312:559–563

Damewood MD, Grochow LB (1986) Prospects for fertility after chemotherapy or radiation for neoplastic disease. Fertil Steril 45:443–459

Dansky LV, Finnell RH (1991) Parental epilepsy, anticonvulsant drugs, and reproductive outcome: epidemiologic and experimental findings spanning three decades. 2. Human studies. Reprod Toxicol 5:301–335

Dansky L, Andermann E, Andermann F (1982) Major congenital malformations in the offspring of epileptic patients: genetic and environmental risk factors. In Janz D, Dam M, Richens A, et al (eds), Epilepsy, pregnancy, and the child. Raven Press, New York, pp 223–234

Davison JM, Hytten FE (1974) Glomerular filtration during and after pregnancy. J Obstet Gynaecol Br Commonwealth 81:588–595

DeStefano F, Annest JL, Kresnow MJ, et al (1987) Automated semen analysis in large epidemiologic studies. J Androl 8:24

Dietrich KN, Krafft KM, Bornschein RL, et al (1987) Low-level fetal lead exposure effect of neurobehavioral development in early infancy. Pediatrics 80:721–730

Dietrich KN, Krafft KM, Bier M, et al (1986) Early effects of fetal lead exposure: neurobehavioral findings at 6 months. Int J Biosoc Res 8:151–168

Droste S (1992) Fetal growth in aneuploid conditions. Clin Obstet Gynecol 35:119–125

Dunlop W (1981) Serial changes in renal haemodynamics during normal human pregnancy. Br J Obstet Gynaecol 88:1–9

Edstein MD, Veenendaal JR, Newman K, Hyslop R (1986) Excretion of chloroquine, dapsone and pyrimethamine in human milk. Br J Clin Pharmacol 22:733–735

English PB, Eskenazi B, Christianson RE (1994) Black–white differences in serum cotinine levels among pregnant women and subsequent effects on infant birthweight. Am J Publ Health 84:1439–1443

Ernhart CB, Morrow-Tlucak M, Wolf AW, et al (1989) Low level lead exposure in the prenatal and early preschool periods: intelligence prior to school entry. Neurotoxicol Teratol 11:161–170

Ernhart CB, Morrow-Tlucak M, Marler MR, Wolf AW (1987a) Low level lead exposure in the prenatal and early preschool periods: early preschool development. Neurotoxicol Teratol 9:259–270

Emhart CB, Sokol RJ, Martier S, et al (1987b) Alcohol teratogenicity in the human: a detailed assessment of specificity, critical period, and threshold. Am J Obstet Gynecol 156:33–39

Emhart CB, Wolf AW, Sokol RJ, et al (1985) Fetal lead exposure: antenatal factors. Environ Res 38:54–66

Everson RB, Randerath E, Santella RM, et al (1988) Quantitative associations between DNA damage in human placenta and maternal smoking and birth weight. J Natl Cancer Inst 80:567–576

Ezimokhai M, Davison JM, Philips PR, Dunlop W (1981) Non-postural serial changes in renal function during the third trimester of normal human pregnancy. 88:465–471

Fergusson DM, Fergusson JE, Horwood LJ, Kinzett NG (1988) A longitudinal study of dentine lead levels, intelligence, school performance and behaviour. Part II. Dentine lead and cognitive ability. J Child Psychiat 29:793–809

Ferris AN, Jensen RG (1984) Lipids in human milk: a review. 1. Sampling, determination and content. J Pediatr Gastro Nutr 3:108–122

Findlay JWA, DeAngelis RL, Kearney MF et al (1981) Analgesic drugs in breast milk and plasma. Clin Pharmacol Ther 29:625–633

Fleishaker JC, Desai N, McNamara PJ (1987) Factors affecting the milk-to-plasma ratio in lactating women: physical interactions with protein and fat. J Pharmaceut Sci 76:189–193

Florey CdV, Taylor DJ (1994) The relation between antenatal care and birth weight. Rev Epidemiol Sante Publ 42:191–197

Fox SH, Koepsell TD, Daling JR (1994) Birth weight and smoking during pregnancy — effect modification by maternal age. Am J Epidemiol 39:1008–1015

Frederiksen MC, Ruo TI, Chow MJ, Atkinson AJ (1986) Theophylline pharmacokinetics in pregnancy. Clin Pharmacol Ther 40:321–328

Fried PA, Watkinson B (1988) 12- and 24-month neurobehavioural follow-up of children prenatally exposed to marihuana, cigarettes and alcohol. Neurotoxicol Teratol 10:305–313

Froescher W, Gugler R, Niesen M, Hoffman F (1984) Protein binding of valproic acid in maternal and umbilical cord serum. Epilepsia 25:244–249

Gabbe SG (1991) Intrauterine growth retardation. In Gabbe SG, Niebyl JR, Simpson JL (eds), Obstetrics: normal and problem pregnancies, 2nd ed. Churchill Livingstone, New York, pp 923–944

Gant NF, Daley GL, Chand S, et al (1973) A study of angiotensin II pressor response throughout primigravid pregnancy. J Clin Invest 52:2682–2689

Gardner MJ, Schatz M, Cousin L, et al (1987) Longitudinal effects of pregnancy on the pharmacokinetics of theophylline. Eur J Clin Pharmacol 31:289–295

Gearhart JM, Clewell HJ III, Crump KS, et al (1995) Pharmacokinetic dose estimates of mercury in children and dose–response curves of performance tests in a large epidemiological study. Water Air Soil Pollut 80:49–58

Gershanik JJ, Brooks GG, Little JA (1974) Blood lead values in pregnant women and their offspring. Am J Obstet Gynecol 119:508–511

Ghai V, Vidyasagar D (1988) Morbidity and mortality factors in twins: an epidemiologic approach. Clin Perinatol 15:123–140

Gibson RA, Kneebone GM (1980) Effect of sampling on fatty acid composition of human colostrum. J Nutr 110:1671–1675

Ginsburg J, Duncan SLB (1967) Peripheral blood flow during normal pregnancy. Cardiovasc Res 1:132–137

Gladstone J, Nulman I, Koren G (1996) Reproductive risks of binge drinking during pregnancy. Reprod Toxicol 10:3–13

Goedde HW, Benkmann HG, Kriese L, et al (1984) Aldehyde dehydrogenase isozyme deficiency and alcohol sensitivity in four different Chinese populations. Hum Hered 34:183–186

Goedde HW, Agarwal DP, Harada S (1983a) Pharmacogenetics of alcohol sensitivity. Pharmacol Biochem Behav 18(Suppl 1):161–166

Goedde HW, Agarwal DP, Harada S (1983b) The role of alcohol dehydrogenase and aldehyde dehydrogenase isozymes in alcohol metabolism, alcohol sensitivity, and alcoholism. Isozymes Curr Top Biol Med Res 8:175–193

Goedde HW, Agarwal DP, Harada S (1980) Genetic studies on alcohol-metabolizing enzymes: detection of isozymes in human hair roots. Enzyme 25:281–286

Grandjean P, Weihe P, Nielsen JB (1994) Methylmercury: significance of intrauterine and postnatal exposures. Clin Chem 40:1395–1400

Greenwood MR (1985) Methylmercury poisoning in Iraq. An epidemiological study of the 1971–1972 outbreak. J Appl Toxicol 5:148–159

Hanson JW, Streissguth AP, Smith DW (1978) The effect of moderate alcohol consumption during pregnancy on fetal growth and morphogenesis. J Pediatr 92:457–460

Hanson JW, Jones KL, Smith DW (1976) Fetal alcohol syndrome. JAMA 235:1458–1460

Harada M (1995) Minamata disease: methylmercury poisoning in Japan caused by environmental pollution. Crit Rev Toxicol 25:1–24

Harada S, Misawa S, Agarwal DP, Goedde HW (1980) Liver alcohol dehydrogenase and aldehyde dehydrogenase in the Japanese: isozyme variation and its possible role in alcohol intoxication. Am J Hum Genet 32:8–15

Hartmann PE, Rattigan S, Saint L, et al (1985) Variation in the yield and composition of human milk. Oxford Rev Reprod Biol 7:118–167

Hattis D, Silver K (1994) Human interindividual variability — a major source of uncertainty in assessing risks for noncancer health effects. Risk Anal 14:421–431

Herngren L, Ernebo M, Boreus LO (1983) Drug binding to plasma proteins during human pregnancy and in the perinatal period. Dev Pharm Ther 6:110–124

Hod M, Merlob P, Friedman S, et al (1992) Prevalence of minor congenital anomalies in newborns of diabetic mothers. Eur J Obstet Gynecol Reprod Biol 44:111–116

Howe G, Westhoff C, Vessey M, Yeates D (1985) Effect of age, cigarette smoking, and other factors on fertility: findings in a large prospective study. Br Med J 290:1697–1700

Hwang S-J, Beaty TH, Panny SR, et al (1995) Association study of transforming growth factor alpha (TGFα) TaqI polymorphism and oral clefts: indication of gene-environment interaction in a population-based sample of infants with birth defects. Am J Epidemiol 141:629–636

Hytten FE (1991) Weight gain in pregnancy. In Hytten FE, Chamberlain G (eds), Clinical physiology in obstetrics, 2nd ed. Blackwell Scientific, Oxford, pp 173

Hytten FE (1980) Nutrition. In Hytten FE, Chamberlain G (eds), Clinical physiology in obstetrics. Blackwell Scientific, Oxford, pp 163–192

Hytten FE, Thomson AM, Taggart N (1966) Total body water in normal pregnancy. J Obstet Gynaecol Br Commonwealth 73:553–561

Institute of Medicine (1990) Nutrition during pregnancy. National Academy Press, Washington, D.C.

International Clearinghouse for Birth Defect Monitoring Systems (1991) Congenital malformations worldwide. Elsevier, New York

Jacobson JL, Jacobson SW, Sokol RJ, et al (1994) Effect of alcohol use, smoking, and illicit drug use on fetal growth in black infants. J Pediatr 124:757–764

Jacobson JL, Jacobson SW, Sokol RJ, et al (1993) Teratogenic effects of alcohol on infant development. Alcohol Clin Exp Res 17:174–183

Jacobson SW, Jacobson JL, Sokol RJ (1994) Effects of fetal alcohol exposure on infant reaction time. Alcohol Clin Exp Res 18:1125–1132

Jacobson SW, Jacobson JL, Sokol RJ, et al (1993) Prenatal alcohol exposure and infant information processing ability. Child Dev 64:1706–1721

Jagoe JR, Washbrook NP, Hudson EA (1986) Morphometry of spermatozoa using semiautomatic image analysis. J Clin Pathol 39:1347–1352

Jick H, Porter J (1977) Relation between smoking and age of natural menopause. Report from the Boston Collaborative Drug Surveillance Program, Boston University Medical Center. Lancet 1:1354–1355

Johannessen SI (1992) Pharmacokinetics of valproate in pregnancy: mother-foetus-newborn. Pharm Week Sci 14:114–117

Johnson L (1982) A re-evaluation of daily sperm output of men. Fertil Steril 37:811–816

Johnson L, Petty CS, Neaves WB (1984a) Influence of age on sperm production and testicular weights in men. J Reprod Fertil 70:211–218

Johnson L, Zane RS, Petty CS, Neaves WB (1984b) Quantification of the human Sertoli cell population: its distribution, relation to germ cell numbers, and age-related decline. Biol Reprod 31:785–795

Juchau MR, Chao ST, Omiecinski CJ (1980) Drug metabolism by the human fetus. Clin Pharmacokinet 5:320–329

Kafetzis D, Siafas CA, Georgakopoulos PA, Papadatos CJ (1981) Passage of cephalosporins and amoxicillin into the breast milk. Acta Paediatr Scand 70:285–288

Katz DF, Overstreet JW, Pelprey J (1981) Integrated assessment of the motility, morphology, and morphometry of human spermatozoa. INSERM 103:97–100

Kemp J, Ilett KF, Booth J, Hackett LP (1985) Excretion of doxepin and N-desmethyldoxepin in human milk. Br J Clin Pharmacol 20:497–499

Kleinman JC, Pierre MB Jr, Madans JH, et al (1988) The effects of maternal smoking on fetal and infant mortality. Am J Epidemiol 127:274–282

Knuth UA, Yeung C-H, Nieschlag E (1987) Computerized semen analysis: objective measurement of semen characteristics is biased by subjective parameter setting. Fertil Steril 48:1118–1124

Koninckx PR, Goddeeris PG, Lauweryns JM, et al (1977) Accuracy of endometrial biopsy dating in relation to the midcycle luteinizing hormone peak. Fertil Steril 28:443–445

Korpela H, Loueniva R, Yrjänheikki E, Kauppila A (1986) Lead and cadmium concentrations in maternal and umbilical cord blood, amniotic fluid, placenta, and amniotic membranes. Am J Obstet Gynecol 155:1086–1089

Kramer MS (1987) Determinants of low birth weight: methodological assessment and meta-analysis. Bull World Health Org 65:663–737

Kuhnz W, Nau H (1983) Differences in *in vitro* binding of diazepam and N-desmethyldiazepam to maternal and fetal plasma protein at birth: relation to free fatty acid concentration and other parameters. Clin Pharmacol Ther 34:220–226

Kyllerman M, Aronson M, Sabel K-G, et al (1985) Children of alcoholic mothers. Growth and motor performance compared to matched controls. Acta Paediatr Scand 74:20–26

Leader A, Wiseman D, Taylor PJ (1985) The prediction of ovulation: a comparison of the basal body temperature graph, cervical mucus score, and real-time pelvic ultrasonography. Fertil Steril 43:385–388

Lehman V, Fabel H (1973) Lungenfunktion untersuchungen an Achwangeren Teil I: Lungenvolumina. Z Geburt Perinatol 177:387

Levy RH, Yerby MS (1985) Effects of pregnancy on antiepileptic drug utilization. Epilepsia 26(Suppl 1):S52–S57

Little RE (1977) Moderate alcohol use during pregnancy and decreased infant birth weight. Am J Public Health 67:1154–1156

Little RE, Asker RL, Sampson PD, Renwick JH (1986) Fetal growth and moderate drinking in early pregnancy. Am J Epidemiol 123:270–278

Luke B, Williams C, Minogue J, Keith L (1993) The changing pattern of infant mortality in the US: the role of prenatal factors and their obstetrical implications. Int J Gynaecol Obstet 40:199–212

Mammen GJ, ed (1990) Clinical pharmacokinetics drug data handbook. ADIS Press, Auckland, New Zealand

Marsh DO, Clarkson TW, Cox G, et al (1987) Fetal methylmercury poisoning. Relationship between concentration in single strands of maternal hair and child effects. Arch Neurol 44:1017–1022

Marsh DO, Clarkson TW, Cox G, et al (1981) Dose–response relationship for human fetal exposure to methylmercury. Clin Toxicol 18:1311–1318

Matheson I, Skjaeraasen J (1988) Milk concentrations of flupenthixol, nortriptyline and zuclopenthixol and between-breast differences in two patients. Eur J Clin Pharmacol 35:217–220

Matheson I, Pande H, Altersen AR (1985) Respiratory depression caused by N-desmethyldoxepin in breast milk. Lancet 2:1124

McKeown-Eyssen GE, Ruedy J, Neims A (1983) Methylmercury exposure in northern Quebec. II. Neurologic findings in children. Am J Epidemiol 118:470–479

McMichael AJ, Baghurst PA, Wigg NR, et al (1988) Port Pirie Cohort Study: environmental exposure to lead and children's abilities at the age of four years. N Engl J Med 319:468–475

Meistrich ML, van Beek MEAB (1990) Radiation sensitivity of the human testis. Adv Radiat Biol 14:227–268

Meistrich ML, Wilson G, Brown BW, et al (1992) Cancer 70:2703–2712

Meistrich ML, Chawla SP, Da Cunha MF, et al (1989) Recovery of sperm production after chemotherapy for osteosarcoma. Cancer 63:2115–2123

Menken J, Trussell J, Larsen U (1986) Age and infertility. Science 233:1389–1394

Mills JL, Knopp RH, Simpson JL, et al (1988a) Lack of relation of increased malformation rates in infants of diabetic mothers to glycemic control during organogenesis. N Engl J Med 318:671–676

Mills JL, Simpson JL, Driscoll SG, et al (1988b) Incidence of spontaneous abortion among normal women and insulin-dependent diabetic women whose pregnancies were identified within 21 days of conception. N Engl J Med 319:1617–1623

Milunsky A, Jick SS, Bruell CL, et al (1989) Predictive values, relative risks, and overall benefits of high and low maternal serum α-fetoprotein screening in singleton pregnancies: new epidemiologic data. Am J Obstet Gynecol 161:291–297

Mongoven M, Dolan-Mullen P, Groff JY, et al (1996) Weight gain associated with prenatal smoking cessation in white, non-Hispanic women. Am J Obstet Gynecol 174:72–77

Morbidity and Mortality Weekly Report (1992) Spina bifida incidence at birth — United States, 1983–1990. 41:497–500

Morriss FH Jr, Brewer ED, Spedale SB, et al (1986) Relationship of human milk pH during course of lactation to concentration of citrate and fatty acids. Pediatrics 78:458–464

Morrow RJ, Knox Ritchie JW, Bull SB (1988) Maternal cigarette smoking: the effects on umbilical and uterine blood flow velocity. Am J Obstet Gynecol 159:1069–1071

Moruzzi JF, Wyrobek AJ, Mayall BH, Gledhill BL (1988) Quantification and classification of human sperm morphology by CAIA. Fertil Steril 50:142–152

Murray FA, Erskine JP, Fielding J (1957) Gastric secretion in pregnancy. J Obstet Gynaecol Br Empire 64:373–381

Neaves WB, Johnson L, Porter JC, et al (1984) Leydig cell numbers, daily sperm production, and serum gonadotropin levels in aging men. J Clin Endocrinol Metab 59:756–763

Neugut RH (1981) Epidemiological appraisal of the literature on the fetal alcohol syndrome in humans. Early Hum Dev 5:411–429

Neville MC, Keller R, Seacat J et al (1988) Studies in human lactation: milk volumes in lactating women during the onset of lactation. Am J Clin Nutr 48:1375–1386

Nevin NC (1979) The importance of screening for genetic diseases. R Soc Health J 99:37–40

Notarianni LJ (1990) Plasma protein binding of drugs in pregnancy and in neonates. Clin Pharmacokinet 18:20–36

Noyes RW, Haman JO (1953) Accuracy of endometrial dating. Fertil Steril 4:504–517

Noyes RW, Hertig AT, Rock J (1950) Dating the endometrial biopsy. Fertil Steril 1:3–25

O'Connor MJ, Brill NJ, Sigman M (1986) Alcohol use in primiparous women older than 30 years of age: relation to infant development. Pediatrics 78:444–450

Parisi VM, Creasy RK (1992) Maternal biologic adaptations to pregnancy. In Reese EA, Hobbins JC, Mahoney MJ, Petrie RH (eds), Medicine of the fetus and mother. Lippincott, Philadelphia, pp 831–848

Pauerstein CJ (1987) Clinical obstetrics. Wiley, New York

Perucca E, Crema A (1982) Plasma protein binding of drugs in pregnancy. Clin Pharmacokinet 7:336–352

Perucca E, Ruprah M, Richens A (1981a) Altered drug binding to serum proteins in pregnant women: therapeutic relevance. J R Soc Med 74:422–426

Perucca E, Ruprah M, Richens A (1981b) Altered drug binding to serum proteins in pregnant women: therapeutic relevance. J R Soc Med 74:422–426

Prager K, Malin H, Spiegler D, et al (1984) Smoking and drinking behavior before and during pregnancy of married mothers of live-born infants and stillborn infants. Public Health Rep 99:117–143

Pritchard JA (1965) Changes in blood volume during pregnancy and delivery. Anesthesiology 26:393–399

Pritchard JA, Cunningham FG, Pritchard SA (1984) The Parkland Memorial Hospital protocol for treatment of eclampsia: evaluation of 245 cases. Am J Obstet Gynecol 148:951–963

Pryzant RM, Meistrich ML, Wilson G, et al (1993) Long-term reduction in sperm count after chemotherapy with and without radiation therapy for non-Hodgkin's lymphomas. J Clin Oncol 11:239–247

Rabinowitz M, Bellinger D, Leviton A, et al (1987) Pregnancy hypertension, blood pressure during labor, and blood lead levels. Hypertension 10:447–451

Raymond EG, Cnattingius S, Kiely JL (1994) Effect of maternal age, parity, and smoking on the risk of stillbirth. Br J Obstet Gynaecol 101:301–306

Rex DK, Bosron WF, Smialek JE, Li TK (1985) Alcohol and aldehyde dehydrogenase isoenzymes in North American Indians. Alcohol Clin Exp Res 9:147–152

Robinson GC, Conry JL, Conry RF (1987) Clinical profile and prevalence of fetal alcohol syndrome in an isolated community in British Columbia. Can Med Assoc J 137:203–207

Robson SC, Hunter S, Boys RJ, Dunlop W (1989) Serial study of factors influencing changes in cardiac output during human pregnancy. Am J Physiol 256:H1061–H1065

Roels H, Hubermont G, Buchet JP, Lauwerys R (1978) Placental transfer of lead, mercury, cadmium, and carbon monoxide in women. III. Factors influencing the accumulation of heavy metals and the relationship between metal concentration in the placenta and in maternal and cord blood. Environ Res 16:236–247

Roman E, Beral V, Zuckerman B (1988) The relation between alcohol consumption and pregnancy outcome in humans. Issues Rev Teratol 4:205–235

Rothenberg SJ, Schnaas L, Cansino-Ortiz S, et al (1989) Neurobehavioral deficits after low level lead exposure in neonates: the Mexico City pilot study. Neurotoxicol Teratol 11:85–93

Rowley MJ, Leach DR, Warner GA, Heller CG (1974) Effect of graded doses of ionizing radiation on the human testis. Radiat Res 59:665–678

Ruprah M, Perucca E, Richens A (1980) Decreased serum protein binding of phenytoin in late pregnancy. Lancet 2:316–317

Sacks PC (1993) The menstrual cycle. In Scialli AR, Zinaman MJ (eds), Reproductive toxicology and infertility. McGraw-Hill, New York, pp 133–185

Schilsky RL, Sherins RJ, Hubbard SM, et al (1981) Long-term follow-up of ovarian function in women treated with MOPP chemotherapy for Hodgkin's disease. Am J Med 71:552–556

Schmassmann A, Mikuz G, Bartsch G, Rohr H (1979) Quantification of human sperm morphology and motility by means of semi-automatic image analysis systems. Microscop Acta 82:163–178

Schrader SM, Turner TW, Simon SD (1991) Longitudinal study of semen quality of unexposed workers. Sperm motility characteristics. J Androl 12:126–131

Schrader SM, Turner TW, Breitenstein MJ, Simon SD (1988) Longitudinal study of semen quality of unexposed workers. I. Study overview. Reprod Toxicol 2:183–190

Schrader SM, Turner TW, Hardin BD, et al (1984) Morphometric analysis of human spermatozoa. J Androl 5:P22

Schwartz D, Laplanche A, Jouannet P, David G (1979) Within-subject variability of human semen in regard to sperm count, volume, total number of spermatozoa and length of abstinence. J Reprod Fertil 57:391–395

Scialli AR (1992) A clinical guide to reproductive and developmental toxicology. CRC Press, Boca Raton, FL

Scott DE (1972) Anemia during pregnancy. Obstet Gynecol Annu 1:219–244

Shaw GM, Lammer EJ (1997) Incorporating molecular genetic variation and environmental exposures into epidemiologic studies of congenital anomalies. Reprod Toxicol 11:275–

Shaw GM, Wasserman CR, Lammer EJ, et al (1996) Orofacial clefts, parental cigarette smoking, and transforming growth factor-alpha gene variants. Am J Hum Genet 58:551–561

Sherins RJ (1995) Are semen quality and male fertility changing? N Engl J Med 322:327–328

Simpson JL, Elias S, Martin AO, et al (1983) Diabetes in pregnancy, Northwestern University Series (1977–1981). Prospective study of anomalies in offspring of mothers with diabetes mellitus. Am J Obstet Gynecol 146:263–270

Smith IE, Coles CD, Lancaster J, et al (1986) The effect of volume and duration of prenatal ethanol exposure on neonatal physical and behavioral development. Neurobehav Toxicol Teratol 8:375–381

Smith KE, Buyalos RP (1996) The profound impact of patient age on pregnancy outcome after early detection of fetal cardiac activity. Fertil Steril 65:35–40

Sokol RJ, Miller SI, Reed G (1980) Alcohol abuse during pregnancy: an epidemiologic study. Alcohol Clin Exp Res 4:135–145

Spetz S, Jansson I (1969) Forearm blood flow during normal pregnancy studied by venous occlusion plethysmography and 133Xenon muscle clearance. Acta Obstet Gynecol Scand 48:285–301

Stillman RJ, Rosenberg MJ, Sachs BP (1986) Smoking and reproduction. Feril Steril 46:545–566

Streissguth AP, Bookstein FL, Sampson PD, Barr HM (1989) Neurobehavioral effects of prenatal alcohol. III. PLS analysis of neuropsychologic tests. Neurotoxicol Teratol 11:493–507

Streissguth AP, Barr HM, Martin DC (1983) Maternal alcohol use and neonatal habituation assessed with the Brazelton scale. Child Dev 54:1109–1118

Streissguth AP, Barr HM, Martin DC, Herman CS (1980) Effects of maternal alcohol, nicotine, and caffeine use during pregnancy on infant mental and motor development at 8 months. Alcohol Clin Exp Res 4:152–164

Streissguth AP, Herman CS, Smith DW (1978) Intelligence, behavior, and dysmorphogenesis in the fetal alcohol syndrome: a report on 20 patients. J Pediatr 92:363–367

Strickler SM, Dansky LV, Miller MA, et al (1985) Genetic predisposition to phenytoin-induced birth defects. Lancet 2:746–749

Todd A, Chettle D (1994) In vitro x-ray fluorescence of lead in bone: review and current issues. Environ Health Perspect 102:172–177

Treloar AE, Boynton RE, Behn BG, Brown BW (1967) Variation of the human menstrual cycle through reproductive life. Int J Fertil 12:77–125

Turner TW, Schrader SM, Simon SD (1988) Sperm head morphometry as measured by three different computer systems. J Androl 9:P45

Ueland K, Novy MJ, Peterson EN, Metcalde J (1969) Maternal cardiovascular dynamics. IV. The influence of gestational age on the maternal cardiovascular response to posture and exercise. Am J Obstet Gynecol 104:856–864

van der Put NMJ, Steegers-Theunissen RPM, Frosst P, et al (1995) Mutated methylenetetrahydrofolate reductase as a risk factor for spina bifida. Lancet 346:1070–1071

van Oppen ACC, Stigter RH, Bruinse HW (1996) Cardiac output in normal pregnancy: a critical review. Obstet Gynecol 87:310–318

Vantman D, Koukoulis G, Dennison L, et al (1988) Computer-assisted semen analysis: evaluation of method and assessment of the influence of sperm concentration on linear velocity determination. Fertil Steril 49:510–515

Van Voorhis BJ, Syrop CH, Hammitt DG, et al (1992) Effects of smoking on ovulation induction for assisted reproductive techniques. Fertil Steril 58:981–985

Vollman RF (1977) The menstrual cycle. Saunders, Philadelphia, pp 19–72

Wibberley DG, Khera AK, Edwards JH, Rushton DI (1977) Lead levels in human placentae from normal and malformed births. J Med Genet 14:339–345

Wigg NR, Vimpani GV, McMichael AJ, et al (1988) Port Pirie Cohort Study: childhood blood lead and neuropsychological development at age two years. J Epidemiol Comm Health 42:213–219

Wolfe HM, Gross TL (1989) Increased risk to the growth retarded fetus. In Gross TL, Sokol RJ (eds), Intrauterine growth retardation. Year Book Medical Publishers, Chicago, p 111

Yip R, Li Z, Cong W (1991) Race and birth weight: the Chinese example. Pediatrics 87:688–693

Yoshida A, Impraim CC, Huang IY (1981) Enzymatic and structural differences between usual and atypical human liver alcohol dehydrogenases. J Biol Chem 256:12430–12436

Zarembski PM, Griffiths PD, Walker J, Goodall HB (1983) Lead in neonates and mothers. Clin Chim Acta 134:35–49

5 Risk Assessment of the Effects of Ozone Exposure on Respiratory Health: Dealing with Variability in Human Responsiveness to Controlled Exposures

Philip A. Bromberg

CONTENTS

INTRODUCTION

This chapter focuses on the scientific basis for risk assessment for the criteria pollutant ozone, whose effects are essentially limited to the respiratory tract. Data obtained from field studies and from controlled (chamber) environmental human exposures are summarized with major emphasis on the short-term impairments of lung function and the large range of individual responses to any given exposure condition. These data are discussed in the context of the primary National Ambient Air Quality Standard (NAAQS) for ozone, which currently is based on a 1-hour averaging time

0-8493-2805-5/99/$0.00+$.50
© 1999 by CRC Press LLC

and set at a level of 0.12 parts per million (ppm), which is not to be exceeded on more than 3 days in 3 consecutive years. Recent statistical approaches to dealing with the wide range of individual responsiveness of lung function to ozone are discussed in the context of risk assessment. Some of the difficulties encountered in defining an adverse health effect on the basis of acute decrements of lung function are noted and the importance of understanding the mechanisms of observed effects is emphasized. The interindividual variability of indices of neutrophilic airways inflammation, a more recently documented short-term effect of human ozone exposure, is described. Finally, the emerging use of large scale time-series studies of the association of acute respiratory illness among whole populations with fluctuations in ambient air pollution is discussed as a valuable complementary approach that avoids some of the problems associated with health risk assessment based on controlled exposure experiments, including the uncertainty associated with interindividual variability in human responses to ozone exposure.

CONTROLLED EXPOSURES: EFFECTS ON LUNG FUNCTION

There is abundant evidence from short-term exposure studies in several animal species that inhalation of high levels of ozone (2 to 10 ppm) causes serious damage to the lower airways and parenchyma. The epithelial cells lining the airways and alveoli bear the brunt of the injury and this can cause lethal pulmonary edema.

Currently, ambient air ozone levels rarely if ever exceed 0.4 ppm (1-hour average) in the United States. Higher concentrations have been documented in Mexico City and may be found in certain work environments. Although some early human controlled exposures used ozone levels as high as 0.75 ppm, few studies since 1980 have exceeded 0.5 ppm. When subjects exercise during exposure to increase their minute ventilation, effects can be elicited at relatively low concentrations of ozone (0.12 to 0.18 ppm) with 2-hour exposures (McDonnell et al. 1983); when exercise is combined with prolonged (6 to 8 hours) exposure, effects are found at levels as low as 0.08 ppm after several hours (Horstman et al. 1990).

The earliest noted effect of such ozone exposure was impairment of lung function, especially the subject's ability to voluntarily make a maximal inspiratory effort. This results in a decrease of inspiratory capacity (IC) and forced vital capacity (FVC) and of other measures dependent on FVC such as the forced expiratory volume in the first second (FEV_1) or the average forced expiratory flow rate over the middle half of the FVC (FEF_{25-75}). Some more subtle alterations in forced expiratory flow rates that are not entirely accounted for by the decreased FVC may also occur (Weinmann et al. 1995), but these are not discussed. The volume of air remaining in the lungs after a complete forced expiration is known as the residual volume (RV). In adults, it is trapped in the lung by diffuse intrapulmonary airways closure. Although RV may be slightly increased after exposure to ozone, almost all the decrease in FVC is attributable to decreased IC and total lung capacity (TLC) (Hazucha et al. 1989). In normal young adults, spirometric indices like FVC and FEV_1 are remarkably reproducible from day to day and have a coefficient of variation (CV) of less than 3% for repeated trials during the same sitting (Cochrane et al. 1977, Pennock et al. 1981) and even for hourly or daily trials (Cochrane et al. 1977). For weekly trials, Cochrane et al. (1977) reported a CV of 3.35% for FEV_1 and 2.9% for FVC. This is because the test maneuver is not highly sensitive to whether the inspiratory and forced expiratory maneuvers were performed at absolutely maximal levels of respiratory muscle effort. Provided that a reasonable fraction of maximal effort is produced by the subject, the pressure-volume and expiratory pressure-flow relationships of the respiratory system are determined by the mechanical properties of the chest wall and lung. These test characteristics contribute to the widespread use of spirometry as a clinical as well as investigative tool. Changes in individual FVC or FEV_1 of 10% or 15% of baseline are generally considered to be clinically significant.

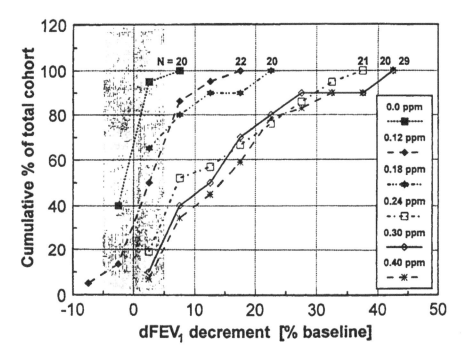

FIGURE 1   With data of McDonnell et al. (1983), the fraction of each exposed cohort showing a FEV₁ decrement (% of baseline FEV₁) no greater than a given amount (abscissa) is shown on the ordinate. Cohorts were exposed to one of six levels of ozone (0.0, 0.12, 0.18, 0.24, 0.30, 0.40 ppm). Each ozone concentration is represented by a different symbol. Cohorts consisted of from 20 to 29 healthy young men (*n* shown at the top of each cumulative curve). The expected range of variability for repeated measures of FEV₁ in normal individuals (±5%) is stippled.

The range of ozone-induced decrements in lung function encountered among similarly exposed healthy young adults is very wide. Those individuals with large responses also note substernal pain and cough, which are provoked by attempted deep inspiration.* Figure 1, derived from data of McDonnell et al. (1983), shows the effect on the spirometric parameter, FEV₁, of exposure to ozone in six cohorts of healthy young white men. The exposures lasted 2 hours with alternating 15-minute periods of heavy exercise [minute ventilation (V_E) ~ 65 L/minute] and rest. Each cohort consisted of 20 to 29 individuals. Individual decrements of FEV₁ are shown on the abscissa as a percentage of the baseline value and grouped into bins at 5% intervals. The reproducibility of spirometry is shown by the fact that sham exposures (0.0 ppm ozone) caused less than a 5% deviation from baseline FEV₁ in 19 of 20 subjects and less than a 10% change in one. The ±5% range is stippled. By contrast, even at 0.12 ppm ozone (the lowest concentration studied) there were three individuals whose FEV₁ decrements exceeded 10%. With progressively higher ozone concentrations (0.18, 0.24, 0.30, 0.40 ppm) more individuals showed decrements >10%, and the group mean decrements were of progressively greater magnitude. However, even at the highest concentration tested, a few individuals showed little, if any, change. The pattern of rightward shift of the cumulative curves with increasing ozone concentration (Figure 1) shows a gap between 0.18 and 0.24 ppm, with

---

* A technical consideration, which may be a source of additional variance ("uncertainty"), also needs to be noted. When, as in the case of ozone exposures, individuals are precluded from making a near-maximal inspiratory air intake effort because of involuntary (or voluntary) inhibition of inspiratory muscle effort, spirometry is less reproducible on repeated trials. The amount of this source of variance does not appear to have been formally investigated but the author's personal impression is that the first trial after ozone exposure tends to elicit a higher FEV₁ and FVC than later trials.

TABLE 1
Group Mean Spirometric Decrements and Response
Variability of Healthy Young Men After Exposure to Ozone

Ozone concentration (ppm)	Baseline FEV$_1$(L) ± SEM	Mean decrement of FEV$_1$(L) ± SD	CV (SD/mean)
0.0	4.42 ± 0.10	0.05 ± 0.09	1.8
0.12	4.64 ± 0.12	0.21 ± 0.23	1.1
0.18	4.50 ± 0.15	0.29 ± 0.36	1.24
0.24	4.10 ± 0.11	0.59 ± 0.46	0.78
0.30	4.41 ± 0.14	0.74 ± 0.50	0.68
0.40	4.46 ± 0.12	0.76 ± 0.49	0.64

relatively little separation between 0.24 and 0.30 ppm and even between 0.24 and 0.40 ppm. This suggests that a sigmoid curve with its steepest slope between 0.18 and 0.24 ppm would provide the best fit to the group mean exposure concentration-response (or even log concentration-response) relationship for this type of exposure protocol.

Interindividual response variability (i.e., standard deviation of mean ozone-induced decrement in FEV$_1$) tends to increase with mean response (see Table 1). The coefficients of variation (CV = SD/mean) of mean responses for the five ozone concentrations tested (Table 1) remain approximately at 1 and decrease only marginally as the mean response increases from 0.21 to 0.76 L.

Spirometric decrements appear by the end of the first hour of strenuous intermittent exercise but increase during the second hour of exposure. After cessation of exposure [or even cessation of exercise while remaining in the ozone atmosphere (Folinsbee et al. 1977)], the observed changes recede markedly over several hours (Folinsbee and Hazucha 1989).* However, in severely affected subjects, small but detectable residual decrements in lung function may persist at 24 hours along with cough (Folinsbee et al. 1994). Nevertheless, it is generally agreed that the effects of acute, controlled ozone exposures on human lung function are entirely reversible. This implies that these reversible ozone-induced changes cannot be taken as a marker of permanent injury. Indeed, were there evidence to the contrary, it would no longer be ethical to expose volunteers to ozone.

Findings of the type reported by McDonnell et al. (1983) were repeatedly demonstrated by various investigators, although there was not complete agreement on the exact position or shape of the exposure concentration vs. mean response curve. Nevertheless, it was clear that FEV$_1$ decrements >10% of baseline would occur in some young normal adults exposed for 1 to 3 hours to <0.2 ppm ozone while performing moderately heavy exercise even when the group mean decrement was much smaller and not significantly different from zero. Although relatively less attention was focused on the mechanism, this decrease was interpreted as an adverse health effect, especially because larger decreases were accompanied by substernal discomfort and cough. It should be noted that these spirometric effects cannot be studied in conscious, spontaneously breathing animals (who cannot be asked to perform voluntary maximal inhalation), although surrogate measures like respiratory frequency and tidal volume might be used.

Substantial resources have been devoted to detailed, descriptive studies involving hundreds (perhaps thousands) of subjects, most of them healthy volunteers. Earlier studies used relatively brief exposure protocols (<3 hours), presumably to mimic the temporal pattern of fluctuation of daily ozone concentrations observed in summertime Los Angeles oxidant pollution. Since 1988,

---

* Thus, protocols in which the subjects rest for 15 to 30 minutes in the chamber after completion of an exercise stint allow some degree of recovery of function before spirometry, and this could produce apparent variability in comparing different studies.

there have been a number of studies with more prolonged exposure protocols (6.6 to 8 hours), inspired by the observation that in the Northeastern megalopolis summertime ozone levels could remain elevated for many hours. In a group of 87 healthy young adults exposed for 6.6 hours to air (control) and to 0.08, 0.10, or 0.12 ppm ozone while performing moderate exercise ($V_E = 39$ L/minute) for 50 minutes of each hour (except for a 35-minute lunch break between hours 3 and 4), the fraction of subjects who exhibited a FEV_1 decrement >10% baseline at the end of exposure was 0% for air and 26%, 31%, and 46%, respectively, for the 0.08, 0.10, and 0.12 ppm ozone exposures. Again, marked heterogeneity of individual responses was observed, even at the 0.08 ppm level (Folinsbee et al. 1991) (see also Table 2).

EXPOSURE TO AMBIENT AIR CONTAINING OZONE: EFFECTS ON LUNG FUNCTION

The effects of exposure to pure ozone diluted with clean air on lung function determined from controlled exposures have been shown to be reasonably predictive of the effects observed from exposures to ambient air containing various levels of naturally occurring ozone. This was first demonstrated in a group of competitive cyclists (Avol et al. 1984) and later in adult joggers (Spektor et al. 1988b) and in children attending summer camps (Spektor et al. 1988a, 1991; Higgins et al. 1990), much of whose day was spent at play out-of-doors. Except for the cyclist study, these studies do not directly compare individual responses to ambient air exposure with the results from controlled exposure. Given that elevated ambient air ozone concentrations are commonly associated with elevated fine particulate matter levels (haze), it may seem almost surprising that the ozone concentration per se appears to be largely responsible for the observed effects. These findings reinforce the significance of the detailed controlled exposure studies and support the use of these studies for risk assessment.

The children's summer camp studies provided impetus for performing controlled prolonged exposures to low ozone concentrations in volunteers. As previously discussed, these studies demonstrated clear decrements of group mean lung function after sufficiently long exposures. When combined with evidence from atmospheric monitoring that ambient air ozone levels could remain elevated above background (0.03 to 0.05 ppm) for many hours without exceeding the 0.12 ppm NAAQS, these data spurred suggestions that an additional daily standard for ozone was needed to protect public health — one lower than 0.12 ppm but averaged over a number of hours. The U.S. Environmental Protection Agency (EPA) has in fact recently proposed an additional primary standard for ozone of 0.08 ppm averaged over an 8-hour period.

INTERINDIVIDUAL RESPONSE VARIABILITY

The wide interindividual range of magnitude of effect has remained largely unexplained. Two studies specifically looked at within-subject reproducibility of response to a given exposure condition and found r values of ~0.9 (McDonnell et al. 1985b, Gliner et al. 1983). Furthermore, although much of the data were not published in detail, two fairly large studies of individual spirometric responses as a function of exposure concentration in young men over ranges of up to 0.25 ppm ozone (Kulle et al. 1985) and 0.32 ppm ozone (Avol et al. 1984), respectively, demonstrate individual monotonic patterns with few exceptions (W. McDonnell, personal communication). This finding, true of weak as well as strong responders, reinforces the notion that degree of responsiveness to ozone is an individual characteristic. Indeed, investigators have taken advantage of within-individual response reproducibility to preselect strong or weak responders for further studies of ozone exposure (e.g., Balmes et al. 1996). An ongoing study of young, adult twins in our laboratory suggests that the degree of ozone responsiveness (high or low) may be more similar within identical twin pairs than within same-sex full sibling pairs. This is consistent with a heritable biologic basis for ozone reactivity. However, examples of discrepant ozone responsiveness among the 24 currently available identical twin pairs in this study suggest that "environmental" factors play a significant role.

Several observations show that other factors can profoundly modify (decrease) individual ozone responsiveness:

1. Striking reduction in response (and in symptoms) is observed in almost all individuals who undergo identical consecutive daily ozone exposures by the 4th day. This phenomenon has been called adaptation, and it recedes over about 7 days after cessation of the daily exposure regimen (Horvath et al. 1981). The phenomenon has been documented repeatedly for relatively short, intense exposures (in which the mean response on day 2 of exposure is usually even greater than on day 1* but then lessens sharply on days 3 and 4, with almost total disappearance on day 5) and more recently for prolonged low-level exposures (in which the daily decline in response is more monotonic) (Folinsbee et al. 1994). One group of investigators in the Los Angeles region (Linn et al. 1988) reported that some vigorous ozone responders tested in the springtime were much less responsive when tested in the fall (after the summer "ozone season"). However, perennial weak responders to ozone did not acquire strong-response character. This suggested that the residents of high oxidant pollution areas who were strong responders might become adapted to ozone over the summer season.

2. Based on cross-sectional studies, healthy men and women older than 50 years appear to include relatively fewer high responders (Drechsler-Parks et al. 1987a, 1987b, 1989; Bedi et al. 1989). In an ongoing study in our laboratory, nonsmoking men and women have been exposed to 0.42 ppm ozone for 2 hours with alternating 15-minute periods of rest and exercise ($V_E \sim 20$ L per minute per m^2 of body surface area). The resulting individual decrements in FEV_1 as a percentage of baseline were evaluated as a function of age. A paucity of large responses among 22 women >35 years old has been noted. The currently available data are more sparse for older men ($n = 9$) (M. J. Hazucha, personal communication). There are no longitudinal observations of sufficient duration to address the issue of whether initially responsive individuals decrease their ozone responsiveness with increasing age. However, increasing age is the only known variable other than inhaled dose of ozone that has a significant (negative) predictive effect on spirometric response in the 18- to 35-year range (McDonnell et al. 1993, 1995).

3. Administration of certain pharmacologic agents (see below) can markedly blunt the effects of ozone exposure on lung function in strong responders. However, no means of enhancing ozone responsiveness has yet been described, although it has been speculated that deficient intake of vitamins with antioxidant functions might do so.

4. Cigarette smokers are probably less responsive than nonsmokers (Emmons and Foster 1991, Kerr et al. 1975). It is not known whether passive exposure to environmental tobacco smoke affects ozone responsiveness.

IMPACT OF RISK ASSESSORS ON THE SCIENTIFIC PROCESS

The pressure from risk assessors on scientists to make experimental exposure protocols "relevant" to real-life conditions is noteworthy. It presumably reflects the conviction that, given sufficiently detailed data from relevant controlled exposures to ozone of a sufficient number of individual subjects representing important susceptible as well as normal subgroups within the total population exposed to ambient air containing ozone, it would be possible to formulate a convincing quantitative short-term health risk assessment capable of supporting air quality standards and policies designed

---

* This effect was not found for two consecutive ozone exposures separated by at least 72 hours (Schonfeld et al. 1989).

to manage at least the short-term aspects of the risk of exposure to ozone-polluted ambient air. The use of controlled exposures to study mechanisms of these effects of ozone exposure on lung function was only a secondary objective. When applied to animal toxicologic studies, this line of reasoning led to insistence on the study of ever-lower ozone concentrations and to a general attitude that the results of investigations at concentrations >1.0 ppm were hardly worth taking into consideration. This attitude fails to take into account the possibility that to achieve tissue doses in resting animals equivalent to those occurring in exercising humans, it might be essential to expose animals to relatively high concentrations of ozone.

Risk assessors therefore were also eager for additional controlled exposure data in identifiable subgroups in the normal population — notably women (Messineo and Adams 1990) and blacks (Seal et al. 1993). Young age (but not old) was also an important consideration because children are likely to be exercising outdoors during summer oxidant pollution episodes. These studies generally failed to show any marked difference in response pattern between men and women or between blacks and whites when ventilation during exposure was normalized for body surface area. Children were studied to a lesser extent than young adults, but they appear to be about as sensitive as the latter at 0.12 ppm (McDonnell et al. 1985a). Thus, with a 2-hour exposure protocol, including an intermittent exercise to produce a V_E of 33 L per minute per m^2 of body surface area, 22 healthy boys (8 to 11 years old) developed a mean decrement of FEV_1 of 3.5% of preexposure baseline (McDonnell et al. 1985a) compared with 4.5% in 20 young adult white males (18 to 30 years old) studied in the same laboratory (McDonnell et al. 1983). The wide range of individual responses found in all groups and unresolved questions on how best to normalize exercise-induced ventilation values (and therefore inhaled dose) for subjects of different sizes increase the difficulty of detecting intergroup differences in pattern of response to ozone, and some studies suggest that women as a group may be more sensitive than men.

MODELS OF DECREMENTS IN LUNG FUNCTION PROVOKED BY OZONE EXPOSURE

LINEAR FUNCTIONS OF OZONE CONCENTRATION (C), VENTILATION (V), AND DURATION (t)

Much effort has been expended on developing dose–response models that take into account the controlled-exposure conditions C, V, and t to predict mean decrements of lung function. (Ambient temperature and relative humidity appear to have little effect.) The measure of V requires some type of normalization if individuals of markedly different body size are to be included in the same models (e.g., women vs. men or children vs. adults). Such models are limited by the very wide range of interindividual variability in lung function response to identical exposure conditions previously mentioned. Thus, the mean decrements in lung function parameters do not adequately reflect the (often skewed) distribution of individual responses. This also makes it more difficult to choose among alternative models or to assign model parameter values with confidence.

Furthermore, the prediction of response as a linear function of the product of concentration, ventilation, and time ($C \times V \times t$) to obtain a cumulative inspired ozone dose as the independent variable is not entirely adequate. In a review and analysis of short (≤ 2 hours) exposures with various levels of exercise, Hazucha (1987) showed that the best data fit required weighting ozone concentration (C) more heavily than the other exposure variables. [This could be done by using $C^{1.3}$ (Larsen et al. 1991)]. In such (intense) 2-hour exposures to higher ozone concentrations, the spirometric decrements at 60 minutes are generally substantially more than 50% of the decrements at 120 minutes (Gliner et al. 1983, McDonnell and Smith 1994). Other data also imply nonlinearity of the cumulative inhaled dose vs. response. Thus, in a study of 8-hour exposures with both a steady ozone concentration (0.12 ppm) and an increasing-decreasing ramp (0.0 → 0.24 ppm → 0.0) in the same subjects, Hazucha et al. (1992) found that (1) there was a response plateau for hours 6 to 8 in the steady 0.12-ppm exposure; (2) the best fit to all the hourly mean data in both protocols was obtained by using a 4-hour running average of cumulative exposure dose ($C \times V \times t$) as the independent variable; (3) in several subjects who were exposed to 0.12 ppm ozone for 10 hours, a tendency toward reduction of effect after the 9th and 10th hours was noted, suggesting possible adaptation.

At low concentrations of ozone (0.08 to 0.12 ppm) several hours of exposure are needed before significant mean decrements of lung function are observed in groups of subjects performing almost continuous moderate exercise ($V_E \sim 40$ L/minute) for 6.6 hours (Horstman et al. 1990). The mean decrements then become progressively more marked with further exposure. No mean response plateau was demonstrated, but this might have occurred [as observed by Hazucha et al. (1992)] had the exposures been extended.

NONLINEAR FUNCTIONS

McDonnell et al. (1993, 1995) and McDonnell and Smith (1994) have used the large body of lung function data resulting from many controlled exposures of young adults in the EPA Human Studies Division (1980–1992) to develop an empirical model of group mean spirometric response as a function of exposure dose for both short and prolonged exposures to steady ozone concentrations (C). By chance, the hourly time-weighted mean ventilation rates (V_E) (including both rest and exercise periods) were similar in both types of protocol (~35 L/minute). Thus, their model is based on an hourly dose rate ($C \times V_E$) that varies only with C. (They did not attempt to apply their model to data sets in which V_E varied.) Tables 1 and 2 in their 1994 paper provide a valuable summary of the group mean percentage decrements in FEV_1 at hourly exposure intervals for both the 2-hour and the 6.6-hour protocols. The sham exposure data ($C = 0.0$ ppm) confirm that FEV_1 is a very reproducible measure (small standard error for the mean change in FEV_1 at all time points). All ozone exposures result in greater variance of group mean FEV_1 decrements as the ozone effect develops, and this variance increases in magnitude with increasing duration of exposure.

Their empirical models are based on the assumption (supported by most available data) that the dose–response relationship is sigmoidal. They investigated the expression

$$\%\Delta FEV_1 = \frac{\alpha\left(1 - e^{-\beta C}\right)}{1 + \gamma e^{-\delta C_t}}$$

where α is the theoretical maximum possible decrement in FEV_1 under any exposure conditions, and $\alpha(1 - e^{-\beta C})$ defines the actual level of the response asymptote for any given dose rate ($C' = C \times V_E$). γ and δ are other constants that define the precise shape of the sigmoid. If one assumes that the level of the response asymptote is not a function of C' (i.e., that $\beta = \infty$), the model equation simplifies to the form

$$\%\Delta FEV_1 = \frac{\alpha}{1 + \gamma e^{-\delta C_t}}$$

This represents a logistic function in which $\%\Delta FEV_1$ is a nonlinear function of $C \times V \times t$.

The total database was randomly divided into an exploratory subset used to compute best-fit values for the adjustable parameters and a confirmatory subset used to evaluate the goodness of these values. Reasonably good fits of all the group mean data were obtained with a single set of values for α, β, γ, and δ with the more complex model. With the simplified model (only three adjustable parameters, α, γ, δ), the fits were still acceptable over the range of exposure conditions. However, there is an indication that exposures longer than 6.6 hours would not be well-fitted by the three-parameter model.

McDonnell et al. (1995) then applied a variant of this simplified model to a more detailed consideration of individual spirometric responses. They were able to successfully predict the proportion of healthy young adults who would be expected to develop lung function changes exceeding a given magnitude (e.g., 5%, 10%, or 15% fall in FEV_1) during prolonged exposures to 0.08, 0.10, and 0.12 ppm ozone as a function of $C \times t$. Again, the ventilation (exercise/rest) pattern was the same for all subjects at all three ozone concentrations. Thus, the model might not work as well if applied to conditions in which a wide range of fluctuating ventilations occur (e.g., children at play out-of-doors). The model was refined by adding another term ($\eta \times$ age) to their simplified 1994 model, resulting in the following equation where P_d is the probability that a randomly selected individual will experience a FEV_1 decrement of at least $d\%$ at time t during exposure to concentration C. Because the hourly mean V_E was invariant, C' was a function only of C.

$$P_d = \frac{\alpha + \eta \times \text{age}}{1 + e^{-\delta(C't - \gamma)}}$$

The inclusion of an age term significantly improved the fit of the model. The coefficient η has a value of –4% of the population per year of age. Thus, the predicted difference between 20- and 30-year-olds is substantial. For example, after a 6.6-hour exposure to 0.12 ppm ozone (under the given exercise/rest conditions) approximately 67% of 20-year-olds but only 27% of 30-year-olds are expected to develop a decrement of $FEV_1 \geq 10\%$ of baseline FEV_1. The model also predicts that essentially all 10-year-olds would develop such a response, but this represents an extrapolation.

Such empirical stochastic models, when combined with other estimates of ambient ozone concentrations, numbers of exposed individuals of the proper age, level of sustained outdoor physical activity, etc., can in principle be used to model the numbers of individuals at risk for a given magnitude of ozone exposure-induced effects on lung function. The outputs from such models could then provide a basis for risk assessment. However, it should be recalled that the McDonnell–Smith model is based on data involving only young adults exposed under conditions in which dose rate ($C \times V_E$) varied only with C and in which C remained constant for each individual during the entire exposure. Fluctuations in C [e.g., Hazucha et al. (1992)] or in V_E might produce more complex response patterns.

MECHANISTIC MODELS

A physiological exposure/response model requires a multistep approach. The tissue dosimetry portion of such a model (see Chapter 8, EPA 1996) would itself be quite complex and involves incompletely understood chemical reactions of inhaled ozone with components of airway surface liquid (e.g., Pryor et al. 1991, 1996; Postlethwait et al. 1994) and secondary reactions of these products with surface cellular and neural elements. Regional uptake of ozone along the airways (Bush et al. 1996, Gerrity et al. 1995) is also dependent on inspiratory airflow rate and on the geometry of the airways (Bush et al. 1996; Hu et al. 1992, 1994; Miller et al. 1985). The activation of cellular and neural elements by ozone products and the duration of such activation (once achieved) are equally complex topics. The richness of airways mucosal innervation by nociceptive fibers and the transmission of their impulses to higher nervous system stations where they are processed and integrated are further identifiable steps in this process. McDonnell and colleagues (personal communication) are currently attempting to formulate a model that would at least separately identify and incorporate a rate-of-dose element and a biological response element. The successful development of a biological-mechanistic model of ozone-induced respiratory effects that predicted individual responses on the basis of independent biological parameters (including genetic markers) would be a substantial achievement in the inhalational toxicology of reactive substances. It might not, however, advance the risk assessment process for ozone much beyond the empirical stochastic model approach so long as the risk to health is evaluated simply on the basis of a given decrement in lung function.

MECHANISM OF DECREASED SPIROMETRIC FUNCTION

What is the precise mechanism of these ozone effects, and how might knowledge of the mechanism affect interpretation of the functional decrements as adverse health effects?

Beckett et al. (1985) (as well as some earlier work) noted that ozone-induced decrements in FVC and FEV_1 were not due to an inability to fully expire — i.e., RV was not increased. By inference, therefore, decreased IC was responsible. The major focus of their paper, however, was on increased airways resistance (R_{aw}) after ozone exposure. They showed that these (generally modest) changes could be prevented by pretreatment with the muscarinic blocker atropine. This implied that the increased R_{aw} is mediated by vagal efferent motor fibers. Furthermore, the increased R_{aw} provoked by ozone exposure could be reversed acutely by administration of inhaled β-adrenergic agonists, implying that airways smooth muscle contraction was the cause. Pertinent to the current discussion, these pharmacologic agents failed either to prevent ozone-induced spirometric decrements or to reverse them after their elicitation. It is of interest that group mean increases in R_{aw} after ozone exposure are relatively small, have a higher ozone concentration threshold than the spirometric effects, and correlate poorly with the latter within individuals (McDonnell et al. 1983). (This is particularly relevant to the subsequent discussion of ozone exposure of asthmatic individuals.)

A systematic investigation of the cause of ozone-induced decrements in lung volume was published by Hazucha et al. (1989). A very small, though statistically significant, increase in RV was found, but the major reason for decreased lung volumes was indeed decreased IC. This in turn accounted for the decreased FVC and TLC. Measurements of inspiratory and expiratory muscle force before and after exposure showed no change. There was a modest increase in lung elastic recoil, but this could not account for the observed lung volume decrements. [Increased lung elastic recoil was suggested to be due to subtle changes in lung mechanics secondary to ozone exposure-induced inhibition of spontaneous periodic deep inflation (inspiration), which is known to be necessary to prevent loss of lung compliance in healthy subjects.] Finally, several of the subjects inhaled aerosolized topical anesthetic (lidocaine), which, in some cases, appeared to reverse ozone-evoked spirometric decrements.

These data indicated that the loss of IC (and thus of FVC and TLC) was due to an involuntary inhibition of inspiration rather than to changes in mechanical properties of the lungs or chest wall or to decreased respiratory muscle strength. The authors suggested that superficially located airway sensory nerve endings were stimulated by ozone inhalation, thus transmitting nociceptive messages via the vagus nerves to higher brain centers and causing premature central neural inhibition of voluntary efforts to inspire maximally. The bronchial C-fiber system was a candidate for this role. Indeed, stimulation of these fibers might also enhance reflex cough, a characteristic symptom of reactive ozone-exposed individuals. The well-known observation that ozone exposure alters breathing to a more shallow but more rapid pattern (in experimental animals as well as in exercising volunteers) can also be accounted for by such a mechanism.

Convincing evidence for stimulation of canine bronchial C-fibers by inhalation of ozone has been published (Schelegle et al. 1993, Coleridge et al. 1993). A recent abstract (Vesely et al. 1995) describes loss of the typical ozone-induced changes of respiratory pattern (increased respiratory frequency and decreased tidal volume) in rats whose C-fiber system had been ablated by neonatal capsaicin administration. Hazbun et al. (1993) described increased levels of a tachykinin, substance P, in lavage of proximal (large) lower airways after ozone exposure of volunteers. C-fiber stimulation characteristically causes secretion of such tachykinins from the nerve endings and a decrease in substance P staining of mucosal nerves has been described in bronchial biopsies from subjects exposed to ozone (Krishna et al. 1996).

Because this nociceptive neural system is modulated by opioid receptors both peripherally and centrally, Passannante et al. (1995) in our laboratory have investigated the short-term effect of intravenous injection of a rapid-acting opioid agonist on the spirometric parameters of volunteers who developed large decrements after exposure to ozone. Injection of saline (control) caused no change but the opioid caused striking and almost complete reversal of the ozone-evoked decrements. Weak responders to ozone were also studied on the hypothesis that release of endogenous opioids (e.g., β-endorphin) might have blocked or blunted an ozone response. However, neither saline (control) nor intravenous naloxone (a potent competitive opioid receptor antagonist) was effective. Thus, the principal effects of ozone on lung function can be acutely reversed by opioid receptor agonists, a finding that strongly supports the bronchial C-fiber hypothesis. A secondary finding is that low responses to ozone are not attributable to secretion of endogenous opioid receptor agonists. It is tempting to speculate that low responders to ozone may have a poorly developed or insensitive bronchial C-fiber system and also that desensitization of these fibers may occur on repeated exposures to ozone. The hypothesis is testable.

The mechanism by which ozone might stimulate bronchial C-fibers likely involves eicosanoid products of the action of cyclooxygenase (COX) on free arachidonate. It is well-established that bronchial epithelial cells liberate increased amounts of arachidonate from membrane phospholipids when exposed to ozone, although the precise mechanisms are not fully known (Madden et al. 1994, Leikauf et al. 1988). These cells also produce increased amounts of the COX product of arachidonate, prostaglandin E_2 (PGE_2) after ozone exposure (McKinnon et al. 1993). After exposure of volunteers to ozone, airways surface liquid sampled by bronchoalveolar lavage (BAL) contains markedly increased PGE_2 concentrations (Koren et al. 1991). Schelegle et al. (1987) first reported that pretreatment of ozone-exposed volunteers with the COX inhibitor indomethacin (a commonly used anti-inflammatory agent) significantly blunted the spirometric responses. This was confirmed by Kreit et al. (1989). Similar findings were reported by Hazucha et al. (1996) with a different anti-inflammatory COX inhibitor, ibuprofen. PGE_2 is known to sensitize bronchial C-fibers and it is therefore a plausible candidate for this role in the response to ozone. Other autacoids (e.g., bradykinin) that stimulate C-fibers might also be involved.

In summary, the mechanism of ozone-induced impairment of inspiratory capacity is very likely to be a prostaglandin-modulated sensitization of C-fiber nerve endings within, or very close to, the epithelium of larger airways. This is basically a reversible, irritant effect rather than an effect on lung mechanics.

IMPACT OF MECHANISTIC INFORMATION ON RISK ASSESSMENT

Specialists in pulmonary medicine are familiar with the clinical use of lung function tests to assess the presence and severity of respiratory disease or to detect and quantify disease exacerbation, progression, or response to therapy. Decrements of 10% to 15% in FVC or FEV_1 would constitute significant events in most clinical contexts because these changes reflect underlying pathophysiological alterations in lung or chest wall mechanics. However, because the major mechanism underlying the reversible ozone-evoked decrements in FVC and FEV_1 has been found to be neurally mediated irritation rather than any change in respiratory mechanics, one may question the weight that should be assigned to this particular phenomenon per se. A risk assessment strategy based heavily on estimates of the fraction of the population in a given region that might develop 10% or 15% reductions in FEV_1 from outdoor exposure (with exercise) occurring at various levels of ambient ozone therefore appears to be open to debate when the mechanism of that effect is taken into consideration.

It is true that ozone-provoked symptoms of substernal pain and irritative cough, which are more likely to be associated with large decrements of lung function, might provide a rationale for basing risk assessment on lung function changes. It is also possible that lung function decrements are a convenient marker for some other effect that is clearly adverse either in the short term (e.g., increased susceptibility to respiratory infection) or after repeated exposures. However, such a marker role for lung function changes remains to be established. In addition, exposure of well-conditioned athletes to ozone-containing air probably causes significant impairment of the ability to sustain very intense exercise at levels approaching 0.12 ppm in at least some individuals (EPA 1996, Chapter 7, pp. 49–53). These observations might also provide some support for using lung function decrements as a surrogate for lifestyle impairment.

This is not to say, of course, that other effects of ozone exposure (e.g., inflammation) may not more directly support a health risk assessment, justifying the current standard. This phenomenon is briefly discussed in the next section.

OZONE-INDUCED RESPIRATORY TRACT INFLAMMATION

Apart from decreased lung function, another important short-term effect of ozone exposure of humans (and animals), neutrophilic airways inflammation, was first found in volunteers by Seltzer et al. (1986) and has been amply confirmed since (Koren et al. 1989a, 1989b; Schelegle et al. 1991; Aris et al. 1993; Devlin et al. 1991, 1996; Hazucha et al. 1996). Tables 2A and 2B list the individual data for degree of neutrophilia in the cells recovered from the lower airways by the BAL technique from healthy, nonsmoking, young adults after ozone exposure and after air (sham) exposure. Neutrophilia is quantified as the percentage of cells of all types present in the total lavage. Most of these cells in healthy individuals after both sham and ozone exposure are alveolar macrophages.* BAL neutrophils are rarely as high as 5% of total cells in healthy, unexposed individuals. This is confirmed in our series of 58 young healthy nonsmokers participating in ozone exposure studies (Table 2). Thus, relatively small degrees of induced neutrophilia are detectable. Five different studies performed in the EPA Human Studies Division research facility in Chapel Hill by EPA and University of North Carolina investigators over the past decade are cited. Table 2A shows data for 2-hour exposures to 0.4 ppm ozone and Table 2B shows data for more prolonged exposures to lower concentrations. The BAL technique used was the same in all these studies, although the time elapsed between termination of exposure and performance of BAL was either 1 or 18 hours in the 0.4 ppm ozone exposures (Table 2A). It should be noted that BAL samples the airway surface rather than the tissue itself. It is assumed that one reflects the other, especially as the primary inflammation-producing events for this highly reactive inhaled toxicant are considered to take place in the airways surface liquid, the subadjacent epithelial cells, and the superficial sensory nerve endings. The time course for onset and regression of BAL neutrophilia after exposure to ozone is slower than for the lung function changes, although it is not yet fully defined (Schelegle et al. 1991, Koren et al. 1991, Devlin et al. 1996). More than 18 hours and perhaps 2 to 3 days are required for resolution of neutrophilia. As is true for the lung function changes (also shown in Table 2A and 2B), there is a wide interindividual range of degree of neutrophilia in BAL liquid. (Some of the apparent interindividual variability in parameters of lower airways inflammation, as revealed in snapshot form by the technique of BAL, may be attributable to individual differences in the time course of the response.) In spite of the wide range of responses, statistically significant group mean BAL neutrophilia was found even after 6.6-hour exposures to as little as 0.08 and 0.10 ppm ozone (Devlin et al. 1991). Thus, the inflammatory response seems about as sensitive to ozone exposure as the lung function changes. However, despite studies in guinea pigs implicating nociceptive nerves in the genesis of ozone-induced neutrophilic lung inflammation (Koto et al. 1995) there appears to be no obvious within-individual positive correlation between indices of inflammatory response and

---

* Other markers of an inflammatory response could also be used, but the neutrophil is probably the best one. The metric used to characterize the degree of neutrophil influx could also reflect the absolute number of neutrophils recovered in the total lavage liquid rather than expressing the data as in Table 2.

lung function decrements (Koren et al. 1989a, 1989b; Devlin et al. 1991, 1996; Koren et al. 1991; Hazucha et al. 1996) (see also Table 2).* It remains to be determined whether more subtle changes in distal airways function may be more closely related to inflammation (Weinmann et al. 1995). Furthermore, it is also possible (indeed likely) that the decrements in spirometric function are incited by events occurring in larger airways, whereas conventional BAL samples the surface of more distal airways and air spaces.

Unlike the situation with lung function decrements, there currently is only one published bronchoscopic study addressing the question whether spontaneous exposure to ambient air oxidant pollution produces such inflammatory changes in the lower respiratory tract (although nasal changes have been described). This study (Kinney et al. 1996) produced equivocal results. A noninvasive procedure (e.g., induced sputum) to detect ozone-induced airways inflammation or epithelial cell damage (e.g., release of ciliary dynein components) would be valuable in this respect as well as for monitoring the time course of inflammation after controlled exposures. An even more difficult problem from the standpoint of risk assessment is what health significance to assign to indices of airway inflammation. Repeated episodes of airways inflammation and epithelial cell damage would seem to this author to constitute a clearer adverse health effect than does the phenomenon of irritant-induced changes in neural regulation of respiration. However, not all would agree with this thesis.

The mechanisms underlying the inflammatory response probably involve ozone-induced generation of potent lipid mediators from phospholipids as well as oxidant-provoked stimulation of epithelial cell production of proinflammatory cytokines and chemokines by upregulation of transcription of the appropriate genes (Devlin et al. 1994). However, the pathways have not been worked out in detail. Nor is it known whether the acute inflammatory process is self-limited and entirely reversible or whether there are residual changes that might lead to longer-term effects in some individuals. Repeated daily ozone exposures appear to produce adaptation (i.e., diminution or disappearance) of some features of the inflammatory response in experimental animals (see EPA 1996, p. 9-26) and in volunteers (Devlin et al. 1997), but the underlying reasons are unknown. Protocol design for long-term ozone exposures in experimental animals should take the adaptation phenomenon adequately into account. Mechanism-oriented studies are desirable to address all these issues before investing heavily in empirically defining dose–response relationships. Nevertheless, it seems obvious that the effects of repeated exposures to ambient ozone at various intervals need to be considered in a risk assessment.

EFFECTS OF CONTROLLED EXPOSURE TO OZONE IN INDIVIDUALS WITH PREEXISTING RESPIRATORY DISEASE

There are several *a priori* reasons to anticipate that patients with established chronic airways disease might constitute a highly susceptible subpopulation for adverse effects of ozone exposure. These include the presence of chronic inflammation, impaired baseline lung function, increased inhaled dose of ozone delivered to the best-ventilated regions of a heterogeneously diseased lung, etc.

CHRONIC OBSTRUCTIVE LUNG DISEASE

A prevalent type of respiratory disease is chronic obstructive lung disease — i.e., patients (usually ex-smokers or current smokers) who have irreversible airflow obstruction usually secondary to combined diffuse small airways inflammation and emphysema. Such individuals are generally older,

---

* A negative correlation is not out of the question, however (see Schelegle et al 1991, Balmes et al. 1996). A speculative explanation would be that strong spirometric responders have absorbed a larger fraction of inhaled ozone in their large airways, thus leaving less ozone to generate more distal effects.

TABLE 2A
Inflammatory Effect of 2-Hour Exposure to 0.4 ppm Ozone (Individual Data)

	Air exposure (sham)	Ozone exposure[a] (0.4 ppm)	
	PMN (% total BAL cells)[b]	PMN (% total BAL cells)[b]	FEV$_1$ decrement (% of baseline FEV$_1$)
Koren et al. (1989)	2.5	13.0	18.6
	0.8	6.0	19.1
	3.0	11.7	4.6
	9.2	21.3	36.3
Healthy men: 18–35 years (n = 11)	1.8	12.0	20.4
	1.0	7.0	12.5
18 hours[a]	0.5	8.3	26.3
	2.3	9.2	38.5
	1.0	8.5	13.3
	0.5	13.4	3.2
	4.2	8.3	31.5
Mean (SD)	2.4 (2.4)	10.8 (4.1)	20.4 (11.3)
Devlin et al. (1996)	0.2	8.9	22.9
	0.7	9.4	0.8
	2.0	8.0	18.8
	1.6	6.9	13.1
Healthy men: 18–35 years (n = 8)	0.3	21.3	24.1
	1.6	7.0	1.7
1 hour[c]	2.1	8.7	15.1
	1.9	9.2	12.4
Mean (SD)	1.3 (0.7)	9.9 (4.4)	13.6 (8.2)
Hazucha et al. (1996)		12.9	13.1
		3.8	8.3
	Not done	7.6	23.5
		7.2	44.5
Healthy men: 20–32 years (n = 10)		14.9	20.6
		2.6	5.7
1 hour[a]		18.2	31.3
		41.5	5.3
		5.0	10.6
		2.0	13.2
Mean (SD)		11.6 (11.2)	17.6 (11.9)

[a] Ozone concentration/duration of exposure: 0.4 ppm/2 hours (15-min rest alternating with 15-min exercise). Ventilation during exercise: V_E 35 L/min/m^2 body surface area.

[b] PMN, polymorphonuclear neutrophil leukocyte.

[c] Time elapsed from end of exposure to BAL.

can sustain only low levels of exercise and ventilation, and often are still smoking. Perhaps for these reasons, experimental exposures of groups of such patients have not revealed any particular sensitivity of lung function to ozone (Kehrl et al. 1985, Linn et al. 1983, Solic et al. 1982), although prolonged exposures remain to be investigated. No studies of the effect of ozone exposure on airways inflammation have been reported in subjects with chronic obstructive lung disease (nor indeed in older normal subjects).

TABLE 2B
Inflammatory Effect of Prolonged Exposure to 0.08, 0.10, and 0.16 ppm Ozone (Individual Data)

	Air exposure (sham)	Ozone exposure (0.08 ppm)		Ozone exposure (0.10 ppm)	
	PMN (% total BAL cells)	PMN (% total BAL cells)	FEV$_1$ decrement (% of baseline FEV$_1$)	PMN (% total BAL cells)	FEV$_1$ decrement (% of baseline FEV$_1$)
Devlin et al. (1991)	3.1	1.3	−2.0	3.8	0.9
0.0, 0.08, 0.10 ppm/6.6 hours	2.7	1.2	2.4	3.8	2.2
(50-min exercise alternating with	2.3	5.8	31.5	3.0	18.6
10-min rest)[a]	0.7	3.2	9.7	5.2	17.4
V_E: 40 L/minute[b]	0.8	3.0	0.0	8.4	4.2
Healthy men: 18–35 years	2.1	1.4	37.0	1.2	45.8
(n = 18 for air/0.08 ppm;10 of	0.2	2.5	6.1	3.1	8.5
these subjects were also exposed	1.8	0.7	2.6	2.6	7.5
to 0.10 ppm)	1.2	1.0	7.2	5.8	8.5
18 hours[c]	0.9	1.8	1.1	0.8	0.2
Subgroup mean (SD): (n = 10)	1.6 (0.9)	2.2 (1.4)	9.6 (12.8)	3.8 (2.1)	11.4 (12.9)
	1.5	3.1	7.7		
	1.4	2.6	3.5		
	0.6	1.3	3.4		
	3.2	1.2	3.4		
	0.9	1.3	−0.8		
	0.2	2.5	18.0		
	4.5	11.4	6.1		
	4.6	7.1	10.4		
Overall mean (SD): (n = 18)	1.8 (1.3)	2.9 (2.6)	8.3 (10.1)		

	Air exposure (sham)	Ozone exposure (0.16 ppm)	
Horstman et al. (1995)	0.5	5.5	
0.0, 0.16 ppm/7.6 hours	0.0	9.0	Individual
(50-min exercise alternating with	0.0	7.0	data not
10-min rest)	0.5	19.0	available
V_E: 15 L/minute/m^2 BSA[b]	0.5	13.5	
Healthy men and women:	0.5	13.0	
18–35 years	0.5	1.5	
(n = 11)	1.5	5.0	
18 hours[c]	0.0	3.5	
	5.0	4.0	
	2.0	1.0	
Mean (SD):	1.0 (1.4)	7.5 (5.4)	8.6 (6.3)
Overall mean (SD) (n = 58)	1.7 (1.7)		

Note: PMN, polymorphonuclear neutrophil leukocyte; BSA, body surface area.

[a] Ozone concentration/duration of exposure; (0.6 hour lunch break (in chamber) between hours 3 and 4).

[b] Ventilation (V_E) during exercise.

[c] Time elapsed from end of exposure to BAL.

ASTHMA

Asthma is a common disease in the United States with a prevalence of about 4% to 8% in various population groups, including children. Asthma is characterized by variable airflow obstruction, chronic diffuse airway inflammation including eosinophils, and hyperreactivity to a variety of bronchoconstrictive stimuli. Air pollution is widely believed to provoke exacerbations of preexisting asthma.* Interestingly, both shorter intense and longer less intense exposures to ozone enhance bronchial reactivity to inhaled bronchoconstrictor drugs, an effect that resolves within 18 to 24 hours after exposure (Folinsbee et al. 1994, Horstman et al. 1990, Folinsbee and Hazucha 1989, Holtzman et al. 1979). Both healthy and asthmatic subjects exhibit this effect (Kreit et al. 1989, Eschenbacher et al. 1989), which is of similar magnitude when expressed in terms of the difference between the pre- and postexposure $\log_{10}$ of the provocative dose (PD) of drug required to achieve a given lung function decrement from baseline (range of $\Delta\log_{10}$ PD, 0.2 to 0.7).**

Despite these *a priori* reasons to expect that asthmatic individuals might prove to be an ozone-susceptible subgroup, and despite suggestive data from early field studies of panels of asthmatic individuals residing in Los Angeles (Whittemore and Korn 1980) and Houston (Holguin et al. 1985), the first cautious controlled-exposure studies (to relatively small inhaled doses of ozone) were negative (Koenig et al. 1985, 1987, 1988; Silverman 1979; Linn et al. 1978). More recently, however, statistically significant excess reactivity (decrements of FEV_1 and FEV_1/FVC ratio and symptoms) of asthmatic individuals has been demonstrated by Horstman et al. (1995) in 7.6-hour exposures to 0.16 ppm ozone while performing mild exercise (V_E = ~27 L/minute) compared with normal volunteers. A similar result was reported by Kreit et al. (1989) in asthmatic subjects with a shorter, more intense exposure protocol. Studies of lung function requiring exercise in asthmatic individuals are complicated by the phenomenon of exercise (or hyperventilation)-induced bronchoconstriction, which occurs even in the absence of any air pollutants in many asthmatic patients. Unlike the spirometric decrements provoked by ozone exposure in normal subjects, which are neither reversed nor prevented by inhaled β-adrenergic agonists, substantial acute reversibility of FEV_1 decrements (and symptoms in some subjects) was found by Horstman et al. (1995) in asthmatic subjects. Furthermore, the ozone exposure produced a mean increase in RV of 0.5 L in the asthmatic group compared with 0.3 L after the sham (air) exposure and (as expected) no change in the normal controls similarly exposed to ozone. These observations suggest that ozone exposure can indeed cause significantly more bronchoconstriction in asthmatic than in normal individuals but that this feature may be masked by the decrements of FVC and FEV_1 occurring in both groups because of the involuntary neural inhibition of inspiration previously discussed.

The short-term effect of ozone exposure on the chronic airways eosinophilic inflammation characteristic of asthma as determined by BAL procedures has been of considerable recent interest (Scannell et al. 1996, Peden et al. 1996, Basha et al. 1994). Whether these effects are qualitatively or merely quantitatively different in asthmatic individuals compared with normal controls is not clear. Some have found that the neutrophil influx is accentuated in asthmatic patients (compared with normal controls) after ozone exposure (Scannell et al. 1996, Basha et al. 1994), whereas one study found an eosinophil influx in asthmatic patients that was not observed in normal volunteers (Peden et al. 1996).

---

* In the case of another air pollutant, SO_2, exercising or hyperventilating subjects with mild to moderate asthma have been clearly shown as a group to exhibit exceptional acute bronchial reactivity to brief exposure (Horstman et al. 1986, Sheppard et al. 1981). Interestingly, there is a wide range of sensitivity to SO_2 exposure even among asthmatic individuals, with some developing bronchoconstriction to SO_2 levels of 0.25 ppm, or even less, whereas others remain unresponsive to 2 ppm or more (Horstman et al. 1986).

** Compared with normal individuals, people with mild to moderate asthma characteristically exhibit baseline airways hyperreactivity in the range of 1 to 2 $\log_{10}$ PD units—i.e., 1/10th to 1/100th the dose of bronchoconstrictor drug is required to achieve similar percentage changes from baseline lung function (Cockcroft et al. 1977). Thus, ozone-induced decrements in $\log_{10}$ PD are modest relative to the effect of the disease itself on baseline $\log_{10}$ PD.

Attention has also focused on the respiratory effects of addition of other likely pollutants to ozone in controlled-exposure studies. Combining ozone with acid aerosols (to mimic summertime oxidant haze) has generally not revealed substantial interactions even when the subjects had mild asthma (Linn et al. 1994, 1995; Utell et al. 1994). Perhaps the most interesting experiments are those in which ozone exposure has been followed by either nasal or bronchial challenge of allergic asthmatic patients with a specific antigen. Several investigators (Molfino et al. 1991, Peden et al. 1995, Jörres et al. 1996), but not all (Ball et al. 1996), have found ozone exposure to enhance later responses to antigen challenge. The potential sensitizing effect of specific antigen exposure on the lung function response of allergic asthmatic patients to subsequent ozone challenge has not been investigated.

In summary, recent controlled studies provide some evidence for enhanced susceptibility of asthmatic subjects to ozone exposure itself and for ozone-induced enhancement of subsequent responses to allergen challenge of the respiratory tract. These findings acquire greater significance when considered in light of recent epidemiologic studies, which are discussed below.

WHITHER RISK ASSESSMENT FOR AMBIENT AIR POLLUTANTS?

The ILSI Workshop on Human Variability has sought to focus on ways in which the uncertainties associated with interindividual variability of human responses to environmental contaminants can be better managed in the context of improved risk assessment.

The published body of research describing the effects on lung function of controlled human exposures to ozone is arguably the largest body of work for any single inhaled toxicant and is complemented by a similarly impressive body of work dealing with responses to short- and long-term exposures in animals. Furthermore, it has been shown that exposures to ambient air containing ozone cause decrements in lung function that are at least as large as those observed in comparable controlled exposures. The range of ozone exposure conditions shown to have a significant effect on human lung function has been extended down to a concentration of only 0.08 ppm and moderate levels of ventilation (albeit for multihour exposure periods). Thus, there is less need for extrapolation in applying these data to real-life situations than is true for many other environmental toxicants. In spite of the interindividual variability of lung function responses to ozone exposure, empirical modeling approaches are promising. However, even when one considers only short-term effects of exposure, the translation of these data into a fully convincing risk assessment remains difficult. Some of the reasons follow:

1. Dosimetry. Modeling the actual population exposures involves many assumptions, including patterns of ambient ozone, number of individuals exposed outdoors, duration of such exposures, individual activity pattern (and ventilation) during exposure, cumulative exposure effects, adjusting exposure for body size or lung size, etc. The separate effects of inspired dose rate and of cumulative inspired dose upon response make effective dose calculations even more uncertain in the spontaneous exposure situation. Furthermore, ambient air ozone pollution is generally associated with other pollutants (especially fine particulate matter), which might interact with ozone.

2. Response. The acute effects of a given exposure may be altered by previous (recent) exposures. Normal individuals, even of the same gender and age, have widely varying lung function decrements in response to similar exposures. This is particularly true of children and young adults, who are the most likely to engage in outdoor exercise. The reason(s) for this intrinsic variability in responsiveness is not currently known. The use of existing empirical dose–response models involves extrapolations and assumptions. Other types of response (e.g., airways inflammation) have been much less extensively studied than lung function changes. Although there again appears to be considerable variability in individual responses, these are not correlated with the lung function

changes. Some fraction of the subpopulation of asthmatic individuals is suspected of being hypersusceptible to the effects of ozone exposure on lung function, but this has been difficult to establish convincingly in controlled exposure studies.

3. Interpretation of effects. The interpretation of a given level of nociceptor-mediated acute decrement in lung function (e.g., 10% to 15% of baseline) as an adverse respiratory health effect is uncertain because the changes do not reflect a change in lung mechanics. (The interpretation of ozone-provoked acute inflammatory effects in the lower airways faces similar uncertainties.)

Is there no approach to risk assessment that would circumvent these problems? Let us consider the time-series analysis approach to the study of environmental respiratory health effects in large, populous regions that has emerged as an important epidemiologic air pollution research tool over the past decade. The independent variables of interest are those of air quality (composition); the dependent health-related variables of interest are relatively crude (but convincing) observations like total mortality, total hospital admissions, or total visits to hospital emergency rooms for respiratory reasons. Both types of variables can be repeatedly quantified, usually on a daily basis. The epidemiologist/statistician then searches for parallelisms in the pattern of day-to-day fluctuation of a dependent variable (measures of adverse health effects) with that of the independent variable (measures of ambient air quality). Fortunately, atmospheric monitoring stations have been measuring air quality with respect to criteria pollutants for a number of years. The clinical endpoints have become increasingly accessible, especially in Canada with its organized health care database.

Such studies do not attempt to estimate individual exposures to air pollutants, but they measure selected health indices (including appropriate controls) of an entire region comprising millions of inhabitants. Thus, there is no question of the importance of the health endpoints and no need to extrapolate these effects from reversible, transient changes observed in small samples of the population exposed under standardized but artificial conditions.

The results of such studies for summertime, ozone-containing air pollution have been briefly summarized and referenced in Bascom et al. (1996) (Table 6, p. 21) and also by Bates (1995) and have been reported in considerable detail in the recent EPA criteria document for ozone (EPA 1996, pp. 7-121–7-143). Reports from Ontario Province, Toronto, Buffalo, New York City, Albany, central New Jersey, and Atlanta have found significant associations between daily respiratory health effects and daily ambient ozone concentrations, which generally did not exceed the current 0.12 ppm NAAQS.* Figure 2 is taken from the paper of Burnett et al. (1994). The total number of daily admissions for respiratory causes in 168 southern Ontario hospitals during the summers of 1983 to 1988 increases as a function of the previous day's maximum 1-hour ozone concentration. The relationship appears linear over the range of 20 to 120 parts per billion (ppb) and has a slope of about 1% per 10 ppb. The ability to replicate such studies in different urban areas (consistency) lends credibility to the observed associations, and multiple studies can be treated more quantitatively by a formal meta-analysis. Different health outcome variables are often related to one another and should therefore exhibit coherent behavior. Unfortunately, ambient respirable particulate matter levels are correlated with ozone concentrations in summertime so that it is often not possible to give precedence to one or the other pollutant in driving the observed association. Indeed, it is possible that both pollutants contribute or even that ozone is merely a surrogate for some other air contaminant. Nor can these studies identify the specific individuals whose respiratory health problems could be attributed to increased air pollutant levels and who might therefore be a sensitive

---

* Burnett et al. (1997) have recently published a time series analysis for the period 1981 to 1991 in 16 Canadian cities. The maximum daily ozone concentration (1-hour average) on the previous day was significantly associated with the number of hospitalizations for respiratory disease from April to December but not in the winter months. Particulate matter levels were also positively associated.

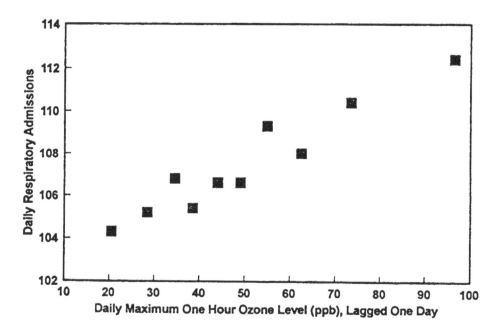

FIGURE 2   Data suggest a linear association in southern Ontario between daily peak (1-hour average) ambient ozone levels over the range 20 to 120 ppb (abscissa) and the 1-day lagged number of hospital admissions for respiratory causes during the summers of 1983–1988. Slope of the relationship is 1 excess admission (1%) per 10 ppb increase in ozone concentration. All points fall within the current 1-hour NAAQS for ozone of 120 ppb. Reproduced with permission from Burnett et al. (1994).

group. Thus, it is not possible to characterize these individuals or solicit their participation in subsequent controlled-exposure studies or in prospective case-control studies. Finally, such associations of phenomena may not have all the qualities required of a causal relationship. The criteria that may increase the weight of an epidemiologically defined association for the purpose of risk assessment are discussed in Appendix B (pp. 72–78) of the booklet published by Federal Focus (1996) and also by Bates (1995).

The association analysis also provides a point estimate and confidence limits for the magnitude of the association and is based on measurements covering the range of spontaneously occurring (i.e., relevant) ambient air pollutant levels. Unusually high levels can be omitted from the analysis to avoid any unduly large influence of such points on the overall association. Admittedly, these estimates could be biased for a number of reasons.

In the case of ambient air particulate matter (PM_{10}) the time-series analysis approach pioneered by Joel Schwartz, Douglas Dockery, and C. Arden Pope, among others, has provided increasingly convincing evidence of acute mortality and morbidity associated with fluctuations of daily PM_{10} levels within the current NAAQS of 150 μg/m³ (Bascom et al. 1996).* A careful independent review of the complex statistical methodology required and its application to very large data sets sponsored by the Health Effects Institute and performed by Samet et al. has recently appeared (1995) and generally supports the analyses of Schwartz, Dockery, Pope, and others.

Unlike the situation for ozone, whose toxic potential was readily demonstrated in controlled-exposure studies of humans and various animal species, previous efforts to demonstrate respiratory (or other) effects of controlled-exposure to various respirable aerosols (relevant to current ambient

---

* One should recall that the initial formal recognition of the adverse effects of air pollution (particulate matter and SO_2) was catalyzed by population studies in London during the health disaster in the winter of 1952.

air particulate matter concentrations in the United States) in the 100-μg/m³ range had rarely produced striking results. The time-series data have led to a renewal of interest on the part of experimental scientists and a number of promising leads are being explored.

It therefore seems (to this author) that in the case of the criteria air pollutants the role of controlled short-term exposure studies of healthy or diseased volunteers, and of animal and *in vitro* studies, will be as follows: (1) to confirm the ability of exposures to specific pollutants to cause biological effects; (2) to define the tissue dose- and exposure dose–response relationships and the underlying toxicologic mechanisms; and (3) to suggest possible susceptible populations. These data then serve to provide plausibility and causal significance to the associations emerging from epidemiologic analyses. Under these circumstances, the task of risk assessment would seem greatly facilitated although some issues concerning thresholds and susceptible individuals would remain. Such an approach would encourage (indeed it would insist on) integration of information obtained from controlled human exposures, acute and chronic animal toxicology, *in vitro* (cultured cells) toxicology, and various types of population (epidemiologic) investigations.*

The descriptive or ecologic or correlational types of epidemiologic studies, which have been very valuable for the risk assessment of air pollutants, are not the only possible types of epidemiologic study. Analytical epidemiologic studies, either of the case-control or cohort type, are also valuable. Indeed, cohort studies have shown or suggested adverse respiratory health effects of long-term residence in areas with chronically elevated levels of air pollution [especially respirable particulate matter (Dockery et al. 1993)], and case-control studies are especially valuable for more rare clinical events like many cancers and other diseases, provided that exposure assessment is reliable. The proceedings of a panel meeting in London in October 1995 on "Principles for Evaluating Epidemiologic Data in Regulatory Risk Assessment" was published by Federal Focus (1996).

SUMMARY AND FUTURE DIRECTIONS

Ozone was identified as a significant summertime ambient oxidant pollutant nearly half a century ago. Health effects attributable to oxidant pollution have been of concern for nearly as long, and the toxicology of ozone itself has been studied intensively in normal and diseased human volunteers, in experimental animals, and in cultured cell systems.

Groups of young healthy volunteers, acutely exposed to relevant levels of ozone develop reduced lung function. This is also seen in children and young adults spontaneously exposed to ozone in ambient air. More severe changes are often accompanied by symptoms of chest pain and cough. The changes are reversible over a period of hours and appear to be largely due to involuntary neurally mediated (airway nociceptive fibers) inhibition of inspiration rather than to altered lung mechanics (which may, however, play a greater role in asthmatic subjects). Although individual subjects have relatively reproducible responses to single ozone exposures, there are large interindividual differences in response to similar exposures. Apart from age (young subjects include a larger number of vigorous responders than do adults >35 years old), no explanation for this variability of response among individuals is currently available.

This response variability has not, however, precluded the development of empirical models of response as a function of exposure conditions (ozone concentration, inspired minute ventilation normalized for body surface area, and duration), which predict the fraction of young individuals who would develop a lung function decrement greater than a given value after a particular exposure. Nevertheless, the use of such models for risk assessment can be questioned because neurally mediated involuntary inhibition of full inspiration, at least of a moderate degree, may not constitute an adverse health effect.

---

* For example, in the case of very short SO_2 exposures, which cause some asthmatics in controlled exposures to develop symptomatic bronchospasm requiring acute inhaled bronchodilator treatment, the absence of population studies showing comparable effect has dampened the support for the need for a short-term SO_2 NAAQS.

Other short-term effects of relevant ozone exposures include a neutrophilic inflammatory response of the airways mucosa. This response is poorly (if at all) correlated with the lung function response. Its within-individual reproducibility is less well-defined and the factors responsible for interindividual response variability are also not known, although the mechanisms responsible for the inflammatory response are partially understood.

Although it is highly desirable to acquire a detailed mechanistic understanding of the adverse health responses and of the basis for differences in individual susceptibility, it seems unlikely to this author that increasingly sophisticated statistical characterization of descriptive studies of human response variability to environmental exposures will be of decisive importance to the risk assessment process. Mechanistic studies can be pursued in human volunteers (in the case of certain toxicants like ozone) as well as in cultured human and animal respiratory cells and animal models.

The author suggests that, from the standpoint of risk assessment, the most profitable approach to ambient air pollution is the use of descriptive epidemiologic (ecologic) studies of large regions that take advantage of accumulated air quality data and of indices of respiratory health to find temporal correlations among these variables. Such associations, when combined with data from controlled studies, acquire sufficient causal weight to furnish a good basis for risk assessment. Other types of epidemiologic research (e.g., prospective cohort studies) may also be valuable.

Efforts to understand the genetic, environmental, and age- or disease-associated sources of unusual susceptibility of some individuals to adverse effects of exposure to air pollutants requires much better understanding of mechanisms. This should lead to enhanced ability to formulate testable hypotheses about the sources of human response variability, ways of identifying susceptible individuals and potential for irreversible effects, and ways to develop targeted prevention strategies for susceptible individuals. Such a level of understanding would be very valuable to risk assessment and to formulation and implementation of cost-effective public health policy.

ACKNOWLEDGMENTS

Support was received from Health Effects Institute Research Contract HEI 91-4 and from EPA Cooperative Agreement CR817643 and CR824915. The author thanks Drs. M. J. Hazucha, L. Folinsbee, J. Raub, and D. Neumann as well as anonymous reviewers for their review and comments on earlier drafts of the manuscript. Dr. William McDonnell has discussed the manuscript at length and made many important suggestions, for which the author is especially grateful. I am also very grateful to Drs. R. Devlin, W. McDonnell, and M. J. Hazucha for sharing unpublished data. The author is, however, solely responsible for the content of the paper. The author is grateful to Dr. Richard Burnett and to the publishers of Environmental Research for permission to reproduce Figure 2. This manuscript was reviewed and approved by the National Health and Environmental Effects Research Laboratory of the EPA but does not represent the official policy of the EPA.

REFERENCES

Aris RM, Christian D, Hearne PQ, et al (1993) Ozone-induced airway inflammation in human subjects as determined by airway lavage and biopsy. Am Rev Respir Dis 148:1363–1372

Avol EL, Linn WS, Venet TG, et al (1984) Comparative respiratory effects of ozone and ambient oxidant pollution exposure during heavy exercise. J Air Pollut Control Assoc 34:804–809

Ball BA, Folinsbee LJ, Peden DB, Kehrl HR (1996) Allergen bronchoprovocation of patients with mild allergic asthma after ozone exposure. J Allergy Clin Immunol 98:563–572

Balmes JR, Chen LL, Scannell C, et al (1996) Ozone-induced decrements in FEV_1 and FVC do not correlate with measures of inflammation. Am J Respir Crit Care Med 153:904–909

Bascom R, Bromberg PA, Costa DA, et al (1996) State of the art: health effects of air pollution (parts I and II). Am J Respir Crit Care Med 153:3–50, 477–498

Basha MA, Gross KB, Gwizdala CJ, et al (1994) Bronchoalveolar lavage neutrophilia in asthmatic and healthy volunteers after controlled exposure to ozone and filtered purified air. Chest 106:1757–1765

Bates DV (1995) Ozone: a review of recent experimental, clinical and epidemiological evidence, with notes on causation (parts I and II) Can Respir J 2:25–31, 161–171

Beckett WS, McDonnell WF, Horstman DH, House DE (1985) Role of the parasympathetic nervous system in acute lung response to ozone. J Appl Physiol 59:1879–1885

Bedi JF, Horvath SM, Drechsler-Parks DM (1989) Adaptation by older individuals repeatedly exposed to 0.45 parts per million ozone for two hours. J Air Pollut Control Assoc 39:194–199

Burnett RT, Brook JR, Yung WT, et al (1997) Association between ozone and hospitalization for respiratory disease in 16 Canadian cities. Environ Res 72:24–31

Burnett RT, Dales RE, Raizenne ME, et al (1994) Effects of low ambient levels of ozone and sulfates on the frequency of respiratory admissions to Ontario hospitals. Environ Res 65:172–194

Bush ML, Asplund PT, Miles KA, et al (1996) Longitudinal distribution of O_3 absorption in the lung: gender differences and intersubject variability. J Appl Physiol 81:1651–1657

Cochrane GM, Prieto F, Clark TJH (1977) Intrasubject variability of maximal expiratory flow volume curve. Thorax 32:171–176

Cockcroft DW, Killian DN, Mellon JJA, Hargreave FE (1977) Bronchial reactivity to inhaled histamine: a method and clinical survey. Clin Allergy 7:235–243

Coleridge JCG, Coleridge HM, Schelegle ES, Green JF (1993) Acute inhalation of ozone stimulates bronchial C-fibers and rapidly adapting receptors in dogs. J Appl Physiol 74:2345–2352

Devlin RB, Folinsbee LJ, Biscardi F, et al (1997) Inflammation and cell damage induced by repeated exposure of humans to ozone. Inhal Toxicol 9:211–235

Devlin RB, McDonnell WF, Becker S, et al (1996) Time-dependent changes in inflammatory mediators in the lungs of humans exposed to 0.4 ppm ozone for 2 hr: a comparison of mediators found in bronchoalveolar lavage fluid 1 and 18 h after exposure. Toxicol Appl Pharmacol 138:176–185

Devlin RB, McKinnon KP, Noah T, et al (1994) Ozone-induced release of cytokines and fibronectin by alveolar macrophages and airway epithelial cells. Am J Physiol 266:L612–L619

Devlin RB, McDonnell WF, Mann R, et al (1991) Exposure of humans to ambient levels of ozone for 6.6 hours causes cellular and biochemical changes in the lung. Am J Respir Cell Mol Biol 4:72–81

Dockery DW, Pope A III, Xu X, et al (1993) An association between air pollution and mortality in six U.S. cities. N Engl J Med 329:1753–1759

Drechsler-Parks DM, Bedi JF, Horvath SM (1989) Pulmonary function responses of young and older adults to mixtures of O_3, NO_2, and PAN. Toxicol Ind Health 5:505–517

Drechsler-Parks DM, Bedi JF, Horvath SM (1987a) Pulmonary function responses of older men and women to ozone exposure. Exp Gerontol 22:91–101

Drechsler-Parks DM, Bedi JF, Horvath SM (1987b) Pulmonary function desensitization on repeated exposures to the combination of peroxyacetyl nitrate and ozone. J Air Pollut Control Assoc 37:1199–1201

Emmons K, Foster WM (1991) Smoking cessation and acute airway response to ozone. Arch Environ Health 46:288–295

Eschenbacher WL, Ying RL, Kreit JW, Gross KB (1989) Ozone-induced lung function changes in normal and asthmatic subjects and the effect of indomethacin. In Schneider T, Lee SD, Wolters GJR, Grant LD (eds), Atmospheric ozone research and its policy implications: proceedings of the 3rd US–Dutch international symposium, May 1988. Elsevier, Nijmegen, The Netherlands, pp 493–499

Federal Focus (1996) Principles for evaluating epidemiologic data in regulatory risk assessment. Federal Focus, Washington, D.C.

Folinsbee LJ, Hazucha MJ (1989) Persistence of ozone-induced changes in lung function and airway responsiveness. In Schneider T, Lee SD, Wolters GJR, Grant LD (eds), Atmospheric ozone research and its policy implications: proceedings of the 3rd US–Dutch international symposium, May 1988. Elsevier, Nijmegen, The Netherlands, pp 483–492

Folinsbee LJ, Horstman DH, Kehrl HR, et al (1994) Respiratory responses to repeated prolonged exposure to 0.12 ppm ozone. Am J Respir Crit Care Med 149:98–105

Folinsbee LJ, Horstman DH, Kehrl HR, et al (1991) Effects of single and repeated prolonged low-level ozone exposure in man. Presented at Annual Meeting of the Society for Occupational and Environmental Health, March, Washington, D.C.

Folinsbee LJ, Silverman F, Shephard RJ (1977) Decrease of maximum work performance following ozone exposure. J Appl Physiol: Respir Environ Exercise Physiol 42:531–536

Gerrity TR, Biscardi F, Strong A, et al (1995) Bronchoscopic determination of ozone uptake in humans. J Appl Physiol 79:852–860

Gliner JA, Horvath SM, Folinsbee LJ (1983) Preexposure to low ozone concentrations does not diminish the pulmonary function response on exposure to higher ozone concentrations. Am Rev Respir Dis 127:51–55

Hazbun ME, Hamilton R, Holian A, Eschenbacher WL (1993) Ozone-induced increases in substance P and 8-epi-prostaglandin $F_{2\alpha}$ in the airways of human subjects. Am J Respir Cell Mol Biol 9:568–572

Hazucha MJ (1987) Relationship between ozone exposure and pulmonary function changes. J Appl Physiol 62:1671–1680

Hazucha MJ, Madden M, Pape G, et al (1996) Effects of cyclo-oxygenase inhibition on ozone-induced respiratory inflammation and lung function changes. Eur J Appl Physiol Occup Med 73:17–27

Hazucha MJ, Folinsbee LJ, Seal E Jr (1992) Effects of steady-state and variable ozone concentration profiles on pulmonary function. Am Rev Respir Dis 146:1487–1493

Hazucha MJ, Bates DV, Bromberg PA (1989) Mechanisms of action of ozone on the human lung. J Appl Physiol 67:1535–1541

Higgins ITT, D'Arcy JB, Gibbons DI, et al (1990) Effect of exposures to ambient ozone on ventilatory lung function in children. Am Rev Respir Dis 141:1136–1146

Holguin AH, Buffer PA, Contant CF, et al (1985) The effects of ozone on asthmatics in the Houston area. In Lee SD (ed), Evaluation of the scientific basis for ozone/oxidants standards: proceedings of an APCA international specialty conference, November 1984, Houston, TX. Air Pollution Control Association, Pittsburgh, PA, pp 262–280

Holtzman MJ, Cunningham JH, Sheller JR, et al (1979) Effect of ozone on bronchial reactivity in atopic and nonatopic subjects. Am Rev Respir Dis 120:1059–1067

Horstman DH, Ball BA, Brown J, et al (1995) Comparison of pulmonary responses of asthmatic and non-asthmatic subjects performing light exercise while exposed to a low level of ozone. Toxicol Ind Health 11:369–385

Horstman DH, Folinsbee LJ, Ives PJ, et al (1990) Ozone concentration and pulmonary response relationships for 6.6 hour exposures with five hours of moderate exercise to 0.08, 0.10, and 0.12 ppm. Am Rev Respir Dis 142:1158–1163

Horstman D, Roger LJ, Kehrl H, Hazucha MJ (1986) Airway sensitivity of asthmatics to sulfur dioxide. Toxicol Ind Health 2:289–298

Horvath SM, Gliner JA, Folinsbee LJ (1981) Adaptation to ozone: duration of effect. Am Rev Respir Dis 123:496–499

Hu SC, Ben-Jebria A, Ultman JS (1994) Longitudinal distribution of ozone absorption in the lung: effects of respiratory flow. J Appl Physiol 77:574–583

Hu SC, Ben-Jebria A, Ultman JS (1992) Longitudinal distribution of ozone absorption in the lung: quiet respiration in healthy subjects. J Appl Physiol 73:1655–1667

Jörres R, Nowak D, Magnussen H (1996) The effect of ozone exposure on allergen responsiveness in subjects with asthma or rhinitis. Am J Respir Crit Care Med 253:56–64

Kehrl HR, Hazucha MJ, Solic JJ, Bromberg PA (1985) Responses of subjects with chronic obstructive pulmonary disease after exposure to 0.3 ppm ozone. Am Rev Respir Dis 131:719–724

Kerr HD, Kulle TJ, McIlhany ML, Swidersky P (1975) Effects of ozone on pulmonary function in normal subjects: an environmental-chamber study. Am Rev Respir Dis 111:763–773

Kinney PL, Nilsen DM, Lippmann M, et al (1996) Biomarkers of lung imflammation in recreational joggers exposed to ozone. Am J Respir Crit Care Med 154:1430–1435

Koenig JQ, Covert DS, Smith MS, et al (1988) The pulmonary effects of ozone and nitrogen dioxide alone and combined in healthy and asthmatic adolescent subjects. Toxicol Ind Health 4:521–532

Koenig JQ, Covert DS, Marshall SG, et al (1987) The effects of ozone and nitrogen dioxide on pulmonary function in healthy and in asthmatic adolescents. Am Rev Respir Dis 136:1152–1157

Koenig JQ, Covert DS, Morgan MS, et al (1985) Acute effect of 0.12 ppm ozone or 0.12 ppm nitrogen dioxide on pulmonary function in healthy and asthmatic adolescents. Am Rev Respir Dis 132:648–651

Koren HS, Devlin RB, Becker S, et al (1991) Time-dependent changes of markers associated with inflammation in the lungs of humans exposed to ambient levels of ozone. Toxicol Pathol 19:406–411

Koren HS, Devlin RB, Graham DE, et al (1989a) Ozone induced inflammation in the lower airways of human subjects. Am Rev Respir Dis 139:407–415

Koren HS, Devlin RB, Graham DE, et al (1989b) The inflammatory response in human lung exposed to ambient levels of ozone. In Schneider T, Lee SD, Wolters GJR, Grant LD (eds), Atmospheric ozone research and its policy implications: proceedings of the 3rd US–Dutch international symposium, May 1988. Elsevier, Nijmegen, The Netherlands, pp 745–753

Koto H, Aizawa H, Takata S, et al (1995) An important role of tachykinins in ozone-induced airway hyper-responsiveness. Am J Respir Crit Care Med 151:1763–1769

Kreit JW, Gross KB, Moore TB, et al (1989) Ozone-induced changes in pulmonary function and bronchial responsiveness in asthmatics. J Appl Physiol 66:217–222

Krishna MT, Springall DR, Meng Q-H, et al (1996) Effects of 0.2 ppm ozone on the sensory nerves in the bronchial mucosa of healthy humans [abstract]. Am J Respir Crit Care Med 153:A700

Kulle TJ, Sauder LR, Hebel JR, Chatham MD (1985) Ozone response relationships in healthy nonsmokers. Am Rev Respir Dis 132:36–41

Larsen RI, McDonnell WF, Horstman DH, Folinsbee LJ (1991) An air quality data analysis system for interrelating effects, standards, and needed source reductions: part 11. A lognormal model relating human lung function decrease to O_3 exposure. J Air Waste Manage Assoc 41:455–459

Leikauf GD, Driscoll KE, Wey HE (1988) Ozone-induced augmentation of eicosanoid metabolism in epithelial cells from bovine trachea. Am Rev Respir Dis 137:435–442

Linn WS, Anderson KR, Shamoo DA, et al (1995) Controlled exposures of young asthmatics to mixed oxidant gases and acid aerosol. Am J Respir Crit Care Med 152:885–891

Linn WS, Shamoo DA, Anderson KR, et al (1994) Effects of prolonged, repeated exposure to ozone, sulfuric acid, and their combination in healthy and asthmatic volunteers. Am J Respir Crit Care Med 150:431–440

Linn WS, Avol EL, Shamoo DA, et al (1988) Repeated laboratory ozone exposures of volunteer Los Angeles residents: an apparent seasonal variation in response. Toxicol Ind Health 4:505–520

Linn WS, Shamoo DA, Venet TG, et al (1983) Response to ozone in volunteers with chronic obstructive pulmonary disease. Arch Environ Health 38:278–283

Linn WS, Buckley RD, Spier CE, et al (1978) Health effects of ozone exposure in asthmatics. Am Rev Respir Dis 117:835–843

Madden MC, Smith JP, Dailey LA, Friedman M (1994) Polarized release of lipid mediators derived from phospholipase A_2 activity in a human bronchial cell line. Prostaglandins 48:197–215

McDonnell WF, Smith MV (1994) Description of acute ozone response as a function of exposure rate and total inhaled dose. J Appl Physiol 76:2776–2784

McDonnell WF, Andreoni S, Smith MV (1995) Proportion of moderately exercising individuals responding to low-level, multi-hour ozone exposure. Am J Respir Crit Care Med 152:589–596

McDonnell WF, Muller KE, Bromberg PA, Shy CM (1993) Predictors of individual differences in acute response to ozone exposure. Am Rev Respir Dis 147:818–825

McDonnell WF, Chapman RS, Leigh MW, et al (1985a) Respiratory responses of vigorously exercising children to 0.12 ppm ozone exposure. Am Rev Respir Dis 132:875–879

McDonnell WF, Horstman DH, Abdul-Salaam S, House DE (1985b) Reproducibility of individual responses to ozone exposure. Am Rev Respir Dis 131:36–40

McDonnell WF, Horstman DH, Hazucha MJ, et al (1983) Pulmonary effects of ozone exposure during exercise: dose–response characteristics. J Appl Physiol: Respir Environ Exercise Physiol 54:1345–1352

McKinnon KP, Madden MC, Noah TL, Devlin RB (1993) In vitro ozone exposure increases release of arachidonic acid products from a human bronchial epithelial cell line. Toxicol Appl Pharmacol 118:215–223

Messineo TD, Adams WC (1990) Ozone inhalation effects in females varying widely in lung size: comparison with males. J Appl Physiol 69:96–103

Miller FJ, Overton JH Jr, Jaskot RH, Menzel DB (1985) A model of the regional uptake of gaseous pollutants in the lung: I. the sensitivity of the uptake of ozone in the human lung to lower respiratory tract secretions and exercise. Toxicol Appl Pharmacol 79:11–27

Molfino NA, Wright SC, Katz I, et al (1991) Effect of low concentrations of ozone on inhaled allergen responses in asthmatic subjects. Lancet 338:199–203

Passannante A, Hazucha MJ, Seal E, et al (1995) Nociceptive mechanisms modulate ozone-induced human lung function decrements [abstract]. Anesth Anal 80:S371

Peden DB, Boehlecke B, Horstman D, Devlin RB (1996) Influx of bronchial neutrophils and eosinophils in asthmatics after prolonged exposure to 0.16 ppm ozone [abstract]. Am J Respir Crit Care Med 153:A700

Peden DB, Setzer RW Jr, Devlin RB (1995) Ozone exposure has both a priming effect on allergen-induced responses and an intrinsic inflammatory action in the nasal airways of perennially allergic asthmatics. Am J Respir Crit Care Med 151:1336–1345

Pennock BE, Rogers RM, McCaffree DR (1981) Changes in measured spirometric indices. What is significant? Chest 80:97–99

Postlethwait EM, Langford SD, Bidani A (1994) Determinants of inhaled ozone in isolated rat lungs. Toxicol Appl Pharmacol 125:77–89

Pryor WA, Bermudez E, Cueto R, Squadrito GL (1996) Detection of aldehydes in bronchoalveolar lavage of rats exposed to ozone. Fundam Appl Toxicol 34:148–156

Pryor WA, Das B, Church DF (1991) The ozonation of unsaturated fatty acids: aldehydes and hydrogen peroxide as products and possible mediators of ozone toxicity. Chem Res Toxicol 4:341–348

Samet J, Zeger S, Berhane K (1995) The association of mortality and particulate air pollution. Health Effects Institute, Particulate Air Pollution and Daily Mortality, August 1995, pp 1–104

Scannell C, Chen L, Aris RM, et al (1996) Greater ozone-induced inflammatory responses in subjects with asthma. Am J Respir Crit Care Med 154:24–29

Schelegle ES, Carl ML, Coleridge HM, et al (1993) Contribution of vagal afferents to respiratory reflexes evoked by acute inhalation of ozone in dogs. J Appl Physiol 74:2338–2344

Schelegle ES, Siefkin AD, McDonald RJ (1991) Time course of ozone-induced neutrophilia in normal humans. Am Rev Respir Dis 143:1353–1358

Schelegle ES, Adams WC, Siefkin AD (1987) Indomethacin pretreatment reduces ozone-induced pulmonary function decrements in human subjects. Am Rev Respir Dis 136:1350–1354

Schonfeld BR, Adams WC, Schelegle ES (1989) Duration of enhanced responsiveness upon re-exposure to ozone. Arch Environ Health 44:229–236

Seal E Jr, McDonnell WF, House DE, et al (1993) The pulmonary response of white and black adults to six concentrations of ozone. Am Rev Respir Dis 147:804–810

Seltzer J, Bigby BG, Stulbarg M, et al (1986) O_3-induced change in bronchial reactivity to methacholine and airway inflammation in humans. J Appl Physiol 60:1321–1326

Sheppard D, Saisho A, Nadel JA, Boushey HA (1981) Exercise increases sulfur dioxide-induced broncho-constriction in asthmatic subjects. Am Rev Respir Dis 123:486–491

Silverman F (1979) Asthma and respiratory irritants (ozone). Environ Health Perspect 29:131–136

Solic JJ, Hazucha MJ, Bromberg PA (1982) The acute effects of 0.2 ppm ozone in patients with chronic obstructive pulmonary disease. Am Rev Respir Dis 125:664–669

Spektor DM, Thurston GD, Mao J, et al (1991) Effects of single and multiday ozone exposures on respiratory function in active normal children. Environ Res 55:107–122

Spektor DM, Lippman M, Lioy PJ, et al (1988a) Effects of ambient ozone on respiratory function in active, normal children. Am Rev Respir Dis 137:313–320

Spektor DM, Lippman M, Thurston GD, et al (1988b) Effects of ambient ozone on respiratory function in healthy adults exercising outdoors. Am Rev Respir Dis 138:821–828

U.S. Environmental Protection Agency (1996) Air quality criteria for ozone and related photochemical oxidants. EPA 600/P-93/004cF, Vol III

Utell MJ, Frampton MW, Morrow PE, et al (1994) Oxidant and acid aerosol exposure in healthy subjects and subjects with asthma. Part II: effects of sequential sulfuric acid and ozone exposures on the pulmonary function of healthy subjects and subjects with asthma. Research Report No. 70. Health Effects Institute, Cambridge, MA, pp 37–39

Vesely KR, Schelegle ES, Hyde DM (1995) Neonatal capsaicin treatment in rats abolishes ozone-induced rapid shallow breathing [abstract]. Am J Respir Crit Care Med 151:A499

Weinmann GG, Weidenbach-Gerbase M, Foster WM, et al (1995) Evidence for ozone-induced small-airway dysfunction: lack of menstrual-cycle and gender effects. Am J Respir Crit Care Med 152:988–996

Whittemore AS, Korn EL (1980) Asthma and air pollution in the Los Angeles area. Am J Publ Health 70:687–696

6 Host-Environment Interactions That Affect Variability in Human Cancer Susceptibility

L. T. Frame, C. B. Ambrosone, F. F. Kadlubar, and N. P. Lang

CONTENTS

INTRODUCTION

Cancers are often classified as "hereditary" or "environmental" based on the perceived cause. However, unless a malignancy results from a known genetic defect or occupational exposure to a carcinogenic agent, the etiology does not easily fall into either of these categories. Therefore, investigators generally rely on epidemiologic studies to deduce what factors lead to variabilities in cancer incidence. When only one factor is isolated, the calculated increased risk is generally small. By contrast, higher levels of risk are assigned when specific combinations of genetic sensitivity and environmental exposure factors act in tandem. Interactions between genetic sensitivity and environmental exposure are not well

0-8493-2805-5/99/$0.00+$.50
© 1999 by CRC Press LLC

understood. Consequently, there is increased interest in better characterizing relevant genetic and environment interactions to lead to identification of at-risk individuals.

This review focuses on the influence of genetic and environmental interactions on interindividual differences in cancer susceptibility in humans. We examine epidemiologic evidence that supports a role for major host-specific factors in cancer risk, including genetics, age, ethnicity, and gender. Within each of those categories, we chose at least two relevant malignancies for discussion because the literature provides an understanding of the interplay of environmental risk factors with host susceptibility. In addition, general comments about the role of diet and physiological variables are addressed briefly. The overall goal of this discussion is to unify the findings of both epidemiologists and molecular biologists in the study of cancer cause and prevention.

CANCER AS A GENETIC DISEASE

For a number of decades, researchers have referred to cancer as a "disease of the genes." The following evidence supports this concept:

- Many chemicals that cause cancer in experimental animals are mutagens.
- Normal cells can be converted to cancer cells by direct introduction of exogenous genetic information (DNA transfection or tumor virus infection) or by loss of genetic information (chromosomal deletions).
- A variety of specific germ-line or somatic cell DNA lesions (point mutations, duplications, rearrangements) are associated with increased risk of cell transformation to a malignant phenotype.
- The cancer-prone genotype can be inherited during the process of cell division.
- Cancer susceptibility can be inherited within families (Watson et al. 1992).

Largely as a consequence of this genetic focus in cancer research, many important molecular targets involved in cancer initiation and progression have now been identified (Table 1). Progress also has been made in characterizing rare germ-line mutations that lead to many hereditary forms of cancer (Table 2). This information has helped build a framework for understanding the role of host defense mechanisms in normal and neoplastic physiology.

CANCER AS A DISEASE OF THE HOST-ENVIRONMENT INTERFACE

Because organisms depend on their immediate surroundings for basic needs, they must be able to monitor changes in the external environment. Individual survival often depends on making the correct systemic adjustment (biochemical, physiological, or behavioral) upon exposure to noxious stimuli. For single-celled organisms, information about the environment is readily accessible, but defense options are limited to short-term changes in the cell cycle, stereotypical avoidance behavior, and biochemical adaptation. Organization of cells into complex multicellular organisms affords more sophisticated and effective defense strategies but creates challenges for disseminating information and coordinating responses. Failure to respond appropriately to internal or environmental stimuli is thought to play a role in the etiology of long-term disease states, such as cancer in higher organisms (Hart and Frame 1996).

Conceptualizing cancer as a disease of the host-environment interface has several important advantages. It provides a vantage point from which to consider genetic, developmental, physiological, environmental, and social factors that, alone or in combination, may alter cancer risk. This perspective is compatible with the concept that cellular and genetic insults and aberrations are a normal consequence of life, with communication between host and environment an important feature. In addition, it is consistent with the idea that there are interacting, opposing, and sometimes

TABLE 1
A Sampling of Molecular Targets That Can Be Modified During Carcinogenesis

Genes coding for	Examples
Receptors	Estrogen receptors (Andersen et al. 1994)
	Progesterone receptors (McKenna et al. 1995)
	Androgen receptors (Ruizeveld de Winter et al. 1994, Suzuki et al. 1993, Lobaccaro et al. 1993)
	Transforming growth factor β (Markowitz et al. 1995)
Tumor suppressors	APC (Thomas and Olschwang 1995, Smith et al. 1994, McKie et al. 1993)
(anti-oncogenes)	DCC (Zhang et al. 1995, Maesawa et al. 1995, Kim et al. 1993)
	p53 (Kondo et al. 1996, Heidenberg et al. 1995, Poremba et al. 1995, Bertorelle et al. 1995, Bosari et al. 1995, Chen et al. 1995, Smith and Fornace 1995, Faille et al. 1994, Ryberg et al. 1994, Berrozpe et al. 1994, Zhang et al. 1994, Hong et al. 1994, Wagner et al. 1994, Ziegler et al. 1994, McManus et al. 1994, Takagi et al. 1994, Enomoto et al. 1993, Honda et al. 1993, Miyamoto et al. 1993, Zou et al. 1993, Smith and Ponder 1993, Suzuki and Tamura 1993, Mazars et al. 1992, Sundaresan et al. 1992, Sakai and Tsuchida 1992, Mitsudomi et al. 1992, Reiss et al. 1992, Tamura et al. 1992, Kihana et al. 1992, Yoshimoto et al. 1992, Evans and Prosser 1992, Bellet 1992, Pierceall et al. 1991b)
	Rb (Miyamoto et al. 1995, Kubota et al. 1995, Pei et al. 1995, Kashii et al. 1994, Phillips et al. 1994, Neubauer et al. 1993, Ozaki et al. 1993, Evans and Prosser 1992, Bellet 1992, Weide et al. 1991)
	p15 (Gombart et al. 1995, de Vos et al. 1995, Komiya et al. 1995, Orlow et al. 1995, Kamb et al. 1994)
	p16 (Hatta et al. 1995, Gombart et al. 1995, Orlow et al. 1995, Kamb et al. 1994)
Oncogenes	Ha-ras (Bittard et al. 1996, Thelu et al. 1993, Pierceall et al. 1991a, Weston et al. 1991, Sugimura et al. 1990a, Ananthaswamy et al. 1988)
	Ki-ras (Malats et al. 1995, Van Laethem et al. 1995, Moerkerk et al. 1994, Scarpa et al. 1993, Chen et al. 1992)
	L-myc (Taylor et al. 1993, Kawashima et al. 1992, Champeme et al. 1992, Kakehi et al. 1991, Tamai et al. 1990)
DNA replication	DNA polymerase β (Matsuzaki et al. 1996)
and repair	DNA repair genes (Boland 1996)
enzymes	

redundant systems for preventing permanent damage; these systems may lose effectiveness with age or disease, or they may be targets for carcinogen action themselves.

Characterizing relevant risk factors may be important for understanding the proximal stages of carcinogenesis as well as for prevention, prognosis, and therapeutic intervention. The functional features of the host-environment interface include the following (Alberts et al. 1989):

1. physical and enzymatic barriers that minimize exposure of critical cell components to potentially damaging chemical or physical agents;
2. receptor-effector pairs linked to second messenger systems and transcriptional control mechanisms that communicate environmental conditions;
3. response mechanisms that act on information about the environment to transiently decrease susceptibility for heritable damage or facilitate long-term adaptation;
4. DNA-, cell-, and tissue-repair mechanisms that reestablish the integrity of the organism if damage has occurred; and
5. behavioral changes that reduce the opportunity for further exposure to damaging agents.

Specific examples are given in Table 3.

Multiple defense systems operate concurrently and interdependently to ensure integrity of the genome. In a simplified scheme, cancer can occur when two or more of these defense mechanisms

TABLE 2
Examples of Rare Hereditary Forms of Cancer

Predisposing factor	Proposed mechanism influencing susceptibility to cancer
Ataxia-telangiectasia (AT)	Chromosome fragility, causing sensitivity to agents that increase genetic recombination. Approximately 10% of AT patients develop neoplasms, 88% of which are lymphoreticular, including Hodgkin's disease and non-Hodgkin's lymphomas (Swift et al. 1993, Taylor 1992, Peterson et al. 1992) . Heterozygous females have increased risk of breast cancer (Peterson et al. 1992) . Autosomal recessive mode of inheritance (Swift et al. 1993, Tadjoedin and Fraser 1965). Incidence: about 1 per 40,000 live births. The gene responsible for AT, *ATM*, was recently cloned and shown to have a phosphatidylinositol 3-kinase-like domain, suggesting a role for ATM protein in signal transduction and cell-cycle regulation (Savitsky et al. 1995). Mutations identified in AT patients primarily result in protein truncations or deletions expected to cause complete inactivation of the ATM Protein (Shiloh 1995).
Basal cell nevus syndrome	Autosomal dominant genetic disorder results in high susceptibility to sunlight-induced cancer, particularly basal cell carcinoma. DNA lesions and repair processes other than the pyrimidine dimer have been implicated (Applegate et al. 1990).
Bloom syndrome	Hypermutability (Sullivan and Willis 1992).
BRCA1	Breast and ovarian cancer (Narod 1994, Hall et al. 1990).
Chediak–Higashi syndrome	Depletion of natural killer cells that combat incipient malignancies.
Down syndrome	Leukemia risk, related to chromosomal abnormality.
Familial polyposis coli	Mutation in *APC* tumor-suppressor gene leads to benign colonic growths that are predisposed to malignant transformation.
Familial testicular cancer	Genetic defect or environmental influence unknown (Forman et al. 1992).
Fanconi anemia (FA)	An autosomal recessive Mendelian trait, associated with an increased incidence of leukemia and solid tumors. Around 15% of children with this disorder develop acute myelogenous leukemia or preleukemia, an incidence about 15,000-fold greater than the general population (Auerbach 1992, Auerbach and Allen 1991). Cells show increased sensitivity to killing with DNA cross-linking agents.
Hereditary non-polyposis colorectal cancer (HNPCC)	High instability of short repeat sequences (microsatellites) because of deficiency in mismatch repair (Radman et al. 1995). Represents one of the most common genetic diseases, affecting as many as 1 in 200 individuals. It segregates as an autosomal dominant form of inheritance.
Hereditary retinoblastoma	Predisposition for retinal cancer because of germ-line mutation of one allele of a tumor suppressor gene, *Rb-1* (Evans and Prosser 1992). The disease affects 1 in 5000 children.
Li–Fraumeni syndrome	Heterozygous germ-line mutation in the p53 tumor-suppressor gene, which predisposes to multiple carcinomas and sarcomas (Li and Fraumeni 1994). Tumors in these patients, however, are homozygous for the mutant p53 allele.
Multiple endocrine neoplasia	Autosomal dominant inherited predisposition to endocrine tumors because of loss of *MEN-1* (a tumor-suppressor candidate on chromosome 11q13) or germ-line mutations in *MEN-2A* (susceptibility maps to *RET*, a protooncogene on 10q11.2) (Padberg et al. 1995, Bordi et al. 1995, Sandelin et al. 1994).
Tyrosinemia	Inherited deficiency in fumarylacetoacetate hydrolase, the enzyme that catalyzes the last step in tyrosine catabolism. If untreated, leads to liver cancer and death in early childhood. The functional impairment is believed to be due to the buildup of heme precursors and their metabolites, some of which may act as reactive oxygen species (Ruppert et al. 1992).
von Hippel–Lindau disease	Loss of tumor-suppressor gene, leading to increased incidence of cerebellar, spinal, and medullary hemangioblastomas, retinal angiomas, renal cell carcinomas, and pheochromocytomas (Martz 1991).
Wilms' tumor	*WT1* polymorphism (Evans and Prosser 1992) but possibly other gene in 11p15 region (Cowell et al. 1993, Schwartz et al. 1991). Somatic mosaicism may be important (Breslow et al. 1993).
Xeroderma pigmentosum (XP)	Inability to repair some kinds of DNA damage, predisposing to skin cancer caused by ultraviolet radiation. XP occurs with an incidence that varies from about 1 in 250,000 in the United States and Europe to as high as 1 in 40,000 in Japan (Takebe et al. 1977, Robbins et al. 1974).

TABLE 3
Features of the Host-Environment Interface

Mechanistic role	Description	Examples/comments	Possible risk factors
Minimizing exposure	Physical barriers (i.e., skin); enzymes that act on endogenous and foreign compounds to minimize exposure to critical cell components: GSH, GST, NAT1 and NAT2, UGT, ST, and some CYPs; antioxidants; SOD; catalase.	Cells of higher organisms are functionally interdependent. Therefore, cells that do not have the capacity for detoxification are dependent on those that do.	Chronic carcinogen exposure; low ability to detoxify relative to bioactivate.
Interpreting extracellular and intracellular conditions	Receptor-effector pairs linked to second messenger systems and growth control; gap junctional communication; immune responses that act on behalf of the whole organism; response elements.	Protooncogenes; immunity; DNA-damage response elements; hormonal responses.	Immune deficiency; oncogene activation; hormonal imbalance.
Decreasing susceptibility for heritable damage	Tumor suppressor genes; stress protein responses (adaptation); cell-cycle check points; apoptosis.	p53: tumor suppressor gene and protein, affects cell-cycle arrest at the G_1 and G_2 checkpoints in response to DNA damage, allowing more time for DNA repair to take place. If repair is not successful, it is thought that p53 may initiate programmed cell death to prevent propagation of genetic defects.	Exposure of multiple stressors possibly acting through common mechanism; tumor-suppressor gene inactivation.
DNA, cell, and tissue repair	DNA-repair enzymes; inflammatory responses; stimulation of growth by autocrine, paracrine, and endocrine-derived factors.	Excision-repair enzymes; mismatch repair genes: *hMSH2, hMLH1, hPMS2, hPMS1*; topoisomerase; telomerase; cytokines; glucocorticoids; hormones; neural-derived factors; angiogenesis factors.	Age-associated decline in function.
Behavioral responses	Immuno/neuro/endocrine communication network coordinating adaptation, sensitization, or avoidance.	Hormones, neurotransmitters, cytokines, sensory receptors.	Exposure to addictive substances, social pressures.

fail. This is more likely to occur when there are critical genetic or physiological deficiencies, or if carcinogen exposure is chronic and accompanied by inadequate DNA repair.

HOST-SPECIFIC SUSCEPTIBILITY FACTORS AND CANCER RISK

Past epidemiologic studies have been made with the assumption that all individuals are equally susceptible to the effects of exposure to a specified substance. Pharmacogenetic studies have shown, however, that individuals vary in their capacity to metabolize drugs. There is also new evidence that this variability extends to metabolism of environmental agents. Therefore, to avoid diluting or masking risk estimates, it is necessary to identify the subsets of the population with increased susceptibility based on polymorphisms in genes involved in metabolism of carcinogenic agents.

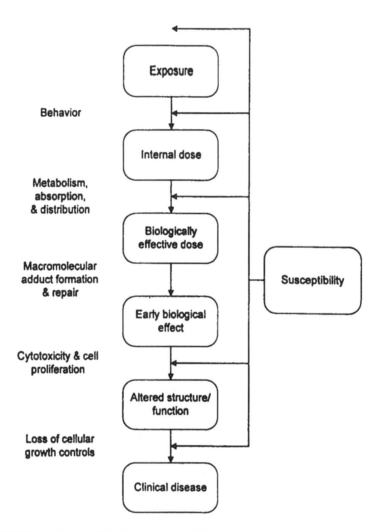

FIGURE 1   Susceptibility factors can modulate numerous stages of carcinogenesis.

Susceptibility factors are divided into two categories. The first category includes factors that confer extremely high risk but have very low incidence in the general population, such as inherited mutations in tumor suppressor genes. The second category consists of factors that confer low-to-moderate risk but that have a relatively high incidence in the population, including normal genetic variability, age, gender, ethnicity, physiological state, concurrent disease, diet, and other lifestyle factors.

As outlined in Figure 1, susceptibility factors modulate risk at any stage of the carcinogenesis process. In some instances, there is a clear association between cancer risk and a marker for exposure or genetic susceptibility. In many cases, however, the relationship is more tenuous.

GENETICS

Cancer-Susceptibility Syndromes

Individuals who inherit susceptibility genes in a Mendelian fashion appear to be at increased risk for some cancers. A diverse set of genes are associated with hereditary forms of cancer. There are diseases associated with defects in carcinogen metabolism (tyrosinemia), DNA repair [basal cell nevus syndrome, hereditary nonpolyposis colorectal cancer (HNPCC), xeroderma pigmentosum

(XP), Bloom syndrome], immune competence (Chediak–Higashi syndrome, X chromosome-linked agammaglobulinemia), cell-cell signaling or cell cycling (ataxia-telangiectasia; *BRCA1* gene), tumor suppression (von Hippel–Lindau disease, familial polyposis coli, hereditary retinoblastoma, Li–Fraumeni syndrome), neutralization of radical species [glutathione reductase deficiency and, possibly, Fanconi's anemia (FA)], and chromosomal integrity (Down syndrome).

However, cancer-susceptibility syndromes are rare. XP, for example, is an autosomal recessive disease with an incidence that ranges from 1 in 250,000 in the United States and Europe to as high as 1 in 40,000 in Japan (Takebe et al. 1977, Robbins et al. 1974). Among affected persons, however, it increases the risk for skin cancer by more than 1500-fold. At the other end of the spectrum, HNPCC represents one of the most common genetic diseases, affecting as many as 1 in 200 persons. HNPCC segregates as an autosomal dominant form of inheritance and accounts for 4% to 13% of all colorectal cancers in the industrial world.

Cancer-susceptibility syndromes may affect multiple physiological systems (see Table 2). Genetic defects that compromise the immune system, for example, may interfere with environmental adaptation, intercellular communication, and immune surveillance and may lead to persistent viral or bacterial infections, which could lead to chronic inflammation and increased risk for carcinogenesis. Consequently, persons with cancer-susceptibility syndromes may experience functional repercussions beyond the occurrence of cancer at the target organ.

Single deficiencies may occur in any number of genes critical for proper functioning of the host-environment interface, resulting in increased cancer risk. Disease states that illustrate this relationship include XP, HNPCC, and FA.

XP is characterized by early onset of severe photosensitivity of the exposed regions of the skin, eyes, and tongue; a high incidence of skin cancers; and frequent neurologic abnormalities (Kraemer et al. 1987). The observation that XP fibroblasts in culture were unable to perform nucleotide-excision repair after exposure to ultraviolet radiation was critical to understanding this rare disease (Cleaver 1969, Setlow et al. 1969, Cleaver 1968). A large interindividual variability in the XP repair defect, measured by unscheduled DNA synthesis, was also discovered. Most of this variability is now attributed to locus heterogeneity. Specifically, XP-associated defects are linked to a number of genes essential for recognition, incision, and repair of damaged DNA. For example, XP mutations have been localized to DNA-binding proteins (XPA), DNA helicases (XPB), transcription factors (XPD), and endonucleases (XPG) (Friedberg et al. 1995). Not surprisingly, the clinical manifestations of XP differ substantially among individuals depending on the specific defect in the nucleotide-excision repair pathway.

HNPCC is characterized by instability of simple repeated DNA sequences, presumably because of defects in mismatch-repair genes. Defective genes from a number of kindred have been characterized (*hMSH1, hMSH2, hPMS1, hPMS2*) that demonstrate polymorphisms at several sites (Marra and Boland 1995, Modrich 1994). Germ-line mutations in *hMSH2* and *hMSH1* are thought to be responsible for approximately 50% and 35% of HNPCC, respectively.

FA is an autosomal recessive Mendelian trait with four known complementary groups. Although the FA phenotype is variable, aplastic pancytopenia is a common feature of the disease. The pathogenesis of FA is unknown, but hypotheses focus on aberrant production of, or responsiveness to, reactive oxygen species. FA cells are very susceptible to damage caused by reactive oxygen species and DNA cross-linking agents, such as nitrogen mustard and mitomycin (Auerbach and Wolman 1976). In addition, in cultured FA lymphocytes, the G_2 phase of the cell cycle is about twice as long as in normal cells, a feature that appears to be oxygen sensitive (Hoehn et al. 1987, Dutrillaux et al. 1982). Increased production of clastogenic factors (Emerit 1994) and tumor necrosis factor by phagocytic cells (Schultz and Shahidi 1993), an aberrant cytokine network (Bagnara et al. 1993, 1992), aberrant endonuclease activities (Lambert et al. 1992, Sakaguchi et al. 1991), increased formation of 8-hydroxydeoxyadenosine possibly associated with catalase deficiency (Takeuchi and Morimoto 1993), and defective damage recognition (Hang et al. 1993) all have been reported (Friedberg et al. 1995). In light of the importance of cell-cycle regulation, it also may be significant that, after exposure to radiation, FA cells demonstrate a marked reduction in apoptosis, as well as p53 induction, compared with normal cells (Rosselli et al. 1995).

Molecular characterizations of these diseases should not be oversimplified in the effort to pinpoint single risk factors. For example, persons with Cockayne's syndrome and trichothiodystrophy (TTD) have DNA-repair defects similar to those that occur with XP and also show an unusual sensitivity to sunlight. However, Cockayne's syndrome does not confer a predisposition for skin cancer (Lehmann 1989, Lehmann et al. 1988). In addition, XP diploid fibroblast cell lines appear to have lower catalase activities compared with TTD-derived lines, suggesting that multiple genetic defects may be necessary for increased cancer risk (Vuillaume et al. 1992).

DNA-repair defects that lead to hypermutability in tumor cells of patients with HNPCC (Liu et al. 1996, 1995) also can be found in normal cells of affected individuals (Parsons et al. 1995). Therefore, investigators should still be mindful that mutagenesis is not synonymous with carcinogenesis and that altered cell microenvironments probably play a role in transformation to the carcinogenic state.

Likewise, the case for cell proliferation in tumorigenesis is often overstated (Farber 1995). Although cellular signals tightly regulate growth in normal tissues, descriptions of this process (particularly as they relate to carcinogenesis) are frequently oversimplified. Many cell types and disease processes (e.g., psoriasis) exhibit high levels of cell turnover without associated cancer risk. Also, many well-known carcinogens inhibit rather than enhance cell population expansion (e.g., carbon tetrachloride, 2-acetylaminofluorene, aflatoxin B1, 7,12-dimethylbenz[a]anthracene, etc.). Cell division is required for fixation of DNA defects, but mutagen- or steroid-mediated modulation of cell-cycle checkpoints also appears to play a critical role in risk (Campbell et al. 1995, Harris 1995, Sicinski et al. 1995, Kamb et al. 1994, Beamish and Lavin 1994, Beamish et al. 1994, Lavin et al. 1994, Arnold et al. 1992, Calabretta and Nicolaides 1992, Fearon 1992, Harris 1991) as do chronic selection pressures, which cause variations in proliferation potential and acquired resistance to apoptosis during tumor promotion (Farber and Cameron 1980).

Single Mutations May Lead to Multiple Risks

A number of important cancer-susceptibility genes have been characterized. Some appear to be malignancy specific, suggesting that some susceptibility genes affect only certain lineages. For example, the *BRCA1* gene has been implicated in familial breast and ovarian cancers, but it may not be important in other common cancers. In other cases, defective genes are initially characterized in one type of tumor but are later found to be associated with other cancers. For example, it now appears that the *DCC* gene, initially linked to colorectal cancer, may play a role in a variety of tumors (Zhang et al. 1995, Milner et al. 1993, Klingelhutz et al. 1993). Mutations in mismatch-repair genes, implicated in HNPCC, may play a role in endometrial and ovarian cancer (Smith and Ponder 1993). Still other genes, such as the tumor suppressor gene p53, appear to be central to cell regulation and are altered in most tumors.

At least 51 types of human cancer are also associated with p53 mutations, the defect found in patients with Li–Fraumeni syndrome (Li and Fraumeni 1994). These patients have a heterozygous p53 defect in their germ line, and tumors from these cancer-prone individuals demonstrate homozygous mutant p53 alleles. The risk of cancer is nearly 100% in affected individuals. The p53 mutation also has been linked to 70% of colorectal cancers, 50% of lung cancers, and 40% of breast cancers. One of the most important functions of the p53 gene product is to regulate the expression of genes whose products inhibit progression through the cell cycle, enhance DNA repair, and trigger apoptosis in cells that sustain DNA damage. The p53 gene product is thus essential for maintaining genomic integrity (Harris 1995).

The significance of the *Rb* gene also has been recognized, initially with respect to the hereditary form of retinoblastoma and now as one of the critical regulators of cell-cycle progression. The normal *Rb* gene encodes a protein that affects reversible growth arrest in the G_1 stage of the cell cycle (Weinberg 1995). A range of different mutations have been characterized that yield an absent or inactive protein in tumor cells and cell lines. However, it is unclear whether additional mutational events are necessary for carcinogenesis. In this regard, it is perhaps significant that tumor cells with inactive *Rb* and p53 genes are refractory to induction of apoptosis (Weinberg 1995).

Multiple Genes May Confer Risks for a Single Cancer

In disentangling the molecular circuitry of critical cell-cycle regulatory proteins, many associated gene products and etiologic agents have emerged as genetic risk factors. For example, human papilloma virus (HPV) may increase the risk of cervical cancer by interfering with mechanisms that mediate DNA damage response via interactions between the E6 and E7 oncoproteins and cellular proteins involved in cell-cycle regulation. By undermining the cell cycle, E6 and E7 protein expression may render HPV-infected cells more susceptible to the genetic lesions required for tumor initiation and progression. Still another level of cell-cycle regulation may be imposed by heterologous cells of the immune system. It has been suggested that interference with the cytokine network may be a necessary precondition for malignant conversion and invasive growth after HPV infection. International studies showing a higher incidence of cervical cancer in tropical and developing countries (de The 1995, Bornstein et al. 1995, Munoz et al. 1994) indicate that factors such as poor nutritional status or concurrent or early exposure to disease or infection may weaken cellular defense mechanisms, thus leading to tumorigenesis.

In the United States, as many as 8.8% of Caucasian patients with breast cancer may be heterozygous for the *AT* gene (Swift et al. 1987). However, in addition to *AT* (11q22–123), other genes are associated with inherited susceptibility to breast cancer, such as *BRCA1*, *BRCA2*, and the androgen receptor gene *AR*. The *BRCA1* gene is estimated to confer a breast cancer risk of about 70% by age 70, and it may account for about 2% of breast cancer incidence overall. This gene accounts for a high proportion of early-onset disease (Ford et al. 1995). Germ-line mutations in p53 (Li–Fraumeni syndrome) are more rare (<1% of all breast cancer), yet 50% of persons with this mutation develop breast cancer by age 50. Germ-line mutations in *AR*, which causes breast cancer in males, may account for one-half of the observed familial clustering of breast cancer (Easton et al. 1993). Other breast cancer susceptibility genes remain unidentified, and the role of somatic mutations (*erbB-2, ras* pathway) in breast cancer progression requires further clarification.

Cancer-susceptibility genes often show allelic variability, with or without corresponding disease. Disease trends characterized by certain mutations may indicate genes that could be useful as biomarkers of carcinogenesis in susceptible individuals (Rodenhiser et al. 1996, Cherpillod and Amstad 1995, Recio and Goldsworthy 1995, Morris et al. 1995, 1994, Jansen et al. 1994, Pourzand and Cerutti 1993, Crespi et al. 1993, Harris 1991).

Genetic Polymorphisms and Enzymatic Expression

Recent research has been aimed at understanding the extent of cancer susceptibility that results from genetic polymorphisms. Polymorphisms arise by gene mutation and, in the absence of strong selection pressure, result in null or variant alleles that become established as stably inherited traits in the population. Many polymorphic enzyme variants have shown no increased cancer risk; however, when combined with certain exposures, there may be a significant increase in cancer risk (Rannug et al. 1995).

The cytochrome P-450 drug-metabolizing enzymes (CYPs) are principally responsible for activating procarcinogens and promutagens (see Table 4) and DNA-reactive electrophiles. Therefore, studies are now examining the relationship between cancer incidence and distribution of polymorphic variants of different CYP isozymes. CYPs show specificity for a variety of substrates, although there is some redundancy among isoforms. Some isoforms (e.g., CYP1A2) appear to be

TABLE 4
Major Carcinogen/Mutagen-Metabolizing CYPs

Enzyme	Tissue	Carcinogen substrates	Possible enzyme inducers
CYP1A1	Lung, placenta, lymphocytes, larynx, kidney, brain, fetal liver, oral mucosa (?)	PAHs	Cigarette smoke, PCBs, TCDD
CYP1A2	Liver, oral mucosa (?)	Arylamines, arylamides	Cigarette smoke, charbroiled foods, cruciferous vegetables, omeprazole, PBBs, TCDD (?)
CYP2A6	Liver, nasal mucosa (?)	Certain nitrosamines (NNN, NNK, DEN)	Barbiturates (?), inflammation
CYP2C9	Liver, larynx, urinary bladder, oral mucosa (?)	Polycyclic hydrocarbons (AHH)	?
CYP2E1	Liver, lymphocytes, larynx, lung, oral mucosa (?)	Numerous low molecular weight suspected carcinogens: dialkylnitrosamines, benzene, styrene, urethane, halomethanes, dihaloalkanes, vinyl halides, vinyl monomers, butadiene	Alcohol, diabetes, isoniazid, obesity
CYP 3A3/4/5/7	Liver (CYP3A3), intestine, larynx, esophagus (CYP3A4), oral mucosa (?)Lung, kidney, liver, pituitary, intestine (CYP3A5)Fetal liver (CYP3A7)	Aflatoxins, polycyclic hydrocarbon dihydrodiols, pyrolizidine alkaloids, tris, arylamines	(CYP3A4): rifampicin, phenytoin, dexamethasone, troleandomycin

Note: AHH, aryl hydrocarbon hydroxylase; NNN, *N*-nitrosonornicotine; NNK, 4-(methylnitrosamino)-1-(3-pyridyl)-1-butanone; DEN, diethylnitrosourea; PBB, polybrominated biphenyl; TCDD, 2,3,7,8-tetrachlorodibenzo-*p*-dioxin.

Data are from Kirby et al. (1996), Carpenter et al. (1996), Murray et al. (1995), Furuya et al. (1995), Raunio et al. (1995), Rannug et al. (1995), Gonzalez (1995a, 1995b), Bereziat et al. (1995), McKinnon et al. (1995), Katoh et al. (1995), H. Yamazaki et al. (1995a, 1995b), Y. Yamazaki et al. (1995), Guengerich (1994), Yamazaki et al. (1994a, 1994b), Sinha et al. (1994), Stern et al. (1993), Guengerich (1993), Camus et al. (1993), Yamazaki et al. (1993), Getchell et al. (1993), Yamazaki et al. (1992a), Rich et al. (1992), Rost et al. (1992), Yamazaki et al. (1992b), Guengerich (1992), Gonzalez et al. (1992), and McLemore et al. (1990).

liver specific, whereas others (e.g., CYP1A1) are widely distributed in extrahepatic tissues (Eaton et al. 1995, Raunio et al. 1995, Guengerich 1993, Gonzalez et al. 1992). Polymorphisms of CYP enzymes can alter gene transcription (CYP1A1, CYP1A2, CYP2E1, CYP3A4), processing (CYP1A2), mRNA stabilization (CYP1A1, CYP2E1, CYP3A4), translation (CYP2E1, CYP1A1, CYP1A2), and enzyme stability (CYP2E1, CYP3A4).

There is wide interindividual variability in the expression of enzymes that detoxify or bioactivate carcinogens and mutagens (Table 5). The toxicologic significance of a deficiency in a single isoform, however, is determined by specific host-environment interactions. An isoform-specific deficiency may be masked if other isoenzymes or detoxification pathways assume an equivalent role. However, a deficiency may be more critical in the context of compounds metabolized by a single isoform, resulting in increased susceptibility to toxicity. Therefore, it is important to determine the pattern of isoform expression as well as any relevant structure-function relationships. In addition, if a compound is a suitable substrate for both bioactivation and detoxification, these pathways may

TABLE 5
Other Enzymes That Demonstrate Interindividual Variability of Expression

Enzyme	Genetic polymorphism	Nature of polymorphism(s)	Reported cancer associations
GSTM1	Yes	Deletion	Lung, bladder, cutaneous tumors
GSTP1	Yes	Single base change	
GSTT1	Yes	Deletion	Astrocytoma and meningioma
NAT1	Yes	Polyadenylation	Bladder, colon
NAT2	Yes	Point mutations	Bladder, colon
Nitric oxide synthase	Yes	RFLP in intron 5	Cancers involved in chronic inflammation
Aromatase (CYP19A1)			Endocrine-dependent tumor promotion
CYP2D6	Yes		Liver, lung, astrocytoma and meningioma
UDP-glucuronosyltransferase	?	Unknown	?
Sulfotransferase TS-ST	Yes	Polygenic	?
CYP2C9	Yes	Single base change	

Note: RFLP = restriction fragment length polymorphism. Data are from Deakin et al. (1996), Haehner et al. (1996), Hakkola et al. (1996), Elexpuru-Camiruaga et al. (1995), Cascorbi et al. (1995), Nakajima et al. (1995), Esumi et al. (1995), Bell et al. (1995), Zimniak et al. (1994), Lang et al. (1994), Cholerton et al. (1994), Kolars et al. (1994), Pemble et al. (1994), Nazar-Stewart et al. (1993), Price et al. (1989), Campbell et al. (1987), and Patel et al. (1995).

compete, resulting in opposing toxicologic outcomes. The balance between bioactivation and detoxification may be the most important parameter for assessment of risk, with relative increases in one or the other affecting the outcome of exposure.

Mechanisms Related to the Carcinogenic Process

Some investigators theorize that cancer predisposition is determined at the cellular level. A carcinogen, or its bioactivated metabolite, initiates a primary lesion in the form of a genetic mutation, strand break, etc., and cells that sustain this lesion arise from a phenotypically diverse initial population (Baron et al. 1986, Ogawa et al. 1980), with the ability to proliferate and expand into focal lesions under appropriate conditions. This phenotype selection response is probably reversible and possibly is transiently advantageous to the organism as an adaptive response. At some point, however, the phenotypic changes become irreversible, and the affected cells become resistant to the cytotoxic actions of carcinogens and hepatotoxins. By overriding the mitoinhibitory effects of the carcinogen, the new phenotype allows the initiated cells to respond to mitogenic stimuli (Farber and Cameron 1980).

Drug-metabolizing enzymes and cancer progression

Experimentally induced alterations of drug-metabolizing enzyme activities in rat hepatocyte nodules suggest that only certain activities contribute to a resistant phenotype. Certain CYP activities are reduced (Tsuda et al. 1980, Cameron et al. 1976), and many Phase 2 enzyme activities are increased. For example, UDP-glucuronosyltransferase activity toward 1-naphthol is increased 5-fold in hepatocyte nodules compared with control tissue, but other isozyme-specific activities (e.g., bilirubin and 4-hydroxybiphenyl glucuronidation) are unchanged or only slightly increased (Bock et al. 1982a, 1982b). Glutathione *S*-transferase (GST) π is strongly expressed in hepatic foci and hepatomas as well as in initiated cells (Sato et al. 1992). Other metabolizing enzymes that appear to be differentially expressed during chemical carcinogenesis include epoxide hydrolase (Kizer et al. 1985, Oesch et al. 1983, Levin et al. 1978), DT-diaphorase (Pickett et al. 1984), Class 3 aldehyde dehydrogenase (Lindahl 1992), and aryl sulfotransferase IV (Yerokun et al. 1992). Genes that control carcinogen transport, such as those involved in multidrug resistance (e.g., *mdr1*), significantly affect cancer progression risk. It is interesting to note that poor prognoses are often associated with certain combinations of factors. For example, *mdr1* and GST π are expressed simultaneously in Epstein–Barr virus-associated recurrent lymphomas (Cheng et al. 1993). Differential expression of gene products during the progression of malignancies affects both prognosis and therapy.

DNA repair capacity

High DNA-repair capacity may lower cancer risk by preventing genotoxic damage, but it may affect cancer susceptibility by indirectly blocking carcinogenic host responses triggered by cytokines (Nishigori et al. 1996, Yarosh and Kripke 1996). There is considerable interindividual variability in DNA-repair capacity among the population. For example, Wei et al. (1993) reported a 2.8-fold increased risk for basal cell carcinoma among individuals reporting more than six severe sunburns in their lifetime. But when the patients were divided into two groups based on DNA-repair capacity of cultured lymphocytes, those with low DNA-repair capacity had a statistically significant 5.3-fold increased risk for this disease. In a study by Wu et al. (1995), an overall 8.8-fold increased risk for lung cancer was associated with smoking; this association was strengthened [$R^2 =$ 12.8; cytotoxicity index (CI) = 4.3 to 38.7] among the subgroup of individuals who were highly susceptible to bleomycin-induced chromosomal damage in an *in vitro* lymphocyte test system.

Interactions with environmental carcinogens

On the cellular level, relevant environmental interactions play a role in malignant transformation. In XP, only a small proportion of repair-deficient cells are ever transformed. However, deficient repair in the presence of DNA damage, combined with continuous cell cycling from activation of protooncogenes or loss of suppressor genes, may tip the scales, leading to development of skin cancer in XP patients (Price et al. 1991).

Table 6 summarizes research of the relationship between inherited predispositions and cancer susceptibility. Many inherited genetic variations show little or no significance except in the context of an environmental stressor. Even after exposure, there may be wide variability of response depending on the candidate gene, suspected carcinogen, and study population. Other factors are also involved in the cancer risk associated with genetic susceptibility and environmental exposures, including immune status, DNA-repair capacity, dietary antioxidant status, and the status of other genes involved in carcinogen metabolism or response. The role of these factors in carcinogenesis, and their variability in at-risk populations, may contribute to some of the conflicting results observed in molecular epidemiologic studies. However, much can be extrapolated from what is known about inherited cancer syndromes:

- Variability in single genes may have multiple consequences (i.e., enzymes that metabolize carcinogens often modulate endocrine or neurotransmitter function) and genetic variants may influence the incidence of more than one cancer type.
- Pleiotropic effects occur when critical defense mechanisms (e.g., immune response, hypothalamic control, stress response, behavior) are compromised by exposure to damaging agents or resulting disease processes. Although compensatory mechanisms may defer any immediate threat to life, corresponding risk may develop in the form of decreased responsiveness to environmental insult.
- Interindividual variability in a single risk factor (e.g., a relevant bioactivation pathway) does not necessarily correspond to a large differential in interindividual risk because compensatory mechanisms are often significant. In other cases, additional factors may be required for full expression of risk. For example, there are important synergistic effects among chemical, physical, and viral carcinogens that multiply risk for certain cancers (Table 7). This reinforces the need for diverse populations to participate in epidemiologic studies specifically aimed at molecular mechanisms of carcinogenesis.

TABLE 6
Examples of Interactions Between Inherited Cancer Predisposition and Environmental Carcinogens in Selected Studies

Candidate gene	Condition	Cancer site	Environmental carcinogens	Odds ratio (95% CI)	Reference
XPAC	Xeroderma pigmentosum	Skin	Sunlight	>1500	Cleaver (1968)
CYP2D6	Extensive-hydroxylator phenotype	Lung	Tobacco smoke	6.1 (2.2–17.1)	Caporaso et al. (1990)
			Asbestos	18.4 (4.6–74)	Caporaso et al. (1989)
			PAH	35.3 (3.9–317)	Caporaso et al. (1989)
CYP1A1	Extensive-metabolic phenotype	Lung	Tobacco smoke	7.3 (2.1–25.1)	Nakachi et al. (1991) Nakachi et al. (1993)
Ha-ras	Restriction fragment length polymorphisms (rare alleles)	Lung	Tobacco smoke	4.2 (1.1–16)	Sugimura et al. (1990a)
NAT2	Slow-acetylator phenotype (recessive inheritance)	Bladder	Aromatic amine dyes	16.7 (2.2–129)	Cartwright et al. (1982)
NAT2	Rapid-acetylator phenotype (dominant inheritance)	Colon	Unknown	1.4 (0.6–3.6) 4.1 (1.7–10.3)	Lang et al. (1986) Ilett et al. (1987)
CYP1A1 and GSTM1	Metabolic balance between activation and detoxification	Lung: squamous cell carcinoma	Aromatic hydrocarbons	9.1 (3.4–24.4)	Hayashi et al. (1992) Also see Caporaso and Rothman, this volume
GSTM1	Metabolic balance between activation and detoxification	Lung	Aromatic hydrocarbons	3.5 (1.1–10.8)	Seidegard et al. (1986)

From National Research Council (1994). Reprinted with permission from National Academy Press.

TABLE 7
Examples of Synergistic Effects Among Chemical, Physical, and Viral Carcinogens in Selected Studies

Cancer	Carcinogens	Odds ratio (95% CI)	Reference
Liver	Hepatitis B virus + aflatoxin B_1 exposure	4.8 (1.2–19.7) 60 (6.4–561.8)	Ross et al. (1992)
Esophagus	Tobacco smoke + alcoholic beverages	5.1 (–) 44.4 (–)	Tuyns et al. (1977)
Mouth	Tobacco smoke + alcoholic beverages	2.4 (–) 15.5 (–)	Rothman and Keller (1972)
Lung	Tobacco smoke + occupational asbestos exposure	8.1 (5.2–12.0) 92.3 (59.2–137.4)	Selikoff and Hammond (1975); Saracci (1977)

From National Research Council (1994). Reprinted with permission from National Academy Press.

- There is no "model" genetic profile that assures defense against disease. In fact, defensive measures against one cancer may significantly increase risk for other cancers. It is hoped that the availability of genetic testing will encourage cancer prevention by targeting at-risk populations and educating them about lifestyle choices that can reduce their cancer risk.

AGE

It is well-documented that cancer incidence and mortality rates increase with age, particularly for tumors originating from epithelial cells (Fraumeni et al. 1993). This trend also occurs in laboratory animals (Sass et al. 1975). Several mechanisms for this association have been suggested, including those related to increased susceptibility to chemical carcinogenesis, weakened immune function (Ershler 1993), impairments in DNA repair, and glutathione deficiency in aging tissue (Richie 1992). It also has been proposed that the association between age and cancer is due to accumulated exposure to carcinogens over a lifetime (Doll 1978, Peto et al. 1975). It is likely that increased cancer risk because of age may be a function of multistage, multifactorial carcinogenesis resulting from accumulated genomic insults over a number of years. Risk estimates reflect not only duration of exposure to carcinogens but also long periods of induction (Doll and Peto 1981).

The interaction of age with genetic and environmental factors occurs in lung and breast carcinogenesis.

Lung Cancer

From an epidemiologic viewpoint, lung cancer is an excellent model for studying chemical carcinogenesis. The incidence is high, and tobacco smoke is a clear risk factor associated with the disease. Most cases of lung cancer occur between 35 and 75 years of age for both males and females, with peak incidence occurring between 55 and 65 years (Minna et al. 1993). It is believed that prolonged exposure to the chemical carcinogens in tobacco smoke leads to an accumulation of genetic lesions in protooncogenes and tumor suppressor genes, ultimately resulting in the transformation of healthy lung cells to malignant cells.

Age of onset
Although tobacco smoke is a recognized carcinogen, not all smokers develop lung cancer. It is likely that some persons are more susceptible to the carcinogenic effects of tobacco smoke. In fact, increased risk for non-lung cancer has been noted among lung cancer patients (Ambrosone et al. 1993, Sellers et al. 1992, McDuffie et al. 1991, McDuffie 1991, Sellers et al. 1987), and numerous studies have shown that relatives of lung cancer patients are at increased risk for lung cancer, regardless of smoking status (Ooi et al. 1986, Heighway et al. 1986, Tokuhata 1976, Tokuhata and Lilienfeld 1963).

Interestingly, some studies of lung cancer report a younger age of onset among individuals with a family history of cancer (Shaw et al. 1991), particularly among patients whose lung cancer is associated with exposure to tobacco smoke (Ambrosone et al. 1993). When allowing for a genetic component, segregation analyses by Ooi and colleagues found that, among lung cancer patients younger than age 50, more than 70% were predicted to have genetic predispositions, compared with only 30% of those over age 70 (Ooi et al. 1986).

The source of genetic variability in lung cancer susceptibility may be linked to polymorphisms in enzymes involved in the biotransformation of chemical carcinogens in tobacco smoke. Although there are inconsistent results among some studies, there is evidence that polymorphisms in CYP1A1, CYP2D6, and GST may increase lung cancer risk (Hamada et al. 1995, Alexandrie et al. 1994, Hirvonen et al. 1993, Nakachi et al. 1991, Caporaso et al. 1990, Seidegard et al. 1986, Sugimura et al. 1990b). Some studies have evaluated the role of these polymorphisms as modifiers of risk related to cigarette smoking, but few have assessed the relationship between polymorphisms and age at onset of disease. Alexandrie and colleagues (1994) found age at onset of disease to be earlier among persons with certain genetic polymorphisms compared with those with wild-type alleles. These associations illustrate the nature of gene-environment interactions in lung cancer risk. Although continued exposure may result in carcinogenesis in one individual, regardless of genetic differences in metabolism of carcinogens, it appears that, among those with a genetic predisposition, fewer years of exposure are needed to result in lung cancer.

Exposure and latency

Lung cancer risk increases with tobacco smoke exposure in a dose–response manner, and it has been shown that the number of years of smoking is more important than number of cigarettes smoked per day (Peto 1986). This suggests a long induction period for lung carcinogenesis. Several studies have reported increased lung cancer risk with earlier age of smoking initiation (IARC 1986, Kahn 1966), although others have been unable to demonstrate this association (Benhamou et al. 1987). After controlling for age and amount of tobacco exposure, Hegmann and colleagues (1993) found that men who began smoking before age 20 had more than twice the risk of lung cancer than those who began smoking at age 20 or older; risk was also greater for women who began smoking at an early age. Additional evidence for a latency effect comes from data showing that smoking cessation reduces risk more significantly for those who quit before age 50 than for smokers quitting at a later age (Sobue et al. 1993, Pathak et al. 1986, Wynder and Stellman 1979). A recent study in China found that childhood exposure to secondhand smoke increased lung cancer risk among nonsmoking women (Wang et al. 1994), with risk being greatest for persons exposed before age 7 (when children spend most of their time in the home and have the most constant exposure). These data suggest that tobacco smoke exposure, even before age 14, may predict future lung cancer risk.

Some of the risk variability associated with early age may be due to differential expression of xenobiotic metabolizing enzymes during development and maturation, such as neonatal deficiencies in CYP1A2 and glucuronidation (Vicellio 1993, Coughtrie et al. 1988). Whether this translates to an age-specific susceptibility to mutagenic agents is unknown, as laboratory animals used for long-term bioassays do not share similar developmental profiles. There is definite cause for concern because children are known to have lower toxicity thresholds for many known substances (Vicellio 1993).

Breast Cancer

The effect of age on risk of breast cancer is evidenced by several factors: (1) there are distinct differences between pre- and postmenopausal breast cancer, (2) women with inherited susceptibility are diagnosed at an earlier age, and (3) the age at which exposures occur may be important in breast cancer etiology.

Pre- and postmenopausal breast cancer

From 1987 to 1989, 6.5% of breast cancers were diagnosed in women under age 40, and 21.8% were diagnosed in women under age 50 (Hankey 1992). Breast cancer incidence increases steadily with age, with a distinct slowing of the rate of increase around age 50, the usual age at menopause (Pike et al. 1993); the increase in incidence levels off after age 50. Pike and colleagues suggest that this pattern of incidence is consistent with hypotheses regarding the role of ovarian hormones in breast cancer risk, which are present in greater amounts in premenopausal women.

It is possible that etiologic pathways differ for pre- and postmenopausal breast cancer. Risk factors differ in their effects depending on menopausal status (Velentgas and Daling 1994, Hislop et al. 1986, Lubin et al. 1985, Janerich and Hoff 1982, de Waard 1979). As reviewed by Hunter and Willett (1993), most breast cancer studies have found that, although lower body mass index (BMI) reduces risk for postmenopausal women, lower BMI increases risk for premenopausal women. It is thought that, among postmenopausal women whose ovaries no longer produce steroid hormones, estrogen is produced in peripheral fat tissue. Thus, postmenopausal women with a larger BMI would be at greater risk for breast cancer. Although there has long been interest in the possibility that pre- and postmenopausal breast cancers have different etiologies, this area of research needs further exploration.

Genetic susceptibility and age at onset

It has long been recognized that there is a familial component to breast cancer; epidemiologic studies have generally found a 2-fold increase in breast cancer risk among women with breast cancer in a first-degree relative. The use of family history as a surrogate for genetic susceptibility may be a crude estimator, however, as family members may have similar reproductive, dietary, and behavior histories that increase breast cancer risk. It also is possible that familial occurrence due to an inherited susceptibility is heterogeneous in mechanism and strength, and genetic susceptibility may reside in more than one gene locus (i.e., in polymorphisms in protooncogenes related to signal transduction and cell-cycle control, in hormone metabolism or responsiveness, in allelic loss in tumor suppressor genes, or in polymorphisms in genes involved in carcinogen metabolism and detoxification, DNA repair, and the immune response).

Reported family history of breast cancer has been used in several studies to evaluate associations between predisposition and age at diagnosis of breast cancer. Lynch and colleagues (1988) found that, whereas a positive family history was present more often in women diagnosed at a younger age, risk for early-onset breast cancer (before age 40) was greatest among women with a first-degree relative also diagnosed at an early age. Similar studies have reported increased likelihood of a family history of breast cancer among women diagnosed at an earlier age (Mettlin et al. 1990, Claus et al. 1990) and greater risk for women whose mother or sister was diagnosed at an early age (Colditz et al. 1993). In the Cancer and Steroid Hormone Study of 4730 women with breast cancer and 4688 controls, researchers fitted genetic models to the data and generated age-specific risk estimates for breast cancer among a high-risk subset of the population, i.e., women with at least one family member with breast cancer (Claus et al. 1993).

Identification of the breast cancer susceptibility genes *BRCA1* and *BRCA2* has enabled researchers to evaluate the effects of genetic predisposition on breast cancer risk, particularly among younger women (Wooster et al. 1994, Miki et al. 1994). These genes are believed to be responsible for many hereditary breast cancers, particularly early-onset breast cancer. Breast cancer associated with *BRCA1* and *BRCA2* has high penetrance, and predisposition is inherited as a dominant genetic trait. This mutation is present in families with hereditary breast cancer; it was also found in 10% of a population-based cohort of women diagnosed with breast cancer before the age of 35 (Langston et al. 1996).

In addition to genetic susceptibility that is based on rare, highly penetrant genes such as *BRCA1* and *BRCA2*, it is possible that some women may be more susceptible to breast cancer as a result of differences in carcinogen and hormone metabolism, DNA repair, and other mechanisms related to the carcinogenic processes. This area of research is in its infancy, and little is known about these possibilities. It is possible that, among women who carry susceptible genotypes, those who are exposed to carcinogenic substrates may develop breast cancer at an earlier age.

Timing of exposures

In human and laboratory studies, it has been demonstrated that timing of carcinogenic exposure is key to breast cancer risk. In rats, carcinogens administered before a first pregnancy result in twice the tumor load of rats exposed after mammary cell differentiation (Russo et al. 1990). As reviewed by Colditz and Frazier (1995), studies of the effects of radiation, alcohol consumption, and cigarette smoking have shown that breast cancer risk is increased by exposures to these environmental factors at an early age. Colditz suggests that genetic damage resulting from exposures before a first pregnancy may be immortalized by cell proliferation during breast development and pregnancy and that the decrease in cell turnover after pregnancy may prevent further genetic damage, thus reducing risk. An understanding of the importance of the age at which exposures occur may be key to cancer prevention strategies.

ETHNICITY

For many cancer types, there is wide variability in incidence by geographic region, race, and ethnicity. Differential patterns of cancer occurrence may be related to polymorphisms in genes that are specific to racial or ethnic groups, which infer susceptibility. It also is possible that ethnic, racial, or regional differences in cancer incidence may be related to variability in exposure to putative risk factors for specific cancers. Finally, it is possible that access to medical care and, therefore, screening, diagnosis, and registry of cancer incidence data vary according to the resources of the individual and the community. Patterns of ethnic variability in cancer risk are discussed in the context of prostate and breast cancer.

Prostate Cancer

Prostate cancer is the most commonly occurring cancer among men in the United States and is second only to lung cancer as a cause of cancer death in men (Boring et al. 1991). The incidence of prostate cancer appears to be increasing, and the widespread use of efficient screening procedures will add to the number of prostate cancers detected. As reviewed by Pienta and Esper (1993), autopsy studies from many countries have shown that approximately 15% to 30% of men older than age 50 have undiagnosed prostate cancer at death. This prevalence increases to 60% to 70% in men 80 years and older. Interestingly, there is little international, ethnic, or racial variability in estimates of the prevalence of these latent tumors (Tulinius 1991, Whittemore et al. 1991, Yatani et al. 1989). However, there is tremendous variability in the incidence patterns of clinical prostate cancer. African-Americans have the highest incidence rates for prostate cancer in the world, followed by Caucasians in the United States, Sweden, and Canada. Incidence rates are lowest for men in Asian countries.

These contrasting patterns for latent vs. clinical prostate cancer, as well as observations in migrant populations, suggest that variability in risk may be related to exogenous exposures related to tumor promotion. Whereas Japanese men in Japan, Hawaii, and the continental United States have a similar incidence of latent prostate cancer, clinical prostate cancer incidence and mortality rates vary markedly for Chinese and Japanese men who migrate to the West (Muir et al. 1991, Shimizu et al. 1991, Carter et al. 1990). Similarly, prostate cancer rates are higher among Polish immigrants in the United States than among those in their native land (Staszewski and Haenszel 1965), and, although the data are scanty and possibly unreliable, it appears that clinical prostate cancer rates are much lower among Africans than African-Americans (Waterhouse et al. 1976). This suggests that, although initiating factors may be fairly similar in different populations, exposure to factors associated with infiltrative tumors may differ significantly.

Genetics and hormones

It has been suggested that variation in prostate cancer risk by race may be due to genetic differences between Caucasians, Asians, and Africans. These differences may be related to metabolism of dietary constituents from meats or other high-fat foods, metabolism of chemical carcinogens, or metabolism of sex hormones. It is known that testosterone and its metabolite, dihydrotestosterone, are involved in prostate growth. It has been hypothesized that prostate cancer could result from excessive hormonal stimulation of prostatic tissue. Case-control studies of plasma or serum concentrations of male sex hormones have yielded inconsistent results (Nomura and Kolonel 1991), but there are numerous methodologic issues related to the study of hormonal factors in a case-control design. Interestingly, Ahluwalia and colleagues (1981) found that patients and controls in Africa had lower levels of testosterone than their African-American counterparts. This is consistent with the lower rates of prostate cancer among Africans compared with African-Americans. A recent study of circulating steroid hormone concentrations in Caucasian and African-American college students in the United States found that, after controlling for age, weight, smoking, and alcohol consumption, African-Americans had 15%

more testosterone than Caucasian students (Ross and Henderson 1994). The authors concluded that these hormonal differences could explain the difference in prostate cancer incidence.

Recently, a polymorphism was identified in the type II steroid 5α-reductase gene (*SRD5A2*), which is involved in the conversion of testosterone to the more active intracellular metabolite, dihydrotestosterone (Davis and Russell 1993). Reichardt and colleagues (1995) have determined the distribution of the polymorphism, a dinucleotide repeat, among low-risk Asian-Americans, high-risk African-Americans, and intermediate-risk non-Hispanic Caucasians. They found some alleles that were specific to African-Americans and suggested that these variants could be related to prostate cancer risk by altering levels of dihydrotestosterone in the prostate. Those variants found primarily among African-Americans could, in part, explain the increased incidence of prostate cancer among these individuals. Because racial variation in prostate cancer incidence is more evident for clinical than for latent prostate cancer, it is likely that additional environmental factors are involved.

Dietary intake

Variations in dietary intake, particularly consumption of red meat and other sources of fat, could be responsible for racial and ethnic disparities in rates of prostate cancer. Support for this comes from observations of differential risk for Japanese living in Japan and Hawaii (Kolonel et al. 1980) as well as the fact that prostate cancer incidence has been increasing in Japan, where a Western diet has become more prevalent over the past three decades (Hirayama 1977). The association between dietary fat intake and prostate cancer has been supported by ecologic data, in which populations with low per capita fat consumption have a low incidence of and low mortality associated with prostate cancer (Blair and Fraumeni 1978, Howell 1974). Additionally, most case-control studies have found that dietary fat is significantly associated with risk of prostate cancer (Nomura et al. 1991), and two recent studies have found a relationship between prostate cancer and consumption of animal fat, particularly red meat (Gann et al. 1994, Giovannucci et al. 1993).

There is some inconsistency among cohort studies. It is possible that genetic differences in response to some dietary agents, such as heterocyclic amines formed in the cooking of meats, could make some individuals more susceptible to the carcinogenic effects of certain agents, and this could account for inconsistencies in studies of diet and prostate cancer.

It is possible that dietary fat alters the production of sex hormones, thus affecting the risk of cancer within the prostate gland (Hutchison 1976). A low-fat, high-fiber diet has been shown to affect metabolism of male sex hormones by decreasing circulating testosterone (Adlercreutz 1990, Hamalainen et al. 1983). Ross and Henderson suggest that diet alters steroid hormone profiles both *in utero* and in adulthood, resulting in higher concentrations of circulating testosterone and altered levels of 5α-reductase (Ross and Henderson 1994, Henderson et al. 1988).

Although the epidemiology of prostate cancer is becoming clearer regarding variability in risk by race, ethnicity, and residence, the reasons for these variations remain unclear. The interaction of dietary and hormonal factors and the role of genetic variability in relationship to these factors is an area that requires more research.

Breast Cancer

As observed for prostate cancer, there is wide ethnic and regional variability in breast cancer incidence. Native Hawaiians and non-Hispanic Caucasian women in the United States have the highest incidence of breast cancer in the world, followed by Northern Europeans, Southern Europeans, Latin Americans, Africans, and Asians (Waterhouse et al. 1976). In assessing geographic and racial variability in incidence patterns, it is important to distinguish between pre- and post-menopausal breast cancer. In countries with a high incidence of breast cancer, such as the United States, rates level off during the menopausal years, followed by a lower rate of increase after menopause (Kelsey et al. 1993). Rates increase even more slowly after menopause among countries with intermediate risk. In low-risk countries, there is a decline in the incidence of breast cancer after menopause. This suggests that, although there is variability in risk by geographic region, factors that influence premenopausal breast cancer may be more universal than those associated with postmenopausal breast cancer. Interestingly, although incidence rates are higher among Caucasians than African-Americans from 40 to 45 years of age, African-American women are at higher risk for premenopausal breast cancer (Sondik 1994, Hankey 1992).

Within the United States, there is marked heterogeneity in breast cancer incidence rates among Hispanic, non-Hispanic Caucasian, and Native American women. In a study of breast cancer in New Mexico, it was found that non-Hispanic Caucasians had the highest rates, followed by Hispanic women (Eidson et al. 1994). Native American women had the lowest breast cancer incidence rates within these groups. The authors conclude that both environmental and genetic factors could influence breast cancer risk within these populations.

Numerous studies of migration of women from low- to high-risk countries have indicated that, within one or two generations, the incidence of breast cancer approaches that of the host country (Ziegler et al. 1993). In a case-control study of breast cancer among Asian-American women, Ziegler and colleagues found that Asian-American women born in the United States had a breast cancer incidence that was 60% higher than that of Asian-American women born in the East. For second generation Asian-Americans, breast cancer incidence was estimated to be higher than in Caucasians living in the same communities.

A number of explanations for ethnic variability in breast cancer risk have been proposed, including reproductive and hormonal patterns, dietary intake, environmental exposures, and genetic factors.

Reproductive and hormonal factors

The finding that breast cancer risk is associated with socioeconomic status may explain the international trends observed in breast cancer incidence. In more urban and developed areas, women of higher socioeconomic groups are more likely to remain childless or to have children at a later age as well as fewer children compared with women of lower socioeconomic groups (Jacobsen and Lund 1990, Kelsey et al. 1981). Because estrogens are believed to be of key importance in breast cancer etiology, these reproductive characteristics may lead to the higher breast cancer incidence among these women. It may also explain the changing disease patterns for women moving from low-risk to high-risk countries.

It also is possible that endogenous hormonal levels may vary between low-risk and high-risk groups. A large serologic survey compared hormone profiles of Chinese women, who have a low incidence of breast cancer, with high-risk British women (Key et al. 1990). It was found that, among premenopausal women, estradiol concentrations were 36% higher in British women than in Chinese women. Other studies comparing Western women with Asians have also found that Asian women have lower concentrations of estrogen.

These ethnic differences in hormone profiles may be closely linked to geographic variability in breast cancer risk. It is possible that endogenous hormonal variability may be related to polymorphisms in genes involved in hormone metabolism and that hormonal concentrations are related to exogenous factors, such as dietary fat, fiber, and micronutrients, such as phytoestrogens.

Dietary intake

The role of dietary fat in breast cancer risk is an area of great debate. Animal studies have found that dietary fat plays a role in the induction of mammary tumors. In support of these laboratory data, ecologic studies also suggest that per capita intake of dietary fat is associated with breast cancer risk. Among women in countries where per capita fat consumption is low, breast cancer rates are also low. A similar association occurs among high-risk groups.

Fat consumption is closely related to socioeconomic status, which, in turn, is also related to breast cancer risk. Additionally, most case-control studies have not found an association between dietary fat consumption and breast cancer risk. Data from several large cohort studies that were pooled and analyzed for dietary factors related to breast cancer risk showed that, among 337,819 women, 4980 of whom had breast cancer, no association was found between dietary fat consumption and breast cancer risk (Hunter et al. 1996).

A high-fat diet also is high in calories and low in fiber, two factors that may be related to breast cancer risk. Dietary fiber increases fecal weight and can decrease the enterohepatic circulation of estrogens (Goldin et al. 1982). It also has been shown that a high-fiber diet is associated with lower concentrations of estradiol (Barbosa et al. 1990, Goldin et al. 1982) and was related to reduced incidence of mammary cancer in an animal study (Cohen et al. 1991). Although, a meta-analysis of case-control studies found dietary fiber to decrease breast cancer risk (Howe et al. 1990), data from the Nurses Health Study, a large cohort study, have not supported this finding (Willett et al. 1992).

Foods that are high in fiber are often sources of phytoestrogens, including lignans and isoflavones, substances found in numerous plants that may reduce breast cancer risk (Adlercreutz et al. 1992). Isoflavones are particularly concentrated in soy products, a staple in many Asian countries. When women were on a controlled diet featuring soy (60 g/day) for 1 month, menstrual cycles were prolonged by 1 to 5 days, follicular phase length increased by 2.5 days, and the midcycle surge in luteinizing and follicle-stimulating hormone peaks was significantly suppressed (Cassidy et al. 1994). Asian women are known to have longer menses, reduced total estrogen exposure, and lower breast cancer risk than their American counterparts. The traditional Japanese diet is based on soy and soy products. It is possible that longer menses and decreased breast cancer risk are directly related to dietary isoflavones.

Of more importance may be the fact that a high-fat diet is high in calories. In many animal models, obesity is a major risk factor for cancer, independent of dietary composition. In epidemiologic studies as well, high BMI often appears as an independent risk factor (Thorling 1996).

Environmental factors

Increased risk for breast cancer exists for women who dwell in urban areas, particularly those in the northeastern United States. In addition to reproductive and dietary habits that may be associated with breast cancer risk in these areas, it is possible that some chemical contaminants, ubiquitous in an urban environment, may be associated with geographic patterns of breast cancer incidence. Polychlorinated biphenyls (PCBs), 2,2-bis(p-chlorophenyl)-1,1,1-trichlorethane (DDT), polycyclic aromatic hydrocarbons (PAHs), and aromatic amines are all putative carcinogens that are more prevalent in an urban environment than in suburban or rural environments. A plot of the geographic distribution of hydrocarbon residues in relationship to the geographic distribution of breast cancer in the United States showed that by-products of hydrocarbon combustion are at consistently higher concentrations in urban than in rural areas, clustering with the distribution of breast cancer cases (Morris and Seifter 1992). Similar associations may exist for other environmental contaminants.

Genetic susceptibility

Susceptibility to numerous dietary and environmental agents may be modified by genetic variability in enzymes involved in their biotransformation. There is a large body of data indicating that, for some genes, the prevalence of the polymorphism varies by ethnicity. N-Acetyltransferase is involved in the metabolism of aromatic and heterocyclic amines (Kadlubar et al. 1992), and several mutations in the gene encoding this enzyme have been found to predict a slow acetylation phenotype. The prevalence of slow acetylators varies markedly by geographic region, with the lowest proportion found among Asians (10% to 20%). Between 40% and 60% of Caucasians are slow acetylators, and some populations descending from the Middle East have a 90% prevalence of the slow phenotype (Blum et al. 1991, Vineis et al. 1990, Karim et al. 1981, Ellard 1976, Harris et al. 1958). There are also alleles specific to Asians and Africans.

Similarly, a polymorphism for CYP1A1, which metabolizes some PAHs and steroid hormones, has been identified that is specific to African-Americans (Crofts et al. 1993), and women with this variant allele have been found to be at increased risk for breast cancer (Taioli et al. 1995). Numerous other genes related to breast cancer risk may be polymorphic, with susceptible genotypes specific to certain populations. The role of these polymorphisms and their distribution among populations is an area of study that warrants further research.

GENDER

Data collected from 1984 to 1988 by the U.S. National Cancer Institute SEER program and the U.S. National Center for Health Statistics indicate that, among males with cancer, the mortality rate is highest for lung cancer, followed by colorectal and prostate cancers. For females, rates are highest for lung and breast cancers, followed by colorectal cancer (Fraumeni et al. 1993).

Incidence rates are higher among men than women for all cancers except that of the breast, gallbladder, and thyroid. The American Cancer Society estimates that 32% of deaths (98,300) in men with cancer are caused by lung cancer. Mortality from prostate cancer ranks second, causing 14% of cancer deaths among males. Among females, the American Cancer Society estimates that cancers of the lung and bronchus (25%), breast (17%), and colon and rectum (10%) will account for more than one-half of all cancer deaths in 1997 (Parker et al. 1997).

It is likely that higher incidence rates for most cancers among men are related to exposure to putative carcinogens. In particular, cigarette smoking, which has been more prevalent among men than women for a number of decades, is thought to account for about 40% of all cancer deaths in men compared with 20% in women. Cancer sites associated with tobacco use include lung, larynx, pharynx, oral cavity, esophagus, pancreas, kidney, bladder, stomach, uterine, and cervix. Leukemia also has been linked to tobacco use. However, a leveling of gender differences in smoking-related cancer rates has begun to occur as the number of women who smoke begins to equal the number of men who smoke. Higher consumption of alcoholic beverages among men also may be related to a higher incidence among males of cancers of the mouth, pharynx, esophagus, and larynx.

Occupational exposures to carcinogenic substances also may explain why men have higher incidence rates for many cancers. Exposures to aromatic amines, asbestos, and herbicides have been found to be associated with cancers of the lung and bladder and with non-Hodgkin's lymphoma (Fraumeni et al. 1993). Dietary factors also may influence the differential cancer rates between males and females. It is possible that gender differences in dietary habits (e.g., increased fruit and vegetable consumption among women, higher meat and fat intake by men) could affect cancer risk at various sites. However, even if exposures to putative carcinogens were similar among men and women, gender still could affect susceptibility to these agents. For both lung and colon cancer, there is evidence that gender-related factors modify the risk associated with carcinogen exposure.

Lung Cancer

There is increasing evidence that gender-based susceptibility plays a role in the etiology of lung cancer, with women being more susceptible to the carcinogens in cigarette smoke than men. A recent large case-control study was conducted (Zang and Wynder 1996) that collected extensive data on smoking habits, including brands of cigarettes smoked and depth of inhalation. Total tar was estimated along with other smoking-related variables. It was found that, at similar levels of smoking, women had an approximately 1.5-fold higher risk of developing all three major histologic types of lung cancer than men. It was also found that the proportion of lung cancer patients who never smoked was more than twice as high in women than in men.

Risch and colleagues (1993) found that the association between smoking and lung cancer was considerably stronger for women than for men in a study of 845 patients and 772 controls. Higher odds ratios for women also were observed for each of the major histologic subtypes of lung cancer at various strata of tobacco exposure. Similar findings have been described in a number of other studies (Zang and Wynder 1996; Begg et al. 1995; Dwyer et al. 1994; Risch et al. 1993; Osann et al. 1993; Brownson et al. 1992; McDuffie et al. 1991, 1987; Lubin and Blot 1984), although some investigators have not found such an association (Taubes 1993, Halpern et al. 1993, Doll and Peto 1976).

A number of methodologic and biologic reasons for enhanced lung cancer risk among women have been proposed. The concern that differential risk may be an artifact of numerous sources of bias was refuted by both Risch and Zang. Women may be more susceptible to carcinogens in tobacco smoke, dose for dose, because of differences in lung and body size (McDuffie 1994). Although lung size was not taken into account, Zang and Wynder adjusted for body weight and size in their study, and the odds ratios remained the same, indicating that body weight and size cannot account for this gender-related risk differential.

Zang and Wynder proposed a number of biologic mechanisms to explain higher lung cancer susceptibility among women related to gender differences in nicotine and carcinogen metabolism by CYP450 enzymes. In a large case-control study in Japan of squamous and small cell carcinomas of the lung, the prevalence of the null *GSTM1* genotype was higher among females with lung cancer than among males with lung cancer (Kihara and Noda 1995) or all controls. Alexandrie and colleagues (1994) observed a similar phenomenon among women with squamous cell lung carcinoma.

Endogenous and exogenous hormones may be related to lung cancer risk. One study found that women with lung cancer had shorter menstrual cycles (Gao et al. 1987). Studies of hormone replacement therapy and lung cancer also have found an increased risk with estrogen use (Taioli and Wynder 1994, Adami et al. 1989). Steroid hormone receptors are present in histologically diverse bronchogenic carcinomas (Jones et al. 1984, Kobayashi et al. 1982, Liu et al. 1980), and steroid hormones are known to regulate lung differentiation (Mendelson et al. 1981, Farrell 1977, Torday et al. 1975). Additionally, Sellers and colleagues (1987) have found that a family history of female reproductive cancer was associated with lung cancer risk. These factors, coupled with the fact that more nonsmoking women are diagnosed with lung cancer than men, support a possible role of steroid hormones in lung cancer susceptibility among women.

Colon Cancer

Colon cancer incidence is similar for men and women overall, although, in highly developed countries, age-adjusted rates are increasing for men. However, there are sex-specific differences associated with the site of the tumor and the age at diagnosis. Males are more likely to be diagnosed with rectal and left-sided colon cancer, particularly at a later age (Devesa and Chow 1993). Women have a higher incidence of right-sided colon cancer, and several studies have noted higher colorectal cancer rates among women under age 55 years than among men of the same age (Zaridze and Filipchenko 1990, Peters et al. 1990, Gerhardsson et al. 1990, Launoy et al. 1989, Mellemgaard et al. 1988, Kune et al. 1986, Eide 1986, McMichael and Potter 1985, Jensen 1984). Differences in tumor site by gender may be related to sex-specific exposure to risk factors. For example, men may have greater exposure to risk factors, such as alcohol consumption and consumption of a high-fat, low-fiber diet, that may be more important in rectal and left-sided colon cancer. The variability in risk by age among women indicates that endogenous and exogenous hormonal factors also may influence risk.

McMichael and Potter (1980) observed that colon cancer shared similar epidemiologic characteristics with hormonally related cancers, such as cancer of the breast, endometrium, and ovary. They also noted that, although women between the ages of 35 and 54 had been at highest risk for colon cancer, this trend was fading, perhaps as reproductive demographic patterns were changing in Western countries. An early study by Fraumeni and colleagues (1969) found that nuns not only had higher rates of cancers of reproductive sites, they also had an excess of colon cancer. Other case-control studies found an increased risk for nulliparous women, with colon cancer risk decreasing with increased numbers of live births (Dales et al. 1979, Bjelke 1974). These data, supported by animal studies showing a protective effect of pregnancy or parity on chemically induced colon cancer (Sjogren 1977), led McMichael and Potter to postulate that female sex hormones influence colon cancer risk. Bile acids and their derivatives have been thought to play a role in colon cancer etiology, and endogenous estrogens affect the hepatic metabolism of cholesterol and increase bile acid production (Cole et al. 1976, Uchida et al. 1970, Nestel et al. 1965). Because progestins, pregnancy, and oral contraceptives reduce bile acid production, it was hypothesized that this process could decrease colon cancer risk.

The lack of uniform agreement on the question of hormonal influences on colon cancer risk may be due to heterogeneity in the study population. As suggested by Devesa (Devesa and Chow 1993), to clearly determine the effects of endogenous and exogenous hormones, it is important to evaluate risk associated with colorectal cancer by site (ascending, transverse, descending, and sigmoid colon, and rectum). Additionally, it is possible that reproductive factors may only modify risk for either pre- or postmenopausal women, as indicated by the study of Davis and colleagues (1989), in which differential effects for parity were observed when women were stratified by age.

Since the hormonally influenced role of bile acid in colon carcinogenesis was proposed, numerous studies have evaluated the role of hormonal and reproductive factors as modifiers of colon cancer risk, with conflicting results. Some studies have found that, compared with nulliparous women, women with children have a lower risk of colon cancer (Kampman et al. 1994, Gerhardsson de Verdier and London 1992, Franceschi et al. 1991, Peters et al. 1990, Wu et al. 1987, McMichael and Potter 1984, Potter and McMichael 1983, Weiss et al. 1981). Others have found no association or an increased risk with parity (Chute et al. 1991, Kvale and Heuch 1991, Furner et al. 1989, Negri et al. 1989, Howe et al. 1985, Plesko et al. 1985, Papadimitriou et al. 1984, Beral 1983, Byers et al. 1982). Similar inconsistencies have been noted in evaluating the associations between age at first full-term live birth and colon cancer risk. Studies also have evaluated the effects of the use of oral contraceptives and hormone replacement therapy on colon cancer risk, with equivocal results (Davis et al. 1993, Gerhardsson de Verdier and London 1992, Chute et al. 1991, Franceschi et al. 1991, Peters et al. 1990, Furner et al. 1989, Wu et al. 1987, Kune et al. 1986, Potter and McMichael 1983, Weiss et al. 1981).

DIET

Approximately 35% of variability in cancer incidence is attributed to dietary factors (Doll and Peto 1981). Research has evaluated the inverse association between cancer risk and the consumption of fruits and vegetables as well as the role of dietary fat in carcinogenesis.

Most studies demonstrate that increased consumption of fruits and vegetables is associated with decreased risk for a number of cancers. Notably, lung cancer risk, although clearly associated with cigarette smoking, may be reduced by a diet high in fruits and vegetables. As reviewed by Steinmetz and Potter (1991), 30 of 32 studies of lung cancer risk showed strong and consistent associations between fruit and vegetable consumption and decreased lung cancer risk, including prospective and case-control studies with questionnaire data as well as plasma studies of circulating antioxidant and carotenoid nutrients. Similarly, 30 of 33 studies of oral, laryngeal, and esophageal cancer, which also are associated with tobacco use, showed that fruit and vegetable consumption decreased risk. Fruit and vegetable consumption also has been consistently associated with decreased risk of breast, colorectal, pancreatic, and cervical cancers, and other cancers to a lesser extent. There is little doubt that a diet rich in fruits and vegetables can modify cancer risk associated with exogenous exposures and endogenous processes.

The role of dietary fat in carcinogenesis remains an issue of discussion. Epidemiologic studies indicate that a high-fat diet increases colon cancer risk (Greenwald and Clifford 1993), particularly when fat is from animal sources (Willett et al. 1990). The role of fat consumption in breast and prostate cancer is less clear.

PHYSIOLOGIC STATE/CONCURRENT DISEASE

Physiologic factors are often associated with cancer risk and may account for variations in the prevalence of common cancers throughout the world. Infections and inflammatory conditions

TABLE 8
Infections and Inflammatory Conditions as Risk Factors in Human Cancers

Infection/inflammation	Cancer site
Viruses	
Hepatitis viruses	Liver (hepatocellular carcinoma)
HPV, herpes simplex virus type 2, cytomegalovirus	Cervix
Epstein–Barr virus	Burkitt's lymphoma, nasopharyngeal carcinoma
Human T-cell leukemia virus	Adult T-cell leukemia
Human immunodeficiency virus	Kaposi's sarcoma, non-Hodgkin's lymphoma
Parasites	
Schistosoma haematobium	Bladder
Schistosoma mansoni	Liver
Schistosoma japonicum	Rectum, liver
Opisthorchis viverrinie	Liver (cholangiocarcinoma)
Clonorchis sinensis	Liver (cholangiocarcinoma)
Malaria	Burkitt's lymphoma
Bacteria	
Helicobacter pylori	Stomach
Urinary infection	Bladder
Other	
Particles (e.g., asbestos, silica dust)	Lung
Hot beverages	Esophagus
Ulcerative colitis	Colon cancer

From Ohshima and Bartsch (1994). Reprinted with permission from *Mutation Research*, Elsevier Science, B. V., Amsterdam.

increase risk of cancer at several tissue sites (Table 8) and may have common molecular mechanisms (Ohshima and Bartsch 1994, Gentile and Gentile 1994, Rosin et al. 1994). Unfortunately, existing toxicologic databases are of little help in explaining differential susceptibility to environmental disease. Rodents used for laboratory testing are inbred for genetic homogeneity and are kept under pathogen-free conditions, with ad libitum purified diets, regimented light/dark cycles, and uncrowded living quarters, all of which minimize interindividual variability. Human population studies tend to exclude individuals with confounding physiological conditions or concurrent disease or are limited to fairly homogeneous study populations. Systematic laboratory-based investigations and broader population studies are required to test the significance of physiological factors as independent variables for cancer risk.

CONCLUSIONS

Carcinogenesis is a complex process, with multiple factors driving the neoplastic cascade. As more studies reveal the complexity of carcinogenesis, the approach of evaluating only single factors in cancer etiology will become increasingly insufficient. Multifactorial models are needed for simultaneously evaluating environmental exposures and host-specific factors that affect individual risk. To this end, reliable biomarkers of both exposure and susceptibility are needed.

There is no doubt that molecular biological techniques have added an important dimension to epidemiologic studies. However, maximum benefits from new methodologies can be achieved only through interdisciplinary approaches. This requires that investigators develop the ability to cross disciplines and apply information from diverse areas of research. The benefits of multidisciplinary approaches will be the development of effective strategies for risk assessment, cancer prevention, prognosis, and therapeutic intervention.

REFERENCES

Adami HO, Persson I, Hoover R, et al (1989) Risk of cancer in women receiving hormone replacement therapy. Int J Cancer 44:833–839

Adlercreutz H (1990) Western diet and Western diseases: some hormonal and biochemical mechanisms and associations [review]. Scand J Clin Lab Invest Suppl 201:3–23

Adlercreutz H, Mousavi Y, Clark J, et al (1992) Dietary phytoestrogens and cancer: *in vitro* and *in vivo* studies. J Steroid Biochem Mol Biol 41:331–337

Ahluwalia B, Jackson MA, Jones GW, et al (1981) Blood hormone profiles in prostate cancer patients in high-risk and low-risk populations. Cancer 48:2267–2273

Alberts B, Bray D, Lewis J, et al, eds (1989) Molecular biology of the cell. Garland Publishing, New York

Alexandrie AK, Sundberg MI, Seidegard J, et al (1994) Genetic susceptibility to lung cancer with special emphasis on *CYP1A1* and *GSTM1*: a study on host factors in relation to age at onset, gender and histological cancer types. Carcinogenesis 15:1785–1790

Ambrosone CB, Rao U, Michalek AM, et al (1993) Lung cancer histologic types and family history of cancer. Analysis of histologic subtypes of 872 patients with primary lung cancer. Cancer 72:1192–1198

Ananthaswamy HN, Price JE, Goldberg LH, Bales ES (1988) Detection and identification of activated oncogenes in human skin cancers occurring on sun-exposed body sites. Cancer Res 48:3341–3346

Andersen TI, Heimdal KR, Skrede M, et al (1994) Oestrogen receptor (ESR) polymorphisms and breast cancer susceptibility. Hum Genet 94:665–670

Applegate LA, Goldberg LH, Ley RD, Ananthaswamy HN (1990) Hypersensitivity of skin fibroblasts from basal cell nevus syndrome patients to killing by ultraviolet B but not by ultraviolet C radiation. Cancer Res 50:637–641

Arnold A, Motokura T, Bloom T, et al (1992) *PRAD1* (cyclin D1): a parathyroid neoplasia gene on 11q13. Henry Ford Hosp Med J 40:177–180

Auerbach AD (1992) Fanconi anemia and leukemia: tracking the genes [review]. Leukemia 1:1–4

Auerbach AD, Allen RG (1991) Leukemia and preleukemia in Fanconi anemia patients. A review of the literature and report of the International Fanconi Anemia Registry [review]. Cancer Genet Cytogenet 51:1–12

Auerbach AD, Wolman SR (1976) Susceptibility of Fanconi's anaemia fibroblasts to chromosome damage by carcinogens. Nature 261:494–496

Bagnara GP, Bonsi L, Strippoli P, et al (1993) Production of interleukin 6, leukemia inhibitory factor and granulocyte-macrophage colony stimulating factor by peripheral blood mononuclear cells in Fanconi's anemia. Stem Cells 2:137–143

Bagnara GP, Strippoli P, Bonsi L, et al (1992) Effect of stem cell factor on colony growth from acquired and constitutional (Fanconi) aplastic anemia. Blood 80:382–387

Barbosa JC, Shultz TD, Filley SJ, Nieman DC (1990) The relationship among adiposity, diet, and hormone concentrations in vegetarian and nonvegetarian postmenopausal women. Am J Clin Nutr 51:798–803

Baron J, Voigt JM, Whitter TB, et al (1986) Identification of intratissue sites for xenobiotic activation and detoxication. Adv Exp Med Biol 197:119–144

Beamish H, Lavin MF (1994) Radiosensitivity in ataxia-telangiectasia: anomalies in radiation-induced cell cycle delay. Int J Radiat Biol 65:175–184

Beamish H, Khanna KK, Lavin MF (1994) Ionizing radiation and cell cycle progression in ataxia telangiectasia. Radiat Res 138:S130–S133

Begg CB, Zhang ZF, Sun M, et al (1995) Methodology for evaluating the incidence of second primary cancers with application to smoking-related cancers from the Surveillance, Epidemiology, and End Results (SEER) program. Am J Epidemiol 142:653–665

Bell DA, Badawi AF, Lang NP, et al (1995) Polymorphism in the *N*-acetyltransferase 1 *(NAT1)* polyadenylation signal: association of *NAT1*10* allele with higher N-acetylation activity in bladder and colon tissue. Cancer Res 55:5226–5229

Bellet D (1992) Tumor-susceptibility markers [review]. Monogr Natl Cancer Inst 12:115–121

Benhamou E, Benhamou S, Flamant R (1987) Lung cancer and women: results of a French case-control study. Br J Cancer 55:91–95

Beral V (1983) Parity and susceptibility to cancer [review]. Ciba Found Symp 96:182–203

Bereziat JC, Raffalli F, Schmezer P, et al (1995) Cytochrome P450 2A of nasal epithelium: regulation and role in carcinogen metabolism. Mol Carcinogen 14:130–139

Berrozpe G, Schaeffer J, Peinado MA, et al (1994) Comparative analysis of mutations in the *p53* and *K-ras* genes in pancreatic cancer. Int J Cancer 58:185–191

Bertorelle R, Esposito G, Del Mistro A, et al (1995) Association of *p53* gene and protein alterations with metastases in colorectal cancer. Am J Surg Pathol 19:463–471

Bittard H, Descotes F, Billerey C, et al (1996) A genotype study of the *c-Ha-ras-1* locus in human bladder tumors. J Urol 155:1083–1088

Bjelke E (1974) Epidemiologic studies of cancer of the stomach, colon, and rectum with special emphasis on the role of diet. Scand J Gastroenterol Suppl 31:1–235

Blair A, Fraumeni JF (1978) Geographic patterns of prostate cancer in the United States. J Natl Cancer Inst 61:1379–1384

Blum M, Demierre A, Grant DM, et al (1991) Molecular mechanism of slow acetylation of drugs and carcinogens in humans. Proc Natl Acad Sci U.S.A. 88:5237–5241

Bock KW, Lilienblum W, Pfeil H (1982a) Functional heterogeneity of UDP-glucuronosyltransferase activities in C57BL/6 and DBA/2 mice. Biochem Pharmacol 31:1273–1277

Bock KW, Lilienblum W, Pfeil H, Eriksson LC (1982b) Increased uridine diphosphate-glucuronyltransferase activity in preneoplastic liver nodules and Morris hepatomas. Cancer Res 42:3747–3752

Boland CR (1996) Roles of the DNA mismatch repair genes in colorectal tumorigenesis [review]. Int J Cancer 69:47–49

Bordi C, Falchetti A, Azzoni C, et al (1995) Lack of allelic loss at the multiple endocrine neoplasia type 1 *(MEN-1)* gene locus in a pancreatic ductal (non-endocrine) adenocarcinoma of a patient with the MEN-1 syndrome. Virch Arch 426:203–208

Boring CC, Squires TS, Tong T (1991) Cancer statistics, 1991. Bol Assoc Med PR 83:225–242

Bornstein J, Rahat MA, Abramovici H (1995) Etiology of cervical cancer: current concepts [review]. Obstet Gynecol Surv 50:146–154

Bosari S, Viale G, Roncalli M, et al (1995) *p53* gene mutations, p53 protein accumulation and compartmentalization in colorectal adenocarcinoma. Am J Pathol 147:790–798

Breslow N, Olshan A, Beckwith JB, Green DM (1993) Epidemiology of Wilms tumor [review]. Med Pediatr Oncol 21:172–181

Brownson RC, Alavanja MC, Hock ET, Loy TS (1992) Passive smoking and lung cancer in nonsmoking women. Am J Publ Health 82:1525–1530

Byers T, Graham S, Swanson M (1982) Parity and colorectal cancer risk in women. J Natl Cancer Inst 69:1059–1062

Calabretta B, Nicolaides NC (1992) *c-myb* and growth control [review]. Crit Rev Eukaryot Gene Expr 2:225–235

Cameron R, Sweeney GD, Jones K, et al (1976) A relative deficiency of cytochrome P-450 and aryl hydrocarbon [benzo(a)pyrene] hydroxylase in hyperplastic nodules induced by 2-acetylaminofluorene in rat liver. Cancer Res 36:3888–3893

Campbell IG, Beynon G, Davis M, Englefield P (1995) LOH and mutation analysis of *CDKN2* in primary human ovarian cancers. Int J Cancer 63:222–225

Campbell NR, Van Loon JA, Sundaram RS, et al (1987) Human and rat liver phenol sulfotransferase: structure-activity relationships for phenolic substrates. Mol Pharmacol 32:813–819

Camus AM, Geneste O, Honkakoski P, et al (1993) High variability of nitrosamine metabolism among individuals: role of cytochromes P450 2A6 and 2E1 in the dealkylation of *N*-nitrosodimethylamine and *N*-nitrosodiethylamine in mice and humans. Mol Carcinogen 7:268–275

Caporaso NE, Tucker MA, Hoover RN, et al (1990) Lung cancer and the debrisoquine metabolic phenotype. J Natl Cancer Inst 82:1264–1272

Caporaso N, Hayes RB, Dosemeci M, et al (1989) Lung cancer risk, occupational exposure, and the debrisoquine metabolic phenotype. Cancer Res 49:3675–3679

Carpenter SP, Lasker JM, Raucy JL (1996) Expression, induction, and catalytic activity of the ethanol-inducible cytochrome P450 (CYP2E1) in human fetal liver and hepatocytes. Mol Pharmacol 49:260–268

Carter HB, Piantadosi S, Isaacs JT (1990) Clinical evidence for and implications of the multistep development of prostate cancer. J Urol 143:742–746

Cartwright RA, Glashan RW, Rogers HJ, et al (1982) Role of *N*-acetyltransferase phenotypes in bladder carcinogenesis: a pharmacogenetic epidemiological approach to bladder cancer. Lancet 2:842–845

Cascorbi I, Drakoulis N, Brockmoller J, et al (1995) Arylamine *N*-acetyltransferase *(NAT2)* mutations and their allelic linkage in unrelated Caucasian individuals: correlation with phenotypic activity. Am J Hum Genet 57:581–592

Cassidy A, Bingham S, Setchell KD (1994) Biological effects of a diet of soy protein rich in isoflavones on the menstrual cycle of premenopausal women. Am J Clin Nutr 60:333–340

Champeme MH, Bieche I, Latil A, et al (1992) Association between restriction fragment length polymorphism of the *L-myc* gene and lung metastasis in human breast cancer. Int J Cancer 50:6–9

Chen J, Compton C, Cheng E, et al (1992) *c-Ki-ras* mutations in dysplastic fields and cancers in ulcerative colitis. Gastroenterology 102:1983–1987

Chen YH, Li CD, Yap EP, McGee JO (1995) Detection of loss of heterozygosity of *p53* gene in paraffin-embedded breast cancers by non-isotopic PCR-SSCP. J Pathol 177:129–134

Cheng AL, Su IJ, Chen YC, et al (1993) Expression of P-glycoprotein and glutathione-S-transferase in recurrent lymphomas: the possible role of Epstein-Barr virus, immunophenotypes, and other predisposing factors. J Clin Oncol 11:109–115

Cherpillod P, Amstad PA (1995) Benzo[a]pyrene-induced mutagenesis of *p53* hot-spot codons 248 and 249 in human hepatocytes. Mol Carcinogen 13:15–20

Cholerton S, Arpanahi A, McCracken N, et al (1994) Poor metabolisers of nicotine and CYP2D6 polymorphism [letter]. Lancet 343:62–63

Chute CG, Willett WC, Colditz GA, et al (1991) A prospective study of reproductive history and exogenous estrogens on the risk of colorectal cancer in women [see comments]. Epidemiology 2:201–207

Claus EB, Risch N, Thompson WD (1993) The calculation of breast cancer risk for women with a first degree family history of ovarian cancer. Breast Cancer Res Treat 28:115–120

Claus EB, Risch NJ, Thompson WD (1990) Age at onset as an indicator of familial risk of breast cancer. Am J Epidemiol 131:961–972

Cleaver JE (1969) Xeroderma pigmentosum: a human disease in which an initial stage of DNA repair is defective. Proc Natl Acad Sci U.S.A. 63:428–435

Cleaver JE (1968) Defective repair replication of DNA in xeroderma pigmentosum. Nature 218:652–656

Cohen LA, Kendall ME, Zang E, et al (1991) Modulation of *N*-nitrosomethylurea-induced mammary tumor promotion by dietary fiber and fat. J Natl Cancer Inst 83:496–501

Colditz GA, Frazier AL (1995) Models of breast cancer show that risk is set by events of early life: prevention efforts must shift focus [review]. Cancer Epidemiol Biomarkers Prev 4:567–571

Colditz GA, Willett WC, Hunter DJ, et al (1993) Family history, age, and risk of breast cancer. Prospective data from the Nurses' Health Study. JAMA 270:338–343 [published erratum (1993) JAMA 270(13):1548]

Cole P, MacMahon B, Brown JB (1976) Oestrogen profiles of parous and nulliparous women. Lancet 2:596–599

Coughtrie MW, Burchell B, Leakey JE, Hume R (1988) The inadequacy of perinatal glucuronidation: immunoblot analysis of the developmental expression of individual UDP-glucuronosyltransferase isoenzymes in rat and human liver microsomes. Mol Pharmacol 34:729–735

Cowell JK, Groves N, Baird P (1993) Loss of heterozygosity at 11p13 in Wilms' tumours does not necessarily involve mutations in the *WT1* gene. Br J Cancer 67:1259–1261

Crespi CL, Penman BW, Gonzalez FJ, et al (1993) Genetic toxicology using human cell lines expressing human P-450. Biochem Soc Trans 21:1023–1028

Crofts F, Cosma GN, Currie D, et al (1993) A novel *CYP1A1* gene polymorphism in African-Americans. Carcinogenesis 14:1729–1731

Dales LG, Friedman GD, Ury HK, et al (1979) A case-control study of relationships of diet and other traits to colorectal cancer in American blacks. Am J Epidemiol 109:132–144

Davis DL, Russell DW (1993) Unusual length polymorphism in human steroid 5 alpha-reductase type 2 gene *(SRD5A2)*. Hum Mol Genet 2:820

Davis DL, Bradlow HL, Wolff M, et al (1993) Medical hypothesis: xenoestrogens as preventable causes of breast cancer [review]. Environ Health Perspect 101:372–377

Davis FG, Furner SE, Persky V, Koch M (1989) The influence of parity and exogenous female hormones on the risk of colorectal cancer. Int J Cancer 43:587–590

de The G (1995) Viruses and human cancers: challenges for preventive strategies. Environ Health Perspect 103:269–273 de Vos S, Miller CW, Takeuchi S, et al (1995) Alterations of *CDKN2* (p16) in non-small cell lung cancer. Genes Chromosomes Cancer 14:164–170

de Waard F (1979) Premenopausal and postmenopausal breast cancer: one disease or two? [review]. J Natl Cancer Inst 63:549–552

Deakin M, Elder J, Hendrickse C, et al (1996) Glutathione S-transferase *GSTT1* genotypes and susceptibility to cancer: studies of interactions with *GSTM1* in lung, oral, gastric and colorectal cancers. Carcinogenesis 17:881–884

Devesa SS, Chow WH (1993) Variation in colorectal cancer incidence in the United States by subsite of origin. Cancer 71:3819–3826

Doll R (1978) An epidemiological perspective of the biology of cancer [review]. Cancer Res 38:3573–3583

Doll R, Peto R (1981) The causes of cancer: quantitative estimates of avoidable risks of cancer in the United States today [review]. J Natl Cancer Inst 66:1191–1308

Doll R, Peto R (1976) Mortality in relation to smoking: 20 years' observations on male British doctors. Br Med J 2:1525–1536

Dutrillaux B, Aurias A, Dutrillaux AM, et al (1982) The cell cycle of lymphocytes in Fanconi anemia. Hum Genet 62:327–332

Dwyer T, Blizzard L, Shugg D, et al (1994) Higher lung cancer rates in young women than young men: Tasmania, 1983 to 1992. Cancer Causes Control 5:351–358

Easton D, Ford D, Peto J (1993) Inherited susceptibility to breast cancer [review]. Cancer Surv 18:95–113

Eaton DL, Gallagher EP, Bammler TK, Kunze KL (1995) Role of cytochrome P4501A2 in chemical carcinogenesis: implications for human variability in expression and enzyme activity [review]. Pharmacogenetics 5:259–274

Eide TJ (1986) The age-, sex-, and site-specific occurrence of adenomas and carcinomas of the large intestine within a defined population. Scand J Gastroenterol 21:1083–1088

Eidson M, Becker TM, Wiggins CL, et al (1994) Breast cancer among Hispanics, American Indians and non-Hispanic whites in New Mexico. Int J Epidemiol 23:231–237

Elexpuru-Camiruaga J, Buxton N, Kandula V, et al (1995) Susceptibility to astrocytoma and meningioma: influence of allelism at glutathione S-transferase *(GSTT1* and *GSTM1)* and cytochrome P-450 (CYP2D6) loci. Cancer Res 55:4237–4239

Ellard GA (1976) Variations between individuals and populations in the acetylation of isoniazid and its significance for the treatment of pulmonary tuberculosis [review]. Clin Pharmacol Ther 19:610–625

Emerit I (1994) Reactive oxygen species, chromosome mutation, and cancer: possible role of clastogenic factors in carcinogenesis [review]. Free Rad Biol Med 16:99–109

Enomoto T, Fujita M, Inoue M, et al (1993) Alterations of the *p53* tumor suppressor gene and its association with activation of the *c-K-ras-2* protooncogene in premalignant and malignant lesions of the human uterine endometrium. Cancer Res 53:1883–1888

Ershler WB (1993) The influence of an aging immune system on cancer incidence and progression [review]. J Gerontol 48:B3–B7

Esumi H, Ogura T, Kurashima Y, et al (1995) Implication of nitric oxide synthase in carcinogenesis: analysis of the human inducible nitric oxide synthase gene. Pharmacogenetics 5 SpecNo:S166–S170

Evans HJ, Prosser J (1992) Tumor-suppressor genes: cardinal factors in inherited predisposition to human cancers [review]. Environ Health Perspect 98:25–37

Faille A, De Cremoux P, Extra JM, et al (1994) *p53* mutations and overexpression in locally advanced breast cancers. Br J Cancer 69:1145–1150

Farber E (1995) Cell proliferation as a major risk factor for cancer: a concept of doubtful validity. Cancer Res 55:3759–3762

Farber E, Cameron R (1980) The sequential analysis of cancer development. Adv Cancer Res 31:125–226

Farrell PM (1977) Fetal lung development and the influence of glucocorticoids on pulmonary surfactant. J Steroid Biochem 8:463–470

Fearon ER (1992) Genetic alterations underlying colorectal tumorigenesis. In Levine AJ (ed), Tumour suppressor genes, the cell cycle and cancer. Cold Spring Harbor Laboratory Press, Plainview, NY, pp 119–136

Ford D, Easton DF, Peto J (1995) Estimates of the gene frequency of *BRCA1* and its contribution to breast and ovarian cancer incidence. Am J Hum Genet 57:1457–1462

Forman D, Newell DG, Fullerton F, et al (1991) Association between infection with *Helicobacter pylori* and risk of gastric cancer: evidence from a prospective investigation. Br Med J 302:1302–1305

Franceschi S, Bidoli E, Talamini R, et al (1991) Colorectal cancer in northeast Italy: reproductive, menstrual and female hormone-related factors. Eur J Cancer 27:604–608

Fraumeni JF, Hoover RN, Devesa SS, Kinlen LJ (1993) Epidemiology of cancer. In DeVita VT, Hellman S, Rosenberg SA (eds), Cancer: principle and practice of oncology, 4th ed. Lippincott, Philadelphia, pp 150–181

Fraumeni JF Jr, Lloyd JW, Smith EM, Wagoner JK (1969) Cancer mortality among nuns: role of marital status in etiology of neoplastic disease in women. J Natl Cancer Inst 42:455–468

Friedberg EC, Walker GC, Siede W (1995) Human hereditary diseases with defective processing of DNA damage. In Friedberg EC, Walker GC, Siede W (eds), DNA repair and mutagenesis. American Society for Microbiology, Washington, D.C., pp 633–685

Furner SE, Davis FG, Nelson RL, Haenszel W (1989) A case-control study of large bowel cancer and hormone exposure in women. Cancer Res 49:4936–4940

Furuya H, Fernandez-Salguero P, Gregory W, et al (1995) Genetic polymorphism of *CYP2C9* and its effect on warfarin maintenance dose requirement in patients undergoing anticoagulation therapy. Pharmacogenetics 5:389–392

Gann PH, Hennekens CH, Sacks FM, et al (1994) Prospective study of plasma fatty acids and risk of prostate cancer. J Natl Cancer Inst 86:281–286

Gao YT, Blot WJ, Zheng W, et al (1987) Lung cancer among Chinese women. Int J Cancer 40:604–609

Gentile JM, Gentile GJ (1994) Implications for the involvement of the immune system in parasite-associated cancers [review]. Mutat Res 305:315–320

Gerhardsson de Verdier M, London S (1992) Reproductive factors, exogenous female hormones, and colorectal cancer by subsite. Cancer Causes Control 3:355–360

Gerhardsson M, Steineck G, Norell SE (1990) Colorectal cancer in Sweden. A descriptive epidemiologic study. Acta Oncol 29:855–861

Getchell ML, Chen Y, Ding X, et al (1993) Immunohistochemical localization of a cytochrome P-450 isozyme in human nasal mucosa: age-related trends. Ann Oto Rhinol Laryngol 102:368–374

Giovannucci E, Rimm EB, Colditz GA, et al (1993) A prospective study of dietary fat and risk of prostate cancer. J Natl Cancer Inst 85:1571–1579

Goldin BR, Adlercreutz H, Gorbach SL, et al (1982) Estrogen excretion patterns and plasma levels in vegetarian and omnivorous women. N Engl J Med 307:1542–1547

Gombart AF, Morosetti R, Miller CW, et al (1995) Deletions of the cyclin-dependent kinase inhibitor genes *p16INK4A* and *p15INK4B* in non-Hodgkin's lymphomas. Blood 86:1534–1539

Gonzalez FJ (1995a) Genetic differences in carcinogen metabolizing enzymes [meeting abstract]. Proc Ann Meet Am Assoc Cancer Res 36:A664

Gonzalez FJ (1995b) Genetic polymorphism and cancer susceptibility: fourteenth Sapporo Cancer Seminar. Cancer Res 55:710–715

Gonzalez FJ, Crespi CL, Czerwinski M, Gelboin HV (1992) Analysis of human cytochrome P450 catalytic activities and expression. Tohoku J Exp Med 168:67–72

Guengerich FP (1994) Catalytic selectivity of human cytochrome P450 enzymes: relevance to drug metabolism and toxicity. Toxicol Lett 70:133–138

Guengerich FP (1993) The 1992 Bernard B. Brodie Award Lecture. Bioactivation and detoxication of toxic and carcinogenic chemicals. Drug Metab Dispos 21:1–6

Guengerich FP (1992) Metabolic activation of carcinogens. Pharmacol Ther 54:17–61

Haehner BD, Gorski JC, Vandenbranden M, et al (1996) Bimodal distribution of renal cytochrome P450 3A activity in humans. Mol Pharmacol 50:52–59

Hakkola J, Pasanen M, Hukkanen J, et al (1996) Expression of xenobiotic-metabolizing cytochrome P450 forms in human full-term placenta. Biochem Pharmacol 51:403–411

Hall JM, Lee MK, Newman B, et al (1990) Linkage of early-onset familial breast cancer to chromosome 17q21. Science 250:1684–1689

Halpern MT, Gillespie BW, Warner KE (1993) Patterns of absolute risk of lung cancer mortality in former smokers. J Natl Cancer Inst 85:457–464

Hamada GS, Sugimura H, Suzuki I, et al (1995) The heme-binding region polymorphism of cytochrome P450IA1 (CypIA1), rather than the RsaI polymorphism of IIE1 (CypIIE1), is associated with lung cancer in Rio de Janeiro. Cancer Epidemiol Biomarkers Prev 4:63–67

Hamalainen EK, Adlercreutz H, Puska P, Pietinen P (1983) Decrease of serum total and free testosterone during a low-fat high-fibre diet. J Steroid Biochem 18:369–370

Hang B, Yeung AT, Lambert MW (1993) A damage-recognition protein which binds to DNA containing interstrand cross-links is absent or defective in Fanconi anemia, complementation group A, cells. Nucleic Acids Res 21:4187–4192

Hankey BF (1992) Breast. In Miller BA, Ries LAG, Hankey BF, et al (eds), Cancer statistics review: 1973-1989. National Cancer Institute, Bethesda, MD, pp IV1–IV20

Harris CC (1995) 1995 Deichmann Lecture — $p53$ tumor suppressor gene: at the crossroads of molecular carcinogenesis, molecular epidemiology and cancer risk assessment. Toxicol Lett 82/83:1–7

Harris CC (1991) Chemical and physical carcinogenesis: advances and perspectives for the 1990s [review]. Cancer Res 51:5023s–5044s

Harris HW, Knight A, Selin MJ (1958) Comparison of isoniazid concentrations in the blood of people of Japanese and European descent — therapeutic and genetic implications. Am Rev Tuberc Pulm Dis 78(6):944–948

Hart RW, Frame LT (1996) Toxicological defense mechanisms and their effect on dose–response relationships. BELLE Newsletter 5:1–8

Hatta Y, Hirama T, Takeuchi S, et al (1995) Alterations of the p16 *(MTS1)* gene in testicular, ovarian, and endometrial malignancies. J Urol 154:1954–1957

Hayashi S, Watanabe J, Kawajiri K (1992) High susceptibility to lung cancer analyzed in terms of combined genotypes of *P450IA1* and *Mu*-class *glutathione S-transferase* genes. Jpn J Cancer Res 83:866–870

Hegmann KT, Fraser AM, Keaney RP, et al (1993) The effect of age at smoking initiation on lung cancer risk. Epidemiology 4:444–448

Heidenberg HB, Sesterhenn IA, Gaddipati JP, et al (1995) Alteration of the tumor suppressor gene *p53* in a high fraction of hormone refractory prostate cancer [review]. J Urol 154:414–421

Heighway J, Thatcher N, Cerny T, Hasleton PS (1986) Genetic predisposition to human lung cancer. Br J Cancer 53:453–457

Henderson BE, Bernstein L, Ross RK, et al (1988) The early *in utero* oestrogen and testosterone environment of blacks and whites: potential effects on male offspring. Br J Cancer 57:216–218

Hirayama T (1977) Changing patterns of cancer in Japan with special reference to the decrease in stomach cancer mortality. In Hiatt HH, Watson JD, Winsten JA (eds), Origins of human cancer. Book A. Cold Spring Harbor Laboratory, Cold Spring Harbor, NY, pp 55–75

Hirvonen A, Husgafvel-Pursiainen K, Anttila S, Vainio H (1993) The *GSTM1* null genotype as a potential risk modifier for squamous cell carcinoma of the lung. Carcinogenesis 14:1479–1481

Hislop TG, Coldman AJ, Elwood JM, et al (1986) Childhood and recent eating patterns and risk of breast cancer. Cancer Detect Prev 9:47–58

Hoehn H, Kubbies M, Schindler D, et al (1987) BrdU-Hoechst flow cytometry links the cell kinetic defect of Fanconi's anemia to oxygen hypersensitivity. In Schroeder-Kurth TM, Auerbach AD, Obe G (eds), Fanconi anemia: clinical, cytogenic and experimental aspects. Springer-Verlag, Heidelberg, Germany, pp 161–173

Honda T, Kato H, Imamura T, et al (1993) Involvement of *p53* gene mutations in human endometrial carcinomas. Int J Cancer 53:963–967

Hong SI, Hong WS, Jang JJ, et al (1994) Alterations of *p53* gene in primary gastric cancer tissues. Anticancer Res 14:1251–1255

Howe GR, Hirohata T, Hislop TG, et al (1990) Dietary factors and risk of breast cancer: combined analysis of 12 case-control studies. J Natl Cancer Inst 82:561–569

Howe GR, Craib KJ, Miller AB (1985) Age at first pregnancy and risk of colorectal cancer: a case-control study. J Natl Cancer Inst 74:1155–1159

Howell MA (1974) Factor analysis of international cancer mortality data and per capita food consumption. Br J Cancer 29:328–336

Hunter DJ, Willett WC (1993) Diet, body size, and breast cancer [review]. Epidemiol Rev 15:110–132

Hunter DJ, Spiegelman D, Adami HO, et al (1996) Cohort studies of fat intake and the risk of breast cancer — a pooled analysis. N Engl J Med 334:356–361

Hutchison GB (1976) Epidemiology of prostatic cancer. Semin Oncol 3:151–159

IARC (1986) International Agency for Research on Cancer. Tobacco smoking. IARC Monogr Eval Carcinogen Risk Chem Hum 38:20–31

Ilett KF, David BM, Detchon P, et al (1987) Acetylation phenotype in colorectal carcinoma. Cancer Res 47:1466–1469

Jacobsen BK, Lund E (1990) Level of education, use of oral contraceptives and reproductive factors: the Tromso Study. Int J Epidemiol 19:967–970

Janerich DT, Hoff MB (1982) Evidence for a crossover in breast cancer risk factors [review]. Am J Epidemiol 116:737–742

Jansen JG, Mohn GR, Vrieling H, et al (1994) Molecular analysis of hprt gene mutations in skin fibroblasts of rats exposed in vivo to N-methyl-N-nitrosourea or N-ethyl-N-nitrosourea. Cancer Res 54:2478–2485

Jensen OM (1984) Different age and sex relationship for cancer of subsites of the large bowel. Br J Cancer 50:825–829

Jones LA, Blocker SH, Rusch VH, Mountain CF (1984) Specific estrogen binding protein in human lung cancer [abstract]. Proc Ann Meet Am Assoc Cancer Res 25:208

Kadlubar FF, Butler MA, Kaderlik KR, et al (1992) Polymorphisms for aromatic amine metabolism in humans: relevance for human carcinogenesis. Environ Health Perspect 98:69–74

Kahn HA (1966) The Dorn study of smoking and mortality among U.S. veterans: report on eight and one-half years of observation. Natl Cancer Inst Monogr 19:1–125

Kakehi Y, Taki Y, Yoshida O (1991) Restriction fragment length polymorphism of the L-myc gene and susceptibility to metastasis in genitourinary cancers. Urologia Internationalis 1:86–89

Kamb A, Shattuck-Eidens D, Eeles R, et al (1994) Analysis of the p16 gene (CDKN2) as a candidate for the chromosome 9p melanoma susceptibility locus. Nature Genet 8:23–26

Kampman E, Bijl AJ, Kok C, van't Veer P (1994) Reproductive and hormonal factors in male and female colon cancer. Eur J Cancer Prev 3:329–336

Karim AK, Elfellah MS, Evans DA (1981) Human acetylator polymorphism: estimate of allele frequency in Libya and details of global distribution. J Med Genet 18:325–330

Kashii T, Mizushima Y, Monno S, et al (1994) Gene analysis of K-, H-ras, p53, and retinoblastoma susceptibility genes in human lung cancer cell lines by the polymerase chain reaction/single-strand conformation polymorphism method. J Cancer Res Clin Oncol 120:143–148

Katoh T, Inatomi H, Nagaoka A, Sugita A (1995) Cytochrome P450IA1 gene polymorphism and homozygous deletion of the glutathione S-transferase M1 gene in urothelial cancer patients. Carcinogenesis 16:655–657

Kawashima K, Nomura S, Hirai H, et al (1992) Correlation of L-myc RFLP with metastasis, prognosis and multiple cancer in lung-cancer patients. Int J Cancer 50:557–561

Kelsey JL, Gammon MD, John EM (1993) Reproductive factors and breast cancer [review]. Epidemiol Rev 15:36–47

Kelsey JL, Fischer DB, Holford TR, et al (1981) Exogenous estrogens and other factors in the epidemiology of breast cancer. J Natl Cancer Inst 67:327–333

Key TJ, Chen J, Wang DY, et al (1990) Sex hormones in women in rural China and in Britain. Br J Cancer 62:631–636

Kihana T, Tsuda H, Teshima S, et al (1992) High incidence of p53 gene mutation in human ovarian cancer and its association with nuclear accumulation of p53 protein and tumor DNA aneuploidy. Jpn J Cancer Res 83:978–984

Kihara M, Noda K (1995) Distribution of GSTM1 null genotype in relation to gender, age and smoking status in Japanese lung cancer patients. Pharmacogenetics 5 Spec No:S74–S79

Kim MS, Li SL, Bertolami CN, et al (1993) State of p53, Rb and DCC tumor suppressor genes in human oral cancer cell lines. Anticancer Res 13:1405–1413

Kirby GM, Batist G, Alpert L, et al (1996) Overexpression of cytochrome P-450 isoforms involved in aflatoxin B-1 bioactivation in human liver with cirrhosis and hepatitis. Toxicol Pathol 24:458–467

Kizer DE, Clouse JA, Ringer DP, et al (1985) Assessment of rat liver microsomal epoxide hydrolase as a marker of hepatocarcinogenesis. Biochem Pharmacol 34:1795–1800

Klingelhutz AJ, Smith PP, Garrett LR, McDougall JK (1993) Alteration of the DCC tumor-suppressor gene in tumorigenic HPV-18 immortalized human keratinocytes transformed by nitrosomethylurea. Oncogene 8:95–99

Kobayashi S, Mizuno T, Tobioka N, et al (1982) Sex steroid receptors in diverse human tumors. Gann 73:439–445

Kolars JC, Lown KS, Schmiedlin-Ren P, et al (1994) CYP3A gene expression in human gut epithelium. Pharmacogenetics 4:247–259

Kolonel LN, Hinds MW, Hankin JH (1980) Cancer patterns among Hawaiians in relation to smoking, drinking, and dietary habits. In Gelboin HV, MacMahon B, Matsushima T (eds), Genetic and environmental factors in experimental and human cancer: Proceedings of the 10th international symposium of the Princess Takamatsu cancer research fund. Japan Scientific Societies Press, Tokyo, pp 307–340

Komiya A, Suzuki H, Aida S, et al (1995) Mutational analysis of *CDKN2 (CDK4I/MTS1)* gene in tissues and cell lines of human prostate cancer. Jpn J Cancer Res 86:622–625

Kondo K, Tsuzuki H, Sasa M, et al (1996) A dose–response relationship between the frequency of *p53* mutations and tobacco consumption in lung cancer patients. J Surg Oncol 61:20–26

Kraemer KH, Lee MM, Scotto J (1987) Xeroderma pigmentosum. Cutaneous, ocular, and neurologic abnormalities in 830 published cases. Arch Dermatol 123:241–250

Kubota Y, Fujinami K, Uemura H, et al (1995) Retinoblastoma gene mutations in primary human prostate cancer. Prostate 27:314–320

Kune S, Kune GA, Watson L (1986) The Melbourne colorectal cancer study: incidence findings by age, sex, site, migrants and religion. Int J Epidemiol 15:483–493

Kvale G, Heuch I (1991) Is the incidence of colorectal cancer related to reproduction? A prospective study of 63,000 women. Int J Cancer 47:390–395

Lambert MW, Tsongalis GJ, Lambert WC, et al (1992) Defective DNA endonuclease activities in Fanconi's anemia cells, complementation groups A and B. Mutat Res 273:57–71

Lang NP, Butler MA, Massengill J, et al (1994) Rapid metabolic phenotypes for acetyltransferase and cytochrome P4501A2 and putative exposure to food-borne heterocyclic amines increase the risk for colorectal cancer or polyps. Cancer Epidemiol Biomarkers Prev 3:675–682

Lang NP, Chu DZ, Hunter CF, et al (1986) Role of aromatic amine acetyltransferase in human colorectal cancer. Arch Surg 121:1259–1261

Langston AA, Malone KE, Thompson JD, et al (1996) *BRCA1* mutations in a population-based sample of young women with breast cancer. N Engl J Med 334:137–142

Launoy G, Pottier D, Gignoux M (1989) [Proximal and distal cancers of the colon: two epidemiologically different cancers] [in French]. Gastroenterol Clin Biol 13:255–259

Lavin MF, Khanna KK, Beamish H, et al (1994) Defect in radiation signal transduction in ataxia-telangiectasia. Int J Radiat Biol 66:S151–S156

Lehmann AR (1989) Trichothiodystrophy and the relationship between DNA repair and cancer [review]. Bioessays 11:168–170

Lehmann AR, Arlett CF, Broughton BC, et al (1988) Trichothiodystrophy, a human DNA repair disorder with heterogeneity in the cellular response to ultraviolet light. Cancer Res 48:6090–6096

Levin W, Lu AY, Thomas PE, et al (1978) Identification of epoxide hydrase as the preneoplastic antigen in rat liver hyperplastic nodules. Proc Natl Acad Sci U.S.A. 75:3240–3243

Li FP, Fraumeni JF Jr (1994) Collaborative interdisciplinary studies of *p53* and other predisposing genes in Li–Fraumeni syndrome. Cancer Epidemiol Biomarkers Prev 3:715–717

Lindahl R (1992) Aldehyde dehydrogenases and their role in carcinogenesis. Crit Rev Biochem Mol Biol 27:283–335

Liu B, Parsons R, Papadopoulos N, et al (1996) Analysis of mismatch repair genes in hereditary non-polyposis colorectal cancer patients. Nature Med 2:169–174

Liu B, Nicolaides NC, Markowitz S, et al (1995) Mismatch repair gene defects in sporadic colorectal cancers with microsatellite instability. Nature Genet 9:48–55

Liu SH, Otal-Brun M, Webb TE (1980) Glucocorticoid receptors in human tumors. Cancer Lett 10:269–275

Lobaccaro JM, Lumbroso S, Belon C, et al (1993) Androgen receptor gene mutation in male breast cancer. Hum Mol Genet 2:1799–1802

Lubin F, Ruder AM, Wax Y, Modan B (1985) Overweight and changes in weight throughout adult life in breast cancer etiology. A case-control study. Am J Epidemiol 122:579–588

Lubin JH, Blot WJ (1984) Assessment of lung cancer risk factors by histologic category. J Natl Cancer Inst 73:383–389

Lynch HT, Watson P, Conway T, et al (1988) Breast cancer family history as a risk factor for early onset breast cancer. Breast Cancer Res Treat 11:263–267

Maesawa C, Tamura G, Suzuki Y, et al (1995) The sequential accumulation of genetic alterations characteristic of the colorectal adenoma-carcinoma sequence does not occur between gastric adenoma and adenocarcinoma. J Pathol 176:249–258

Malats N, Porta M, Pinol JL, et al (1995) *Ki-ras* mutations as a prognostic factor in extrahepatic bile system cancer. PANK-ras I Project Investigators. J Clin Oncol 13:1679–1686

Markowitz S, Wang J, Myeroff L, et al (1995) Inactivation of the type II TGF-beta receptor in colon cancer cells with microsatellite instability. Science 268:1336–1338

Marra G, Boland CR (1995) Hereditary nonpolyposis colorectal cancer: the syndrome, the genes, and historical perspectives [review]. J Natl Cancer Inst 87:1114–1125

Martz CH (1991) von Hippel–Lindau disease: a genetic condition predisposing tumor formation [review]. Oncol Nurs Forum 18:545–551

Matsuzaki J, Dobashi Y, Miyamoto H, et al (1996) DNA polymerase beta gene mutations in human bladder cancer. Mol Carcinogen 15:38–43

Mazars R, Spinardi L, BenCheikh M, et al (1992) *p53* mutations occur in aggressive breast cancer. Cancer Res 52:3918–3923

McDuffie HH (1994) Re: "Are female smokers at higher risk for lung cancer than male smokers? A case-control analysis by histologic type". Am J Epidemiol 140:185–186; discussion 187–188

McDuffie HH (1991) Clustering of cancer in families of patients with primary lung cancer. J Clin Epidemiol 44:69–76

McDuffie HH, Klaassen DJ, Dosman JA (1991) Men, women and primary lung cancer — a Saskatchewan personal interview study. J Clin Epidemiol 44:537–544

McDuffie HH, Klaassen DJ, Dosman JA (1987) Female-male differences in patients with primary lung cancer. Cancer 59:1825–1830

McKenna NJ, Kieback DG, Carney DN, et al (1995) A germline TaqI restriction fragment length polymorphism in the progesterone receptor gene in ovarian carcinoma. Br J Cancer 71:451–455

McKie AB, Filipe MI, Lemoine NR (1993) Abnormalities affecting the APC and MCC tumour suppressor gene loci on chromosome 5q occur frequently in gastric cancer but not in pancreatic cancer. Int J Cancer 55:598–603

McKinnon RA, Burgess WM, Hall PM, et al (1995) Characterisation of *CYP3A* gene subfamily expression in human gastrointestinal tissues. Gut 36:259–267

McLemore TL, Adelberg S, Liu MC, et al (1990) Expression of *CYP1A1* gene in patients with lung cancer: evidence for cigarette smoke-induced gene expression in normal lung tissue and for altered gene regulation in primary pulmonary carcinomas. J Natl Cancer Inst 82:1333–1339

McManus DT, Yap EP, Maxwell P, et al (1994) *p53* expression, mutation, and allelic deletion in ovarian cancer. J Pathol 174:159–168

McMichael AJ, Potter JD (1985) Diet and colon cancer: integration of the descriptive, analytic, and metabolic epidemiology. Natl Cancer Inst Monogr 69:223–228

McMichael AJ, Potter JD (1984) Parity and death from colon cancer in women: a case-control study. Commun Health Stud 8:19–25

McMichael AJ, Potter JD (1980) Reproduction, endogenous and exogenous sex hormones, and colon cancer: a review and hypothesis [review]. J Natl Cancer Inst 65:1201–1207

Mellemgaard A, Engholm G, Lynge E (1988) High and low risk groups for cancer of colon and rectum in Denmark: multiplicative Poisson models applied to register linkage data. J Epidemiol Commun Health 42:249–256

Mendelson CR, Johnston JM, MacDonald PC, Snyder JM (1981) Multihormonal regulation of surfactant synthesis by human fetal lung *in vitro*. J Clin Endocrinol Metab 53:307–317

Mettlin C, Croghan I, Natarajan N, Lane W (1990) The association of age and familial risk in a case-control study of breast cancer. Am J Epidemiol 131:973–983

Miki Y, Swensen J, Shattuck-Eidens D, et al (1994) A strong candidate for the breast and ovarian cancer susceptibility gene *BRCA1*. Science 266:66–71

Milner BJ, Allan LA, Kelly KF, et al (1993) Linkage studies with 17q and 18q markers in a breast/ovarian cancer family. Am J Hum Genet 52:761–766

Minna JD, Pass H, Glatstein E, Ihde DC (1993) Cancer of the lung. In DeVita VT, Hellman S, Rosenberg SA (eds), Cancer: principle and practice of oncology. Lippincott, Philadelphia, pp 591–705

Mitsudomi T, Steinberg SM, Nau MM, et al (1992) *p53* gene mutations in non-small-cell lung cancer cell lines and their correlation with the presence of *ras* mutations and clinical features. Oncogene 7:171–180

Miyamoto H, Shuin T, Torigoe S, et al (1995) Retinoblastoma gene mutations in primary human bladder cancer. Br J Cancer 71:831–835

Miyamoto H, Kubota Y, Shuin T, et al (1993) Analyses of *p53* gene mutations in primary human bladder cancer. Oncol Res 5:245–249

Modrich P (1994) Mismatch repair, genetic stability, and cancer [review]. Science 266:1959–1960

Moerkerk P, Arends JW, van Driel M, et al (1994) Type and number of *Ki-ras* point mutations relate to stage of human colorectal cancer. Cancer Res 54:3376–3378

Morris JJ, Seifter E (1992) The role of aromatic hydrocarbons in the genesis of breast cancer [review]. Med Hypoth 38:177–184

Morris SM, Domon OE, McGarrity LJ, et al (1995) Programmed cell death and mutation induction in AHH-1 human lymphoblastoid cells exposed to m-amsa. Mutat Res 329:79–96

Morris SM, Domon OE, Delclos KB, et al (1994) Induction of mutations at the hypoxanthine phosphoribosyltransferase (HPRT) locus in AHH-1 human lymphoblastoid cells. Mutat Res 310:45–54

Muir CS, Nectoux J, Staszewski J (1991) The epidemiology of prostatic cancer. Geographical distribution and time-trends [review]. Acta Oncol 30:133–140

Munoz N, Bosch FX, de Sanjose S, Shah KV (1994) The role of HPV in the etiology of cervical cancer. Mutat Res 305:293–301

Murray GI, Pritchard S, Melvin WT, Burke MD (1995) Cytochrome P450 CYP3A5 in the human anterior pituitary gland. FEBS Lett 364:79–82

Nakachi K, Imai K, Hayashi S, Kawajiri K (1993) Polymorphisms of the *CYP1A1* and *glutathione S-transferase* genes associated with susceptibility to lung cancer in relation to cigarette dose in a Japanese population. Cancer Res 53:2994–2999

Nakachi K, Imai K, Hayashi S, et al (1991) Genetic susceptibility to squamous cell carcinoma of the lung in relation to cigarette smoking dose. Cancer Res 51:5177–5180

Nakajima T, Elovaara E, Anttila S, et al (1995) Expression and polymorphism of *glutathione S-transferase* in human lungs: risk factors in smoking-related lung cancer. Carcinogenesis 16:707–711

Narod SA (1994) Genetics of breast and ovarian cancer [review]. Br Med Bull 50:656–676

National Research Council (1994) Science and judgment in risk assessment. Appendix H-2: individual susceptibility factors. National Academy Press, Washington, D.C., pp 505–514

Nazar-Stewart V, Motulsky AG, Eaton DL, et al (1993) The *glutathione S-transferase mu* polymorphism as a marker for susceptibility to lung carcinoma. Cancer Res 53:2313–2318

Negri E, La Vecchia C, Parazzini F, et al (1989) Reproductive and menstrual factors and risk of colorectal cancer. Cancer Res 49:7158–7161

Nestel PJ, Hirsch EZ, Couzens EA (1965) The effect of chlorophenoxyisobutyric acid and ethinyl estradiol on cholesterol turnover. J Clin Invest 44:891–896

Neubauer A, de Kant E, Rochlitz C, et al (1993) Altered expression of the retinoblastoma susceptibility gene in chronic lymphocytic leukaemia. Br J Haematol 85:498–503

Nishigori C, Yarosh DB, Ullrich SE, et al (1996) Evidence that DNA damage triggers interleukin 10 cytokine production in UV-irradiated murine keratinocytes. Proc Natl Acad Sci U.S.A. 93:10354–10359

Nomura AM, Kolonel LN (1991) Prostate cancer: a current perspective [review]. Epidemiol Rev 13:200–227

Nomura AM, Stemmermann GN, Chyou PH (1991) Prospective study of serum cholesterol levels and large-bowel cancer. J Natl Cancer Inst 83:1403–1407

Oesch F, Vogel-Bindel U, Guenthner TM, et al (1983) Characterization of microsomal epoxide hydrolase in hyperplastic liver nodules of rats. Cancer Res 43:313–319

Ogawa K, Solt DB, Farber E (1980) Phenotypic diversity as an early property of putative preneoplastic hepatocyte populations in liver carcinogenesis. Cancer Res 40:725–733

Ohshima H, Bartsch H (1994) Chronic infections and inflammatory processes as cancer risk factors: possible role of nitric oxide in carcinogenesis [review]. Mutat Res 305:253–264

Ooi WL, Elston RC, Chen VW, et al (1986) Increased familial risk for lung cancer. J Natl Cancer Inst 76:217–222

Orlow I, Lacombe L, Hannon GJ, et al (1995) Deletion of the *p16* and *p15* genes in human bladder tumors. J Natl Cancer Inst 87:1524–1529

Osann KE, Anton-Culver H, Kurosaki T, Taylor T (1993) Sex differences in lung-cancer risk associated with cigarette smoking. Int J Cancer 54:44–48

Ozaki T, Ikeda S, Kawai A, et al (1993) Alterations of retinoblastoma susceptible gene accompanied by *c-myc* amplification in human bone and soft tissue tumors. Cell Mol Biol 39:235–242

Padberg B, Schroder S, Capella C, et al (1995) Multiple endocrine neoplasia type 1 (MEN 1) revisited [review]. Virch Arch 426:541–548

Papadimitriou C, Day N, Tzonou A, et al (1984) Biosocial correlates of colorectal cancer in Greece. Int J Epidemiol 13:155–159

Parker SL, Tong T, Bolden S, Wingo PA (1997) Cancer statistics, 1997. CA Cancer J Clin 47(1):5–27

Parsons R, Li GM, Longley M, et al (1995) Mismatch repair deficiency in phenotypically normal human cells. Science 268:738–740

Patel M, Tang BK, Grant DM, Kalow W (1995) Interindividual variability in the glucuronidation of (S)-oxazepam contrasted with that of (R)-oxazepam. Pharmacogenetics 5:287–297

Pathak DR, Samet JM, Humble CG, Skipper BJ (1986) Determinants of lung cancer risk in cigarette smokers in New Mexico. J Natl Cancer Inst 76:597–604

Pei L, Melmed S, Scheithauer B, et al (1995) Frequent loss of heterozygosity at the retinoblastoma susceptibility gene (RB) locus in aggressive pituitary tumors: evidence for a chromosome 13 tumor suppressor gene other than RB. Cancer Res 55:1613–1616

Pemble S, Schroeder KR, Spencer SR, et al (1994) Human glutathione S-transferase theta (GSTT1): cDNA cloning and the characterization of a genetic polymorphism. Biochem J 300:271–276

Perera FP (1996) Uncovering new clues to cancer risk. Sci Am 274(5):54–62

Peters RK, Pike MC, Chang WW, Mack TM (1990) Reproductive factors and colon cancers. Br J Cancer 61:741–748

Peterson RD, Funkhouser JD, Tuck-Muller CM, Gatti RA (1992) Cancer susceptibility in ataxia-telangiectasia [review]. Leukemia 1:8–13

Peto R (1986) Influence of dose and duration of smoking on lung cancer rates. IARC Sci Publ 74:23–33

Peto R, Roe FJ, Lee PN, et al (1975) Cancer and ageing in mice and men. Br J Cancer 32:411–426

Phillips SM, Barton CM, Lee SJ, et al (1994) Loss of the retinoblastoma susceptibility gene (RB1) is a frequent and early event in prostatic tumorigenesis. Br J Cancer 70:1252–1257

Pickett CB, Williams JB, Lu AY, Cameron RG (1984) Regulation of glutathione transferase and DT-diaphorase mRNAs in persistent hepatocyte nodules during chemical hepatocarcinogenesis. Proc Natl Acad Sci U.S.A. 81:5091–5095

Pienta KJ, Esper PS (1993) Risk factors for prostate cancer. Ann Intern Med 118:793–803

Pierceall WE, Goldberg LH, Tainsky MA, et al (1991a) Ras gene mutation and amplification in human nonmelanoma skin cancers. Mol Carcinogen 4:196–202

Pierceall WE, Mukhopadhyay T, Goldberg LH, Ananthaswamy HN (1991b) Mutations in the p53 tumor suppressor gene in human cutaneous squamous cell carcinomas. Mol Carcinogen 4:445–449

Pike MC, Spicer DV, Dahmoush L, Press MF (1993) Estrogens, progestogens, normal breast cell proliferation, and breast cancer risk [review]. Epidemiol Rev 15:17–35

Plesko I, Preston-Martin S, Day NE, et al (1985) Parity and cancer risk in Slovakia. Int J Cancer 36:529–533

Poremba C, Yandell DW, Huang Q, et al (1995) Frequency and spectrum of p53 mutations in gastric cancer — a molecular genetic and immunohistochemical study. Virch Arch 426:447–455

Potter JD, McMichael AJ (1983) Large bowel cancer in women in relation to reproductive and hormonal factors: a case-control study. J Natl Cancer Inst 71:703–709

Pourzand C, Cerutti P (1993) Mutagenesis of H-ras codons 11 and 12 in human fibroblasts by N-ethyl-N-nitrosourea. Carcinogenesis 14:2193–2196

Price FM, Parshad R, Tarone RE, Sanford KK (1991) Radiation-induced chromatid aberrations in Cockayne syndrome and xeroderma pigmentosum group C fibroblasts in relation to cancer predisposition. Cancer Genet Cytogenet 57:1–10

Price RA, Spielman RS, Lucena AL, et al (1989) Genetic polymorphism for human platelet thermostable phenol sulfotransferase (TS PST) activity. Genetics 122:905–914

Radman M, Matic I, Halliday JA, Taddei F (1995) Editing DNA replication and recombination by mismatch repair: from bacterial genetics to mechanisms of predisposition to cancer in humans [review]. Philos Trans R Soc London B Biol Sci 347:97–103

Rannug A, Alexandrie AK, Persson I, Ingelman-Sundberg M (1995) Genetic polymorphism of cytochromes P450 1A1, 2D6 and 2E1: regulation and toxicological significance [review]. J Occup Environ Med 37:25–36

Raunio H, Husgafvel-Pursiainen K, Anttila S, et al (1995) Diagnosis of polymorphisms in carcinogen-activating and inactivating enzymes and cancer susceptibility — a review [review]. Gene 159:113–121

Recio L, Goldsworthy TL (1995) The use of transgenic mice for studying mutagenicity induced by 1,3-butadiene. Toxicol Lett 83:607–612

Reichardt JK, Makridakis N, Henderson BE, et al (1995) Genetic variability of the human *SRD5A2* gene: implications for prostate cancer risk. Cancer Res 55:3973–3975

Reiss M, Brash DE, Munoz-Antonia T, et al (1992) Status of the *p53* tumor suppressor gene in human squamous carcinoma cell lines. Oncol Res 4:349–357

Rich KJ, Murray BP, Lewis I, et al (1992) *N*-hydroxy-MeIQx is the major microsomal oxidation product of the dietary carcinogen MeIQx with human liver. Carcinogenesis 13:2221–2226

Richie JP Jr (1992) The role of glutathione in aging and cancer [review]. Exp Gerontol 27:615–626

Risch HA, Howe GR, Jain M, et al (1993) Are female smokers at higher risk for lung cancer than male smokers? A case-control analysis by histologic type. Am J Epidemiol 138:281–293

Robbins JH, Kraemer KH, Lutzner MA, et al (1974) Xeroderma pigmentosum. An inherited diseases with sun sensitivity, multiple cutaneous neoplasms, and abnormal DNA repair. Ann Intern Med 80:221–248

Rodenhiser D, Chakraborty P, Andrews J, et al (1996) Heterogeneous point mutations in the *BRCA1* breast cancer susceptibility gene occur in high frequency at the site of homonucleotide tracts, short repeats and methylatable cpg/cpnpg motifs. Oncogene 12:2623–2629

Ross RK, Henderson BE (1994) Do diet and androgens alter prostate cancer risk via a common etiologic pathway? [editorial]. J Natl Cancer Inst 86:252–254

Ross RK, Yuan JM, Yu MC, et al (1992) Urinary aflatoxin biomarkers and risk of hepatocellular carcinoma. Lancet 339:943–946

Rosselli F, Ridet A, Soussi T, et al (1995) *p53*-dependent pathway of radio-induced apoptosis is altered in Fanconi anemia. Oncogene 10:9–17

Rost KL, Brosicke H, Brockmoller J, et al (1992) Increase of cytochrome P450IA2 activity by omeprazole: evidence by the ^{13}C-[*N*-3-methyl]-caffeine breath test in poor and extensive metabolizers of S-mephenytoin. Clin Pharmacol Ther 52:170–180

Rothman K, Keller A (1972) The effect of joint exposure to alcohol and tobacco on risk of cancer of the mouth and pharynx. J Chronic Dis 25:711–716

Ruizeveld de Winter JA, Janssen PJ, Sleddens HM, et al (1994) Androgen receptor status in localized and locally progressive hormone refractory human prostate cancer. Am J Pathol 144:735–746

Ruppert S, Kelsey G, Schedl A, et al (1992) Deficiency of an enzyme of tyrosine metabolism underlies altered gene expression in newborn liver of lethal albino mice. Genes Dev 6:1430–1443

Russo J, Gusterson BA, Rogers AE, et al (1990) Comparative study of human and rat mammary tumorigenesis [review]. Lab Invest 62:244–278

Ryberg D, Kure E, Lystad S, et al (1994) *p53* mutations in lung tumors: relationship to putative susceptibility markers for cancer. Cancer Res 54:1551–1555

Sakaguchi K, Harris PV, Ryan C, et al (1991) Alteration of a nuclease in Fanconi anemia. Mutat Res 255:31–38

Sakai E, Tsuchida N (1992) Most human squamous cell carcinomas in the oral cavity contain mutated *p53* tumor-suppressor genes. Oncogene 7:927–933

Sandelin K, Larsson C, Decker RA (1994) Genetic aspects of multiple endocrine neoplasia types 1 and 2 [review]. Curr Opin Gen Surg 1994: 60–68

Saracci R (1977) Asbestos and lung cancer: an analysis of the epidemiological evidence on the asbestos-smoking interaction. Int J Cancer 20:323–331

Sass B, Rabstein LS, Madison R, et al (1975) Incidence of spontaneous neoplasms in F344 rats throughout the natural life-span. J Natl Cancer Inst 54:1449–1456

Sato K, Satoh K, Tsuchida S, et al (1992) Specific expression of glutathione S-transferase Pi forms in (pre)neoplastic tissues: their properties and functions [review]. Tohoku J Exp Med 168:97–103

Savitsky K, Sfez S, Tagle DA, et al (1995) The complete sequence of the coding region of the *ATM* gene reveals similarity to cell cycle regulators in different species. Hum Mol Genet 4:2025–2032

Scarpa A, Capelli P, Mukai K, et al (1993) Pancreatic adenocarcinomas frequently show *p53* gene mutations. Am J Pathol 142:1534–1543

Schultz JC, Shahidi NT (1993) Tumor necrosis factor-alpha overproduction in Fanconi's anemia. Am J Hematol 42:196–201

Schwartz CE, Haber DA, Stanton VP, et al (1991) Familial predisposition to Wilms tumor does not segregate with the *WT1* gene. Genomics 10:927–930

Seidegard J, Pero RW, Miller DG, Beattie EJ (1986) A glutathione transferase in human leukocytes as a marker for the susceptibility to lung cancer. Carcinogenesis 7:751–753

Selikoff IJ, Hammond EC (1975) Multiple risk factors in environmental cancer. In Fraumeni JF (ed), Persons at high risk of cancer: an approach to cancer etiology and control. Academic Press, New York, pp 467–483

Sellers TA, Bailey-Wilson JE, Potter JD, et al (1992) Effect of cohort differences in smoking prevalence on models of lung cancer susceptibility. Genet Epidemiol 9:261–271

Sellers TA, Ooi WL, Elston RC, et al (1987) Increased familial risk for non-lung cancer among relatives of lung cancer patients. Am J Epidemiol 126:237–246

Setlow RB, Regan JD, German J, Carrier WL (1969) Evidence that xeroderma pigmentosum cells do not perform the first step in the repair of ultraviolet damage to their DNA. Proc Natl Acad Sci U.S.A. 64:1035–1041

Shaw GL, Falk RT, Pickle LW, et al (1991) Lung cancer risk associated with cancer in relatives. J Clin Epidemiol 44:429–437

Shiloh Y (1995) Ataxia-telangiectasia: closer to unraveling the mystery [review]. Eur J Hum Genet 3:116–138

Shimizu H, Ross RK, Bernstein L, et al (1991) Cancers of the prostate and breast among Japanese and white immigrants in Los Angeles County. Br J Cancer 63:963–966

Sicinski P, Donaher JL, Parker SB, et al (1995) Cyclin D1 provides a link between development and onco-genesis in the retina and breast. Cell 82:621–630

Sinha R, Rothman N, Brown ED, et al (1994) Pan-fried meat containing high levels of heterocyclic aromatic amines but low levels of polycyclic aromatic hydrocarbons induces cytochrome P4501A2 activity in humans. Cancer Res 54:6154–6159

Smith AJ, Stern HS, Penner M, et al (1994) Somatic *APC* and *K-ras* codon 12 mutations in aberrant crypt foci from human colons. Cancer Res 54:5527–5530

Smith ML, Fornace AJ Jr (1995) Genomic instability and the role of *p53* mutations in cancer cells [review]. Curr Opin Oncol 7:69–75

Smith SA, Ponder BA (1993) Predisposing genes in breast and ovarian cancer: an overview [review]. Tum-origenesis 79:291–296

Sobue T, Yamaguchi N, Suzuki T, et al (1993) Lung cancer incidence rate for male ex-smokers according to age at cessation of smoking. Jpn J Cancer Res 84:601–607

Sondik EJ (1994) Breast cancer trends: incidence, mortality, and survival. Cancer 74:995–999

Staszewski J, Haenszel W (1965) Cancer mortality among the Polish-born in the United States. J Natl Cancer Inst 35:291–297

Steinmetz KA, Potter JD (1991) Vegetables, fruit, and cancer. I. Epidemiology. Cancer Causes Control 2:325–357

Stern SJ, Degawa M, Martin MV, et al (1993) Metabolic activation, DNA adducts, and H-ras mutations in human neoplastic and non-neoplastic laryngeal tissue. J Cell Biochem Suppl 17F:129–137

Sugimura H, Caporaso NE, Modali RV, et al (1990a) Association of rare alleles of the Harvey ras protoonco-gene locus with lung cancer. Cancer Res 50:1857–1862

Sugimura H, Caporaso NE, Shaw GL, et al (1990b) Human debrisoquine hydroxylase gene polymorphisms in cancer patients and controls. Carcinogenesis 11:1527–1530

Sullivan NF, Willis AE (1992) Cancer predisposition in Bloom's syndrome [review]. Bioessays 14:333–336

Sundaresan V, Ganly P, Hasleton P, et al (1992) *p53* and chromosome 3 abnormalities, characteristic of malignant lung tumours, are detectable in preinvasive lesions of the bronchus. Oncogene 7:1989–1997

Suzuki H, Sato N, Watabe Y, et al (1993) Androgen receptor gene mutations in human prostate cancer. J Steroid Biochem Mol Biol 46:759–765

Suzuki Y, Tamura G (1993) Mutations of the *p53* gene in carcinomas of the urinary system. Acta Pathol Jpn 43:745–750

Swift M, Heim RA, Lench NJ (1993) Genetic aspects of ataxia telangiectasia [review]. Adv Neurol 61:115–125

Swift M, Reitnauer PJ, Morrell D, Chase CL (1987) Breast and other cancers in families with ataxia-telangiectasia. N Engl J Med 316:1289–1294

Tadjoedin MK, Fraser EC (1965) Heredity of ataxia telangiectasia (Louis-Barr syndrome). Am J Dis Child 110:64–68

Taioli E, Wynder EL (1994) Re: Endocrine factors and adenocarcinoma of the lung in women [letter]. J Natl Cancer Inst 86:869–870

Taioli E, Crofts F, Trachman J, et al (1995) A specific African-American *CYP1A1* polymorphism is associated with adenocarcinoma of the lung. Cancer Res 55:472–473

Takagi S, Naito E, Yamanouchi H, et al (1994) Mutation of the *p53* gene in gallbladder cancer. Tohoku J Exp Med 172:283–289

Takebe H, Miki Y, Kozuka T, et al (1977) DNA repair characteristics and skin cancers of xeroderma pigmentosum patients in Japan. Cancer Res 37:490–495

Takeuchi T, Morimoto K (1993) Increased formation of 8-hydroxydeoxyguanosine, an oxidative DNA damage, in lymphoblasts from Fanconi's anemia patients due to possible catalase deficiency. Carcinogenesis 14:1115–1120

Tamai S, Sugimura H, Caporaso NE, et al (1990) Restriction fragment length polymorphism analysis of the L-myc gene locus in a case-control study of lung cancer. Int J Cancer 46:411–415

Tamura G, Maesawa C, Suzuki Y, et al (1992) *p53* gene mutations in esophageal cancer detected by polymerase chain reaction single-strand conformation polymorphism analysis. Jpn J Cancer Res 83:559–562

Taubes G (1993) Claim of higher risk for women smokers attacked [news]. Science 262:1375

Taylor AM (1992) Ataxia telangiectasia genes and predisposition to leukaemia, lymphoma and breast cancer [editorial] [review]. Br J Cancer 66:5–9

Taylor JA, Bell DA, Nagorney D (1993) *L-myc* proto-oncogene alleles and susceptibility to hepatocellular carcinoma. Int J Cancer 54:927–930

Thelu MA, Zarski JP, Froissart B, et al (1993) *c-Ha-ras* polymorphism in patients with hepatocellular carcinoma. Comparison of healthy subjects and alcoholic patients with cirrhosis. Gastroenterol Clin Biol 17:903–907

Thomas G, Olschwang S (1995) Genetic predispositions to colorectal cancer [review]. Pathol Biol 43:159–164

Thorling EB (1996) Obesity, fat intake, energy balance, exercise and cancer risk, a review [review]. Nutr Res 16:315–368

Tokuhata GK (1976) Cancer of the lung: host and environmental interaction. In Lynch H (ed), Cancer genetics. Thomas, Springfield, IL, pp 213–232

Tokuhata GK, Lilienfeld AM (1963) Familial aggregation of lung cancer in humans. J Natl Cancer Inst 30:289–312

Torday JS, Smith BT, Giroud CJ (1975) The rabbit fetal lung as a glucocorticoid target tissue. Endocrinology 96:1462–1467

Tsuda H, Lee G, Farber E (1980) Induction of resistant hepatocytes as a new principle for a possible short-term *in vivo* test for carcinogens. Cancer Res 40:1157–1164

Tulinius H (1991) Latent malignancies at autopsy: a little used source of information on cancer biology [review]. IARC Sci Publ 112:253–261

Tuyns AJ, Pequignot G, Jensen OM (1977) [Esophageal cancer in Ille-et-Vilaine in relation to levels of alcohol and tobacco consumption. Risks are multiplying] [in French]. Bull Cancer 64:45–60

Uchida K, Nomura Y, Kadowaki M, et al (1970) Effects of estradiol, dietary cholesterol and 1-thyroxine on biliary bile acid composition and secretory rate, and on plasma, liver and bile cholesterol levels in rats. Endocrinol Jpn 17:107–121

Van Laethem JL, Vertongen P, Deviere J, et al (1995) Detection of *c-Ki-ras* gene codon 12 mutations from pancreatic duct brushings in the diagnosis of pancreatic tumours. Gut 36:781–787

Velentgas P, Daling JR (1994) Risk factors for breast cancer in younger women [review]. Monogr Natl Cancer Inst 16:15–24

Vicellio P (1993) Handbook of medical toxicology. Little Brown, Boston, pp 36–37

Vineis P, Caporaso N, Tannenbaum SR, et al (1990) Acetylation phenotype, carcinogen-hemoglobin adducts, and cigarette smoking. Cancer Res 50:3002–3004

Vuillaume M, Daya-Grosjean L, Vincens P, et al (1992) Striking differences in cellular catalase activity between two DNA repair-deficient diseases: xeroderma pigmentosum and trichothiodystrophy. Carcinogenesis 13:321–328

Wagner J, Portwine C, Rabin K, et al (1994) High frequency of germline *p53* mutations in childhood adrenocortical cancer. J Natl Cancer Inst 86:1707–1710

Wang FL, Love EJ, Liu N, Dai XD (1994) Childhood and adolescent passive smoking and the risk of female lung cancer. Int J Epidemiol 23:223–230

Waterhouse J, Muir C, Correa P (1976) Cancer incidence in five continents. IARC, Lyons, France

Watson JD, Gilman M, Witkowski J, Zoller M, eds (1992). Recombinant DNA. Scientific American Books. Freeman, New York

Wei Q, Matanoski GM, Farmer ER, et al (1993) DNA repair and aging in basal cell carcinoma: a molecular epidemiology study. Proc Natl Acad Sci U.S.A. 90:1614–1618

Weide R, Dowding C, Sucai B, et al (1991) Inactivation of the retinoblastoma susceptibility gene in a human high grade non-Hodgkin's lymphoma cell line. Br J Haematol 78:500–505

Weinberg RA (1995) The retinoblastoma protein and cell cycle control [review]. Cell 81:323–330

Weiss NS, Daling JR, Chow WH (1981) Incidence of cancer of the large bowel in women in relation to reproductive and hormonal factors. J Natl Cancer Inst 67:57–60

Weston A, Vineis P, Caporaso NE, et al (1991) Racial variation in the distribution of *Ha-ras-1* alleles. Mol Carcinogen 4:265–268

Whittemore AS, Keller JB, Betensky R (1991) Low-grade, latent prostate cancer volume: predictor of clinical cancer incidence? J Natl Cancer Inst 83:1231–1235

Willett WC, Hunter DJ, Stampfer MJ, et al (1992) Dietary fat and fiber in relation to risk of breast cancer. An 8-year follow-up. JAMA 268:2037–2044

Willett WC, Stampfer MJ, Colditz GA, et al (1990) Relation of meat, fat, and fiber intake to the risk of colon cancer in a prospective study among women. N Engl J Med 323:1664–1672

Wooster R, Neuhausen SL, Mangion J, et al (1994) Localization of a breast cancer susceptibility gene, *BRCA2*, to chromosome 13q12–13. Science 265:2088–2090

Wu AH, Paganini-Hill A, Ross RK, Henderson BE (1987) Alcohol, physical activity and other risk factors for colorectal cancer: a prospective study. Br J Cancer 55:687–694

Wu X, Hsu TC, Anneegers JF, et al (1995) A case-control study of nonrandom distribution of bleomycin-induced chromatid breaks in lymphocytes of lung cancer cases. Cancer Res 55:557–561

Wynder EL, Stellman SD (1979) Impact of long-term filter cigarette usage on lung and larynx cancer risk: a case-control study. J Natl Cancer Inst 62:471–477

Yamazaki H, Guo Z, Guengerich FP (1995a) Selectivity of cytochrome P4502E1 in chlorzoxazone 6-hydroxylation. Drug Metab Dispos 23:438–440

Yamazaki H, Inui Y, Wrighton SA, et al (1995b) Procarcinogen activation by cytochrome P450 3A4 and 3A5 expressed in *Escherichia coli* and by human liver microsomes. Carcinogenesis 16:2167–2170

Yamazaki H, Guo Z, Persmark M, et al (1994a) Bufuralol hydroxylation by cytochrome P450 2D6 and 1A2 enzymes in human liver microsomes. Mol Pharmacol 46:568–577

Yamazaki H, Mimura M, Oda Y, et al (1994b) Activation of *trans*-1,2-dihydro-1,2-dihydroxy-6-aminochrysene to genotoxic metabolites by rat and human cytochromes P450. Carcinogenesis 15:465–470

Yamazaki H, Mimura M, Oda Y, et al (1993) Roles of different forms of cytochrome P450 in the activation of the promutagen 6-aminochrysene to genotoxic metabolites in human liver microsomes. Carcinogenesis 14:1271–1278

Yamazaki H, Inui Y, Yun CH, et al (1992a) Cytochrome P450 2E1 and 2A6 enzymes as major catalysts for metabolic activation of N-nitrosodialkylamines and tobacco-related nitrosamines in human liver microsomes. Carcinogenesis 13:1789–1794

Yamazaki H, Oda Y, Funae Y, et al (1992b) Participation of rat liver cytochrome P450 2E1 in the activation of N-nitrosodimethylamine and N-nitrosodiethylamine to products genotoxic in an acetyltransferase-overexpressing *Salmonella typhimurium* strain (NM2009). Carcinogenesis 13:979–985

Yamazaki Y, Ogawa Y, Afify AS, et al (1995) Difference between cancer cells and the corresponding normal tissue in view of stereoselective hydrolysis of synthetic esters. Biochim Biophys Acta 1243:300–308

Yarosh DB, Kripke ML (1996) DNA repair and cytokines in antimutagenesis and anticarcinogenesis. Mutat Res Fundam Mol Mech Mutagen 350:255–260

Yatani R, Kusano I, Shiraishi T, et al (1989) Latent prostatic carcinoma: pathological and epidemiological aspects [review]. Jpn J Clin Oncol 19:319–326

Yerokun T, Etheredge JL, Norton TR, et al (1992) Characterization of a complementary DNA for rat liver aryl sulfotransferase IV and use in evaluating the hepatic gene transcript levels of rats at various stages of 2-acetylaminofluorene-induced hepatocarcinogenesis. Cancer Res 52:4779–4786

Yoshimoto K, Iwahana H, Fukuda A, et al (1992) Role of *p53* mutations in endocrine tumorigenesis: mutation detection by polymerase chain reaction-single strand conformation polymorphism. Cancer Res 52:5061–5064

Zang EA, Wynder EL (1996) Difference in lung cancer risk between men and women: examination of the evidence. J Natl Cancer Inst 88:183–192

Zaridze DG, Filipchenko VV (1990) Incidence of colo-rectal cancer in Moscow. Int J Cancer 45:583–585

Zhang J, Ding F, Wang X (1995) [Expression and loss of heterozygosity of *DCC* gene in human lung cancer] [in Chinese]. Chin Med J 75:211–213, 254–255

Zhang LF, Hemminki K, Szyfter K, et al (1994) *p53* mutations in larynx cancer [review]. Carcinogenesis 15:2949–2951

Ziegler A, Jonason AS, Leffell DJ, et al (1994) Sunburn and p53 in the onset of skin cancer. Nature 372:773–776

Ziegler RG, Hoover RN, Pike MC, et al (1993) Migration patterns and breast cancer risk in Asian-American women. J Natl Cancer Inst 85:1819–1827

Zimniak P, Nanduri B, Pikula S, et al (1994) Naturally occurring human glutathione S-transferase *GSTP1-1* isoforms with isoleucine and valine in position 104 differ in enzymatic properties. Eur J Biochem 224:893–899

Zou M, Shi Y, Farid NR (1993) *p53* mutations in all stages of thyroid carcinomas. J Clin Endocrinol Metab 77:1054–1058

7 Genetic Susceptibility and Cancer Risk

Neil Caporaso and Nathaniel Rothman

CONTENTS

CHEMICAL ENVIRONMENT AND THE NATURE OF GENETIC SUSCEPTIBILITY

Humans are exposed to a wide variety of environmental carcinogens in life, yet some individuals develop cancer and others do not. Heritable susceptibility factors for cancer based on interindividual variation in xenobiotic metabolizing enzymes have been postulated to explain some of this difference. Hereditary factors may cause human cancer through single genes that largely determine disease [i.e., retinoblastoma, Li–Fraumeni syndrome, and hereditary polyposis coli (Evans 1993)], or, as is the case with the more common complex diseases, act to alter the probability of a condition given a specific exposure to a medication, dietary agent, toxicant, or other xenobiotic. This latter category of susceptibility genes includes glucose-6-phosphate dehydrogenase (G6PD) deficiency [hemolytic anemia after exposure to primaquine (medication) or fava beans (diet)], α_1-antitrypsin deficiency (accelerated emphysema in tobacco smokers), and hepatitis in slow acetylators after administration of isoniazid. Other hereditary factors are implicated in cancer etiology because they

0-8493-2805-5/99/$0.00+$.50
© 1999 by CRC Press LLC

encode enzymes that metabolize (activate or eliminate) procarcinogenic substrates (Caporaso et al. 1991).

ACTIVATION/DEACTIVATION

A basic tenet of chemical carcinogenesis is that carcinogens must be activated before they can initiate carcinogenesis. This process generally requires enzyme activity. A promutagen or procarcinogen may react with a hydroxyl, methyl, or other functional group and then bind to DNA, subsequently altering the genetic structure or damaging other biologically important entities. Many of the enzymes that control activation or inactivation/elimination of these compounds are polymorphic, i.e., individuals inherit different forms that function well, poorly, or not at all. The metabolism of procarcinogens is classified into two general categories that loosely correspond with the activation/detoxification mechanisms.

Phase 1 enzymes (i.e., cytochrome P-450 isozymes such as *CYP2D6* and *CYP1A1*) oxidize chemicals to active electrophilic intermediates that may bind to biologically important macromolecules. Phase 2 enzymes [i.e., *N*-acetyltransferases and glutathione *S*-transferases (GST)] attach a bulky moiety to the substrate and increase solubility and thereby enhance elimination (Nelson 1982, Daly et al. 1993b) in the urine or bile. Both Phase 1 (e.g., *CYP1A1*, *CYP2D6*, and *CYP2E1*) and Phase 2 [e.g., acetylation phenotype (*NAT2*) and GST mu null phenotype (*GSTM1*)] enzymes have been studied in relationship to risk of tobacco-associated cancers.

NATURE OF GENETIC SUSCEPTIBILITY

Hereditary risk factors for cancer can be loosely classified in two groups. Traditional single genes such as retinoblastoma (*Rb*) cause cancers in a largely independent manner. Though rare in the population, these genes exhibit high penetrance and familial aggregation. They account for only a small amount of the cancer in the population.

The other group, defined here as susceptibility genes, are common (i.e., the minor allele frequency is >0.01), are typically associated with low-to-modest relative risks (under 10), and do not exhibit familial patterns of inheritance. These genes exert a cancer-promoting effect in concert with exposure to carcinogens (Caporaso and Goldstein 1995). Because these genes are common and act in conjunction with carcinogenic exposure, they are more relevant to a discussion of risk in the population. Some contrasts between single and susceptibility genes are offered in Table 1. Four examples of genetic susceptibility and exposure susceptibility relationships are presented to illustrate these points.

CYP1A2, HETEROCYCLIC AROMATIC AMINES (HAAs), AND WELL-COOKED MEAT

Cooking meat at high temperatures can result in the formation of carcinogenic compounds such as HAAs (Adamson 1990). The cancer risk posed to humans by exposure to HAAs in the diet may depend on the extent to which the compounds are activated *in vivo* (Kato 1986, Snyderwine and Battula 1989, Degawa et al. 1989). Two processes involved in the activation of these compounds are N-oxidation by liver cytochrome *P4501A2* and N-acetylation (Flammang and Kadlubar 1986). Interindividual variability in acetyltransferase activity segregates into slow and rapid *NAT2* phenotypes. Interindividual variability in *P4501A2* activity is known to be in part due to induction by environmental exposure, such as cigarette smoking, and by dietary exposure to polycyclic aromatic hydrocarbons (PAHs), although there may be a genetic component as well (Kadlubar et al. 1992). Both of these enzymes can be evaluated by measuring the excretion of caffeine metabolites in urine after consumption of caffeine (Butler et al. 1992), which can distinguish between slow and rapid acetylators and N-oxidizers. On the basis of metabolic studies in animals, it is expected that subjects

TABLE 1
Characteristics of Genes in Relationship to Cancer Etiology

Characteristic	Classic (single gene) ("causes" cancer)	Susceptibility gene (alters risk)
Genetic		
Genes and associated cancer	*MEN1:* pancreas, pituitary	*CYP1A1:* lung (Kawajiri et al. 1990)
	BRCA1: breast	*CYP2D6:* lung (Ayesh et al. 1984)
	Rb: retinoblastoma	*GSTM1:* bladder (Bell et al. 1993)
	(see Evans 1993)	
Gene frequency	Generally <1%	Often >1%
Disease prevalence (specific cancer)	Rare, but may be subset of common cancer	Often common
Familiality	Typical	No
Starting point for investigation	Cancer family	Mechanistic hypothesis and candidate gene
Epidemiologic		
Strength of association, i.e., relative risk (odds ratio)	Very high, relative risks often >100	Modest, relative risks <10
Absolute risk	High	Low to moderate
Population attributable risk	Low	Potentially high
Gene-environment interaction	Variable	Primary and implicit
Role of environmental "exposure"	Variable	Critical
Clinical/laboratory		
Convincing demonstration in human cancer	Yes	Increasing evidence
Role of genetic counseling	Established for many disorders, but less so for cancer	Investigational
Clinical strategies	Gene therapy	High-risk screening in populations (theoretical)
	Genetic counseling within high-risk families	Avoidance of relevant exposure
	Avoid relevant exposures	
Methodologic		
Typical study design	Genetic: linkage (positional cloning)	Epidemiologic: i.e., case-control or cohort with biologic component (molecular epidemiology)
Anticipated methodologic advances relevant to study	Improved DNA methodology, more numerous and informative markers	Genotyping supplanting phenotyping
	Smaller requirements for DNA	Improved methods to collect DNA, e.g., buccal cells
	Improved computational (hardware and software) approaches: multipoint linkage, incorporation of covariate information, nonparametric linkage approaches	Larger studies of improved design
		Better treatment of effect modification (interaction)
Application in cancer screening (investigational)	In cancer families to identify at-risk family members	In groups at high-risk due to relevant exposure

who both rapidly acetylate and have rapid *P4501A2* activity may be at higher risk of having colon neoplasms, particularly given exposure to HAAs. Lang et al. (1994) evaluated the risk of colon neoplasia as a result of HAA intake (using preference for meat "doneness" as a surrogate) in a case-control study of colon polyps and cancer. The overall risk from eating well-done meat was 2.08 (adjusted for age and metabolic phenotypes). Compared with subjects with slow *NAT2* and slow *P4501A2* activity who consumed meat cooked rare or medium, subjects with slow/slow phenotypes who ate well-done meat had a 2.06-fold increased risk; subjects with rapid/rapid phenotypes who ate well-done meat had a 6.45-fold increased risk (tests for interaction were not significant, however).

BLADDER CANCER, NAT2, AND OCCUPATIONAL EXPOSURE TO AROMATIC AMINES

Many epidemiologic studies over the past decade have examined the hypothesis that the slow acetylator phenotype (present in approximately 50% of individuals in Western countries), encoded by the NAT2 gene, confers excess risk of bladder cancer in workers with occupational exposure to carcinogenic aromatic amines. Although mixed findings have been reported in bladder cancer studies in the general population (Hein 1988), more convincing associations have been observed in studies of people with occupational exposure to arylamines. This is consistent with biochemical and animal data indicating that slow acetylators are less able to detoxify this class of carcinogenic substrate. Although the acetylation reaction is generally thought to enhance detoxification of arylamines, the role of NAT2 may not always be deactivating. The slow acetylator trait may not confer increased risk for specific exposures, i.e., bladder cancer in populations exposed to benzidine (Hayes et al. 1993). This finding was supported by observing that the predominant adduct in exfoliated urothelial cells of exposed workers was N-acetylated (Rothman et al. 1996).

Cigarette smokers have a 2- to 3-fold excess risk of bladder cancer. A dose–response (risk related to both intensity and duration of smoking) correlation with per capita cigarette sales and a decrease in risk with cessation of smoking have been demonstrated. People who smoke black tobacco (i.e., air-cured tobacco with a high concentration of aromatic amines) exhibit excess risk of bladder cancer compared with people who smoke blond tobacco (i.e., flue-cured tobacco, with a lower concentration of aromatic amines), a difference that has been related to the higher aromatic amine content of black tobacco. In a cross-sectional study, 4-aminobiphenyl hemoglobin adducts have been shown to be related both to type of tobacco smoked and to the acetylator phenotype, a finding that supports the idea that the genetic trait influences the effective dose of a carcinogen (Vineis et al. 1990).

OTHER POLYMORPHISMS AND CANCER: DNA-REPAIR

The ability to repair damaged DNA has also been evaluated as a potential modifier of risk for smoking-associated tumors. For example, assays have been developed to detect the *in vitro* sensitivity of cultured peripheral blood lymphocytes to mutagen-induced chromosomal aberrations, which provides an indirect assessment of DNA-repair capability. In particular, Hsu et al. (1989) developed an assay that quantifies bleomycin-induced chromatid breakage by using cultured lymphocytes from study subjects. Recently, Wu et al. (1995) used this assay in a case-control study of lung cancer. There was an overall risk of lung cancer for ever-smoking individuals of 8.8 (unadjusted), which was strengthened among the subgroup of people who were susceptible to bleomycin-induced chromosomal damage and weakened among subjects who were not susceptible. For example, the mean risk of lung cancer among smokers who did not form breaks on chromosome 2 was 4.5 (95% confidence interval (CI), 1.6 to 12.4) (compared with nonsmokers who did not form breaks). Risk was 12.8 (95% CI, 4.3 to 38.7) among smokers who did form breaks. Further, a multiplicative interaction was suggested (but was not statistically significant) between smoking and susceptibility for breaks on chromosome 4; lung cancer risks (adjusted for age, sex, ethnicity) were 8.1 among smokers who did not have breaks on chromosome 4, 3.9 among nonsmokers who had breaks, and 128.8 for subjects who both smoked and were susceptible for forming breaks (Wu et al. 1995). This assay also has been used to study susceptibility to malignancies of the upper aerodigestive tract (Spitz et al. 1993) and for second malignancies of the head and neck (Spitz et al. 1994).

SKIN CANCER AND DNA-REPAIR

Although the relationship between sun exposure and risk of skin cancer is well-established, it has afforded the opportunity to test potential genetic susceptibility markers. For example, Athas et al. (1991) developed an assay that evaluates the capability of subjects to repair ultraviolet-damaged DNA by using cultured lymphocytes. Wei et al. (1993) applied this assay to a case-control study of basal cell carcinoma. There was a 2.8-fold increased risk (unadjusted) for basal cell carcinoma given a history of ≥6 severe sunburns. This partitioned into a nonsignificant 1.9-fold increased risk among subjects with ≥6 severe sunburns with high repair capability (compared with subjects with a smaller number of severe sunburns and high repair capacity) and a statistically significant 5.3-fold increased risk (95% CI, 2.04 to 12.9, adjusted for age and sex) among subjects with ≥6 severe sunburns with low repair capacity.

GST

Cytosolic GST enzymes are important detoxifying species that catalyze the reaction of reduced glutathione (GSH) with a variety of electrophilic compounds (Board 1981, Ketterer 1988). Inert chemicals may undergo biotransformation in the body to highly reactive chemicals that may damage DNA and cause cell death or transformation. The formation of GSH conjugates, catalyzed by these enzymes, facilitates the formation of nontoxic excretory products, and it is thought that these enzyme systems are important in preventing certain mutagenic electrophiles from becoming carcinogenic. GSH is a free radical trap, because it donates a hydrogen atom to free radicals (hydroxy or carbon) and thereby quenches the free radical OH·. GST also binds nonsubstrate organic anions with uncertain consequences. Although it is a valid generalization that for most substrates Phase II enzymes are detoxifying, under certain circumstances GST may make a compound more toxic (see Hayes et al. 1993). There are four classes of dimeric GSTs based on isoelectric point: α (basic), μ (neutral), π (acidic), and θ. All are located on different chromosomes (Mannervik et al. 1992). The enzymes are ubiquitous, with the greatest activity found in liver, kidney, intestine, testis, and adrenal gland (Strange et al. 1984). The *GSTM1* isozyme detoxifies *trans*-stilbene oxide (alkene epoxides), a compound that can be assayed in red blood cells to assign the phenotype. Compounds relevant to carcinogenesis with known isoenzyme specificity for *GSTM1* include aflatoxin-8,9,-oxide, benzo[a]pyrene and other PAHs, 1-nitropyrene, and other endogenous electrophiles — products of lipid peroxidation (alkene epoxides). van Poppel et al. (1992) have shown sister chromatid exchange increase in *GSTM1*-deficient individuals. The GST μ class (*GSTM1*) has interindividual variation in expression with about 50% of Caucasians lacking enzyme activity (Lin et al. 1994). The deficient enzyme phenotype is due to a deletion of both paternal and maternal *GSTM1* alleles (Seidegard et al. 1988). In addition to the postulated role in detoxification, simple binding of GST to ligands may reduce carcinogenic potency. Another speculative mechanism involves GSTs as peroxidases, i.e., having the capacity to remove organic peroxides generated from lipids (Wolf 1990).

GSTM1 PHENOTYPING AND GENOTYPING

An individual's *GSTM1* status was determined in early studies by measuring phenotypic expression of GST activity toward a known substrate (i.e., *trans*-stilbene oxide) in an *in vitro* assay system. It is now routine to determine an individual's genotype by polymerase chain reaction (PCR) technology. Using a small sample of DNA obtained from blood cells, PCR-based analysis corresponds with phenotypic enzyme expression assays and is a simple, straightforward method for characterizing individuals.

EPIDEMIOLOGIC STUDIES OF *GSTM1* GENOTYPE/PHENOTYPE

Seven case-control studies have investigated the relationship between *GSTM1* expression and the risk of bladder cancer. Five of these studies have shown a statistically significant increased risk (95% CI, 1.4 to 3.8) for developing bladder cancer in individuals who do not express the *GSTM1* genotype, whereas two studies have not found a statistically significant difference. Early studies measured the phenotypic expression (Heckbert et al. 1992, LaFuente et al. 1993) of GST activity toward a known substrate (i.e., *trans*-stilbene oxide), and later studies (Zhong et al. 1993, Bell et al. 1993, Daly et al. 1993a, Brockmoller et al. 1994, Lin et al. 1994) and follow-up analyses (Brockmoller et al. 1996) used PCR technology to determine *GSTM1* genotype. There is a good correlation between the phenotype and genotype (Bell et al. 1992, Brockmoller et al. 1994), with the genotype assay generally preferred because it is inexpensive and simple. The studies are briefly described below.

INDIVIDUAL GST STUDIES OF BLADDER CANCER

Heckbert et al. (1992) conducted a case-control study of *GSTM1* phenotype in 113 incident smoking-related cancers (lung, oral pharyngeal, and bladder), 50 other cancers (prostate and non-Hodgkin's lymphoma), and 120 population controls in Washington state. Overall, there was no difference in the distribution of *GSTM1* phenotype in smoking-related cases compared with controls. In the heaviest smokers, the odds ratio for developing cancer in *GSTM1* null compared with the gene present was 1.7 (95% CI, 0.9 to 3.3). No separate results for bladder cancer were presented. The study concluded that there was a suggestion of a moderate protective effect associated with high or intermediate enzyme activity (i.e., the gene-deficient group was at higher risk) among persons heavily exposed to tobacco, although the results were not statistically significant.

In a study from Spain, Lafuente et al. (1993) studied the *GSTM1* phenotype in 75 smokers with bladder cancer and in 78 smokers with laryngeal cancer compared with 127 outpatient controls matched for age and smoking history. Leukocytic *GSTM1* was measured by an enzyme-linked immunosorbent assay using affinity-purified rabbit polyclonal antibody to human *GSTM1*. The overall odds ratio of persons with the *GSTM1* null phenotype developing bladder cancer was 2.4 (95% CI, 1.75 to 3.06) compared with those with the *GSTM1* phenotype. It was also noted that a larger proportion of smokers in the bladder cancer group had the null phenotype compared with controls (66.7% vs. 45.4%; $p < 0.01$). Presumably these were incident cases because the patients were interviewed at the time of definitive surgery.

A case-control study in the United Kingdom of 53 patients with transitional cell bladder cancer and controls from a urology service ($n = 52$) and from a healthy population ($n = 58$) found an odds ratio of 4.4 (range, 1.8 to 11.1) for the risk of bladder cancer in those with the *GSTM1* null genotype compared with those with the gene present (Daly et al. 1993a). In the bladder cancer patients, no difference was found in *GSTM1* expression in smokers vs. nonsmokers. It was not specified whether these were incidence or prevalence cases.

Bell et al. (1993) described a large case-control study of transitional cell carcinoma of the bladder in the United States. Two hundred twenty-nine incident and prevalent cases and 211 hospital urology service controls were studied. Among Caucasians, patients were more likely to carry the *GSTM1* null phenotype than controls (odds ratio, 1.7; 95% CI, 1.1 to 2.5). The odds ratio was 2.0 (95% CI, 0.4 to 9.5) in African-Americans. Genotype was not affected by whether the cases were incident or prevalent. Overall, the odds ratio for bladder cancer associated with the *GSTM1* null genotype in smokers was higher (1.8; 95% CI, 1.2 to 3.0) compared with nonsmokers (1.3; 95% CI, 0.6 to 2.7). More detailed analysis of tobacco use revealed a greater risk of bladder cancer in those with the *GSTM1* null phenotype and a greater than 50-pack-year smoking history (odds ratio, 5.9; 95% CI, 2.6 to 13). Although tests for additive or multiplicative interaction were not significant, this study suggests there may be higher smoking-associated bladder cancer risks for certain subpopulations. Evidence supporting the biologic plausibility of this association has recently been provided in a cross-sectional study, which showed that smokers with the null *GSTM1* genotype had higher mutagenic activity in the urine than subjects with heterozygous (*GSTM1* +/–) or wild-type (*GSTM1* +/+) genotypes (Hirvonen et al. 1994).

Brockmoller et al. (1994) studied 296 patients with bladder cancer and 400 controls; 59% of the patients were *GSTM1* null compared with 51% of the controls (odds ratio, 1.40; 95% confidence interval 1.0–1.9). A follow-up analysis by Brockmoller in a slightly larger group confirmed that deficiency in *GSTM1* (and *NAT2*) were bladder cancer risk factors. *GSTM1* deficiency was a risk factor independent of smoking and occupation (Brockmoller et al. 1996).

In contrast to the above studies, two studies have found only a small and non-statistically significant difference in *GSTM1* genotype in bladder cancer patients vs. controls. Zhong et al. (1993) studied the *GSTM1* genotype in breast, colon, and bladder cancer patients (*n* = 97) and in 225 volunteer controls and found only a minimal difference (odds ratio, 1.1; *p* = 0.79). There was no analysis by smoking status or occupation.

In a study from California, Lin et al. (1994) used PCR to study the *GSTM1* genotype in archived tissues from 114 bladder cancer patients and 1473 healthy individuals. There was a non-statistically significant difference (odds ratio, 1.4; 95% CI, 0.9 to 2.1), but the controls were younger and had a different gender distribution. The estimate was not adjusted for age or sex, and there was no information on smoking or occupation.

GST MECHANISM

The presence of a gene-environment interaction between smoking and the *GSTM1* genotype has also been suggested in a small study (Hirvonen et al. 1994). The mutagenicity of urine from smokers with the *GSTM1* null genotype was greater than the mutagenicity of smokers with the *GSTM1* genotype. The two largest, well-designed studies (Bell et al. 1993, Brockmoller et al. 1994) also examined the interaction between *GSTM1* genotype and smoking on the risk of bladder cancer. Bell found that persons with a 50-pack-year tobacco history and a deletion of the *GSTM1* gene were nearly six times more likely to develop bladder cancer (odds ratio, 5.9; 95% CI, 2.6 to 13.0) than nonsmokers without the gene deletion. On the other hand, no relationship between pack-years of smoking and *GSTM1* genotype was observed by Brockmoller.

GST AND OTHER SMOKING-RELATED CANCERS

There have been a number of studies of *GSTM1* and lung cancer. A review of eight case-control studies (Zhong et al. 1991; Brockmoller et al. 1993; Hayashi et al. 1992; Nazar-Stewart 1993; Heckbert et al. 1992; Kihara et al. 1994; Hirvonen et al. 1993; Seidegard et al. 1986, 1990; Alexandrie 1994) finds a 1.3 (95% CI, 1.1 to 1.5) risk for null subjects (no heterogeneity), suggesting that, as with bladder cancer, there is a significant association of increased risk with the null genotype. Overall there were 53% null individuals among the patients and 48% among the controls. This type of summary ignores important potential sources of bias; for example, ethnic variation, the possibility that negative data have not been published, and the inability to adjust for confounding effects. In spite of these limitations, these data are consistent with an association of the *GSTM1* null genotype and major smoking-related cancers. With regard to non-smoking-related cancers [i.e., pituitary adenomas (Fryer et al. 1993), skin tumors (LaFuente et al. 1995, Heagerty et al. 1994), colon and breast cancer (Zhong et al. 1993), and hepatocellular cancer (Yu et al. 1995)], the data are negative or too sparse to draw firm conclusions.

CONSENSUS OF EPIDEMIOLOGIC FINDINGS

The studies in the aggregate suggest that the *GSTM1* null genotype confers a minor increment of risk for both smoking-related cancers. Although not all studies are positive, the confidence bounds for the studies fall within a narrow range and null individuals are at significantly increased risk.

Negative studies may be due to inadequate power (small size) or possibly control group selection. The test for heterogeneity is not significant. The point estimate for lung cancer is 1.3 (95% CI, 1.1 to 1.5), whereas a similar estimate for bladder cancer is slightly higher. This suggests overall a 10% to 50% increased risk of these malignancies in *GSTM1* null individuals. The precise manner in which exposure and genes act together is much less clear. It is not known precisely how the degree of smoking alters *GSTM1* null genotype risk; nor is it understood whether there is increased risk for other smoking-related cancers (i.e., larynx) or whether the risk extends to all smoking-related conditions (i.e., bronchitis, heart disease, stroke, etc.). The precise biologic mechanism for GST in the respective malignancies remains controversial. Finally, a role for GST in other malignancies, such as breast and colon cancer, has been proposed but at this time has not been adequately studied.

NEED FOR BETTER UNDERSTANDING OF GENE ACTION IN THE CONTEXT OF CANCER RESEARCH

Numerous environmental carcinogens have been identified through epidemiologic study and this knowledge has directly led to approaches to reduce the occurrence of cancer. This approach has been most successful for relatively potent environmental carcinogens associated with risks of about 2-fold or greater (Rothman and Hayes 1995). There are distinct limitations to identifying cancer risks due to environmental risk factors through studies in human populations. Very large studies are required to detect small increments of risk. In addition, because of the observational character of epidemiologic studies, precisely determining cause-effect relationships is problematic and the goal of identifying the environmental origins of cancer has become increasingly difficult. The causes of certain tumors, including prostate, brain, and leukemia, remain obscure and others are incompletely understood in spite of decades of study. If new carcinogens could be identified, new avenues for cancer control may result. Understanding gene-environment interactions is important because it may allow cancer-control measures to be focused on a smaller group of genetically susceptible individuals. If a substantial multiplicative interaction is present, there exists the theoretical possibility of blocking the disease process by interrupting either of the factors interacting in disease causation. Further, these studies may provide insight into the mechanism of carcinogenesis, which can help establish the biologic plausibility of an exposure-cancer relationship. Paradoxically, the identification of a genetic susceptibility factor for a particular malignancy may enhance understanding of the environmental component, because it will focus attention on the possible substrate(s).

Table 2 demonstrates a hypothetical scenario in which a particular exposure is associated with a 1.5-fold increased risk of lung cancer in an exposed cohort compared with an unexposed cohort. If 10% of the exposed cohort carry a susceptibility gene responsible for the entire cancer burden, in combination with the exposure, then their relative risk increases to 6.0-fold. Although it is unlikely that single factors will be found that contribute so strongly to excess cancer risk for any particular tumor type in the general population, this example demonstrates the potential to uncover stronger exposure-disease relationships that may be diluted if potential susceptibility factors are not taken into account.

OTHER TYPES OF SUSCEPTIBILITY MARKERS

Potential sources of susceptibility for cancer risk include interindividual variation in enzymes that activate and detoxify procarcinogens and carcinogens [e.g., cytochrome P-450 enzymes (Phase I enzymes), which catalyze oxidation, reduction, and hydrolysis reactions, and *N*-acetyltransferase and GST (Phase II enzymes)], which catalyze conjugation and synthetic reactions], variation in DNA repair, and variation in sensitivity for mutagen-induced chromosomal damage. Table 3 includes several examples of these processes.

TABLE 2
Hypothetical Example of Lung Cancer Risk, Exposure Status, and Susceptibility

A. Hypothetical risk of lung cancer in exposed compared
with unexposed cohort

Population		Cancer rate	Relative risk of exposure
Cases: Exposed cohort:	15 1000 }	0.015	0.015/0.01 = 1.5
Cases: Unexposed cohort:	10 1000 }	0.01	

B. Risk of lung cancer in susceptible (10%) and nonsusceptible (90%)
subjects in an exposed cohort compared with unexposed cohort

Exposed cohort		Cancer rate	Relative risk of exposure
Cases: Susceptible subcohort:	6 100 }	0.06	Exposed susceptibles/unexposed: 0.06/0.01 = 6.0
Cases: Nonsusceptible subcohort:	9 900 }	0.01	Exposed nonsusceptibles/unexposed: 0.01/0.01 = 1.0

TABLE 3
Examples of Susceptibility Biomarkers and Potentially Relevant Chemical Exposures

Susceptibility factor	Exposure	Refs.
Metabolism		
Phase I enzymes		
CYP1A1	PAHs	Kawajiri et al. 1993
CYP1A2	PAHs, HAAs	Snyderwine et al. 1989, Flammang and Kadlubar 1986
CYP2E1	Benzene, methylene chloride, N-nitrosamines	Guengerich et al.1991
Phase II enzymes		
NAT2	HAAs, aromatic amines	Hein 1998, Flammang and Kadlubar 1986
GSTM1	PAHs, aflatoxin	Ketterer 1988
GSTT1	Butadiene, methylene chloride	Hallier et al. 1993, Robertson et al. 1986
DNA repair	Sunlight (ultraviolet exposure)	Wei et al. 1993

EPIDEMIOLOGIC APPROACHES TO GENETIC EPIDEMIOLOGY STUDY

There will be many opportunities to explore the interaction of genetic susceptibility factors and environmental exposures for cancer risk as more information about the human genome is accumulated over the next decade. The case-control design is very efficient for examining the role of genetic susceptibility markers, particularly when high-quality exposure data are available from questionnaires or environmental monitoring. One limitation is that study size must be substantial (often >1000 subjects), particularly if the gene-environment effect is modest, or if either the gene or exposure involved is uncommon.

GENOTYPE VERSUS PHENOTYPE

If the basis of observed interindividual variation in a potential susceptibility factor has been characterized at the DNA level, it can be readily studied with DNA collected from peripheral blood samples or from less-invasive sources of DNA (e.g., buccal swabs, which collect small numbers of exfoliated epithelial cells). If a susceptibility factor has not been characterized at the DNA level, then it is necessary to assay the phenotypic expression of that variation, keeping in mind the potential for disease status to influence such assays. Regardless of which assay is used, reliability and accuracy are critical to any epidemiologic investigation. For example, Rothman et al. (1993) have demonstrated how even small degrees of genotype or phenotype misclassification may have a substantial impact on risk estimates, particularly when the prevalence of the at-risk alleles is either very low or high.

FACTORS THAT COMPLICATE USE OF GENETIC INFORMATION TO INFER RISK

The results of studies that show consistent environment-gene interactions may be applied to the risk assessment process, because a stronger overall risk, or a clearer dose–response relationship, may be evident for a particular exposure only among certain subgroups. Although it is tempting to consider using susceptibility markers to screen particular high-risk subgroups as part of primary or secondary prevention efforts, adequate data are not available at this time to support such activities. To use these markers for screening purposes, we must be able to estimate the cumulative probability that an individual, given a certain age and possessing relevant biologic and nonbiologic risk factors, will develop cancer over a defined period, along with an estimate of uncertainty (Gail et al. 1989; Benichou and Gail, 1995). At this time, we are far from using any of the markers discussed here for screening purposes. Even if such data are eventually accumulated, the ethical considerations of genetic screening must be carefully evaluated.

RATIONALE FOR THE STUDY OF SUSCEPTIBILITY

The primary rationale for using susceptibility biomarkers in cancer epidemiology at this time is to gain better insight into exposure-disease relationships so that the workplace and general environment can be made safer for all individuals. Although the impact of genetic factors for cancer risk in the general population, excluding familial cancers, is unclear, the broader literature as well as the examples cited here support the concept that interindividual variation in some metabolic processes, particularly in combination with particular patterns and levels of exposure to potential carcinogens, may contribute to the cancer burden. Evaluating these factors will remain a goal in future epidemiologic studies of cancer etiology.

RISK ASSESSMENT AND GENETICS

The purpose of this discussion is to reflect on the implications of genetic susceptibility as it is understood in the context of the examples described and risk assessment. Given the interindividual variability evident by the data presented, how can a regulatory agency deal with this information? Because of the paucity of information, detailed recommendations are unrealistic.

First, the type of genetic factors that will be of interest to the regulator are susceptibility genes, as we have defined them. This is the case because single genes generally exert their deleterious effects largely (but never completely) independent of environmental factors. Susceptibility genes, in contrast, require exposures (substrates) before they act. In addition, in the general population, single genes involved in cancer are rare. Although the risk to the individuals that bear these deleterious genes is high (both relative and absolute), both the numbers of subjects involved and the environmental component are small, and therefore the attributable risk for the population is small. In contrast, susceptibility genes confer modest increases of relative and absolute risk to relatively large numbers of subjects, and environmental exposure plays a crucial role. This results in a large attributable risk and therefore the potential for important public health implications.

Second, although gene-environment effects are implicit in the schema for the action of susceptibility genes, the precise substrates, and the dose at which exposures become relevant to disease (i.e., threshold, linear, and sub- or supralinear models), are unknown or at best sketchy. For long-term diseases such as cancer, exposures act over extended periods of time. Currently, data can be cited to support a role for genetic factors at both low exposure [Vineis et al. 1994 (acetylation and bladder cancer)] and high exposure [Kihara et al. 1994 (*GST* and smoking), Caporaso et al. 1995 (*CYP2D6* and smoking)]. Both possibilities may be correct depending on the enzyme kinetics and the specific substrate. From an epidemiologic perspective, much larger studies are required to resolve these issues.

Third, ethnic differences in gene frequency are common. Certain genes are present only in particular ethnic groups, such as the trait conferring G6PD deficiency, which is observed in Mediterranean and African populations but rarely in others.

Fourth, larger, well-designed interdisciplinary studies are required to resolve these issues. Studies of susceptibility markers must include adequate numbers, appropriate design (especially with regard to control selection), and careful assessment of exposure.

Fifth, genetic factors cannot be applied as a rationale to loosen restrictions on chemicals/xenobiotics in the environment because (1) multiple mechanisms may influence the effect of a particular chemical, i.e., genetic and nongenetic; (2) the same gene product may both activate and deactivate potential carcinogens — in the absence of complete knowledge of all potentially harmful substrates, action would not be rational; (3) altered risk may be present only within certain bounds of substrate concentration; (4) specific genes may not be relevant to other types of risk, i.e., to the environment, ecosystems, nonhuman species, etc.; (5) many types of effects may result from an exposure — a reduction in risk of cancer might be offset by increased pulmonary toxicity; (6) environmental exposures are likely to be complex, including mixtures of compounds that can have qualitatively and quantitatively variable relationships to a particular xenobiotic-metabolizing enzyme.

Sixth, other nongenetic sources of variability, such as extremes of age and body size or changes in metabolic status due to diet, illness, medications, and pregnancy, all contribute to human variation.

The study of these issues is in its infancy. The metabolism of a number of human carcinogens clearly exhibits interindividual variability (aromatic amines and *NAT2*, cyclophosphamide and *ADH*, benzene and *CYP2E1*, etc.) and is the subject of intense study. Currently, we are able to recognize the existence of a genetic (hereditary) impact on exposure-driven cancers, but we lack the specific understanding to use this information in the service of public health.

REFERENCES

Adamson RH (1990) Mutagens and carcinogens formed during cooking of foods and methods to minimize their formation. In DeVita VT, Hellman S, Rosenberg SA (eds), Cancer Prevention. Lippincott, Philadelphia, pp 1–7

Alexandrie AK, Sundberg MI, Seidegaard J, et al (1994) Genetic susceptibility to lung cancer with special emphasis of CYP1A1 and GSTM1: a study of host factors in relation to age at onset, gender, and histological cancer type. Carcinogenesis 15(9):1785–1790

Ayesh R, Idle JR, Ritchie JC, et al (1984) Metabolic oxidation phenotypes as markers for susceptibility to lung cancer. Nature 312:169–170

Bell DA, Taylor JA, Paulson DF, et al (1993) Genetic risk and carcinogen exposure: a common inherited defect of the carcinogen-metabolism gene glutathione S-transferase M1 (*GSTM1*) that increases susceptibility to bladder cancer. J Natl Cancer Inst 85:1159–1164

Bell DA, Thompson CL, Taylor J, et al (1992) Genetic monitoring of human polymorphic cancer susceptibility genes by polymerase chain reaction: applications to glutathione transferase mu. Environ Health Persp 98:113–117

Benichou J, Gail MH (1995) Methods of inference for estimates of absolute risk derived from population-based case-control studies. Biometrics, 51:182–194

Brockmoller J, Cascorbi C, Kerb R, Roots I (1996) Combined analysis of inherited polymorphisms in arylamine N-acetyltransferase 2, glutathione S-transferase M1 and T1, microsomal epoxide hydrolase, and cytochrome P450 enzymes as modulators of bladder cancer risk. Cancer Res 56:3915–3925

Brockmoller J, Kerb R, Drakoulis N, et al (1994) Glutathione S-transferase M1 and its variants A and B as host factors of bladder cancer susceptibility: a case-control study. Cancer Res 54:4103–4111

Brockmoller J, Kerb R, Drakoulis N, et al (1993) Genotype and phenotype of glutathione s-transferase class μ isozymes μ and θ in lung cancer patients and controls. Cancer Res 53:1004–1011

Butler MA, Lang NP, Young JF, et al (1992) Determination of *CYP1A2* and *NAT2* phenotypes in human populations by analysis of caffeine urinary metabolites. Pharmacogenetics 2:116–127

Caporaso N, Goldstein A (1995) Cancer genes: single and susceptibility: exposing the difference. Pharmacogenetics 5:59–63

Caporaso N, DeBaun MR, Rothman N (1995) Lung cancer and CYP2D6 (the debrisoquine polymorphism): sources of heterogeneity in the proposed association. Parmacogenetics 5:S129–S134

Caporaso N, Landi MT, Vineis P (1991) Relevance of the metabolic polymorphisms to human carcinogenesis: evaluation of epidemiologic evidence. Pharmacogenetics 1:4–19

Daly AK, Thomas DJ, Cooper J, et al (1993a) Homozygous deletion of gene for glutathione S-transferase M1 in bladder cancer. Br Med J 307:481–482

Daly AK, Cholerton S, Gregory W, Idle JR (1993b) Metabolic polymorphisms. Pharmacol Ther 57:129–160

Degawa M, Tanimura S, Agatsuma T, et al (1989) Hepatocarcinogenic heterocyclic aromatic amines that induce cytochrome P-488H (P-450IA2), responsible for mutagenic activation of the carcinogens in rat liver. Carcinogenesis 10:1119–1122

Evans HJ (1993) Molecular genetic aspects of human cancer: the 1993 Frank Rose Lecture. Br J Cancer 68:1051–1060

Flammang TJ, Kadlubar FF (1986) Acetyl coenzyme A-dependent metabolic activation of N-hydroxy-3,2'-dimethyl-4-aminobiphenyl and several carcinogenic N-hydroxy arylamines in relation to tissue and species differences, other acyl donors, and arylhydroxamic acid-dependent acyltransferases. Carcinogenesis 7:919–926

Fryer AA, Zhao L, Alldersea J, et al (1993) The glutathione S-transferases: polymerase chain reaction studies and the frequency of the *GSTM1* 0 genotype in patients with pituitary adenomas. Carcinogenesis 14(4):563–566

Gail MH, Brinton LA, Byar DP, et al (1989) Projecting individualized probabilities of developing breast cancer for white females who are being examined annually. J Natl Cancer Inst 81:1879–1886

Guengerich FP (1992) Metabolic activation of carcinogens. Pharmacol Ther 54:17–61

Guengerich FP, Kim D-H, Iwasaki M (1991) Role of human cytochrome P-450 IIE1 in the oxidation of many low molecular weight cancer suspects. Chem Res Toxicol 4:168–179

Hallier E, Langhof T, Dannappel D, et al (1993) Polymorphism of glutathione conjugation of methyl bromide, ethylene glycol and dichloromethane in human blood: influence on the induction of sister chromatid exchanges in lymphocytes. Arch Toxicol 67:173–178

Hayashi S, Watanabe J, Kawajiri K (1992) High susceptibility to lung cancer analyzed in terms of combined genotypes of P450IA1 and mu-class glutathione S-transferase genes. Jpn J Cancer Res 83:866–870

Hayes RB, Bi W, Rothman N, et al (1993) N-acetylation phenotype and genotype and risk of bladder cancer, in benzidine-exposed workers. Carcinogenesis 14:675–678

Heagerty A, Fitzgerald D, Smith A, et al (1994) Glutathione S-transferase *GSTM1* phenotypes and protection against cutaneous tumors. Lancet 343:266–268

Heckbert SR, Weiss NS, Hornung SK, et al (1992) Glutathione S-transferase and epoxide hydrolase activity in human leukocytes in relation to risk of lung cancer and other smoking-related cancers. J Natl Cancer Inst 84:414–422

Hein DW (1988) Acetylator genotype and arylamine-induced carcinogenesis. Biochim Biophys Acta 948:37–66

Hirvonen A, Nylund L, Kociba P, et al (1994) Modulation of urinary mutagenicity by genetically determined carcinogen metabolism in smokers. Carcinogenesis 15:813–815

Hirvonen A, Husgafvel-Pursiainen K, Anttila S, Vainio H (1993) The *GSTM1* null genotype as a potential risk modifier for squamous cell carcinoma of the lung. Carcinogenesis 14:1479–1481

Hsu TC, Johnston DA, Cherry LM, et al (1989) Sensitivity to genotoxical effects of bleomycin in humans: possible relationship to environmental carcinogenesis. Int J Cancer 43:403–409

Kadlubar FF, Butler MA, Kaderlik KR, et al (1992) Polymorphisms for aromatic amine metabolism in humans: relevance for human carcinogenesis. Environ Health Persp 98:69–74

Kato R (1986) Metabolic activation of mutagenic heterocyclic aromatic amines from protein pyrolysate. CRC Crit Rev Toxicol 16:307–348

Kawajiri K, Nakachi K, Imai K, et al (1993) The *CYP1A1* gene and cancer susceptibility. Crit Rev Oncol Hematol 14:77–87

Kawajiri K, Nakachi K, Imai K, et al (1990) Identification of genetically high risk individuals to lung cancer by DNA polymorphisms of the cytochrome *P4501A1* gene. FEBS Lett 263:131–133

Ketterer B (1988) Protective role of glutathione and glutathione transferases in mutagenesis and carcinogenesis. Mutat Res 202:343–361

Kihara M, Kihara M, Noda K (1994) Lung cancer risk of *GSTM1* null genotype is dependent on the extent of tobacco smoke exposure. Carcinogenesis 15:415–418

Lafuente A, Molina R, Palou J, et al Phenotype of the glutathione S-transferase Mu (GSTM1) and susceptibility to malignant melanoma. Br J Cancer 72(2):324–326

Lafuente A, Pujol F, Carretero P, et al (1993) Human glutathione S-transferase mu (GSTmu) deficiency as a marker for the susceptibility to bladder and larynx cancer among smokers. Cancer Lett 68:49–54

Lang NP, Butler MA, Massengill J, et al (1994) Rapid metabolic phenotypes for acetyltransferase and cytochrome *P4501A2* and putative exposure to food-borne heterocyclic amines increase the risk for colorectal cancer or polyps. Cancer Epidemiol Biomarkers Prev 3:675–682

Lin HJ, Han C-Y, Bernstein DA, et al (1994) Ethnic distribution of the glutathione transferase Mu 1-1 (*GSTM1*) null genotype in 1473 individuals and application to bladder cancer susceptibility. Carcinogenesis 15:1077–1081

Mannervik B, Awasthi YC, Board PG, et al (1992) Nomenclature for human glutathione transferases. Biochem J 282:305–306

Nazar-Stewart V, Motulsky AG, Eaton DL, et al (1993) The glutathione S-transferase μ polymorphism as a marker for susceptibility to lung cancer. Cancer Res 53:2313–2318

Robertson IGC, Guthenberg C, Mannervik B, Jernstrom B (1986) Differences in stereoselectivity and catalytic efficiency of three human glutathione transferases in the conjugation of glutathione with 7b,8a,-dihydroxy-9a,10a-oxy-7,8,9,10-tetrahydrobenzo(a)pyrene. Cancer Res 46:2220–2224

Rothman N, Hayes RB (1995) Using biomarkers of genetic susceptibility to enhance the study of cancer etiology. Environ Health Persp 103(8):291–294

Rothman N, Bhatnagar VK, Hayes RB, et al (1996) The impact of interindividual variation in NAT2 activity on benzidine urinary metabolites and urothelial DNA adducts in exposed workers. Proc Natl Acad Sci U.S.A. 93:5084–5089

Rothman N, Stewart W, Caporaso NE, Hayes RB (1993) Misclassification of genetic susceptibility markers in case-control studies of cancer: implications for cross-population comparisons and study design. Cancer Epidemiol Biomarkers Prev 2:299–303

Seidegard J, Pero RW, Markowitz MM, et al (1990) Isoenzyme(s) of glutathione transferase (class μ) as a marker for the susceptibility to lung cancer: a follow up study. Carcinogenesis 11:33–36

Seidegard J, Vorachek WR, Pero RW, Pearson WR (1988) Hereditary differences in the expression of human glutathione transferase activity on *trans*-stilbene oxide are due to a gene deletion. Proc Natl Acad Sci U.S.A. 85:7293–7297

Seidegard J, Pero RW, Miller DG, Beattie EJ (1986) A glutathione transferase in human leukocytes as a marker for the susceptibility to lung cancer. Carcinogenesis 7:751–753

Snyderwine EG, Battula N (1989) Selective mutagenic activation by cytochrome P_3-450 of carcinogenic arylamines found in foods. J Natl Cancer Inst 81:223–227

Spitz MR, Fueger JJ, Halabi S, et al (1993) Mutagen sensitivity in upper aerodigestive tract cancer: a case-control analysis. Cancer Epidemiol Biomarkers Prev 2:329–333

Strange RC, Faulder CG, Davis BA, et al (1984) The human glutathione S-transferase: studies on the tissue distribution and genetic variation of GST1, GST2, and GST3 isozymes. Ann Hum Genet 48:11–20

van Poppel G, de Vogel N, van Balderen PJ, Kok FJ (1992) Increased cytogenetic damage in smokers deficient in glutathione S-transferase isozyme mu. Carcinogenesis 13(2):303–305

Vineis P, Caporaso N, Tannenbaum SR, Skipper PL, Glogowski J, Bartsch H, Coda M, Talaska G, Kadlubar F (1990) Acetylatin phenotype, carcinogen-hemoglobin adducts, and cigarette smoking. Cancer Res 50:3002–3005.

Vineis P, Bartsch H, Caporaso N, et al (1994) Genetically based N-acetyltransferase metabolic polymorphism and low-level environmental exposure to carcinogens. Nature 369:154–156

Wei Q, Matanoski GM, Farmer ER, et al (1993) DNA repair and aging in basal cell carcinoma: a molecular epidemiology study. Proc Natl Acad Sci U.S.A. 90:1614–1618

Wiencke JK, Spitz MR (1994) *In vitro* chromosomal assays of mutagen sensitivity and human cancer risk. Cancer Bull 46:238–246

Wolf CR (1990) Metabolic factors in cancer susceptibility. Cancer Surv 9(3):437–474

Wu X, Hsu TC, Annegers JF, et al (1995) A case-control study of nonrandom distribution of bleomycin-induced chromatid breaks in lymphocytes of lung cancer cases. Cancer Res 55:557–561

Yu MW, Gladek-Yarborough A, Chiamprasert S, et al (1995) Cytochrome P450 2E1 and glutathione S-transferase M1 polymorphisms and susceptibility to hepatocellular carcinoma. Gastroenterology 109(4):1266–1273

Zhong S, Wyllie AH, Barnes D, et al (1993) Relationship between the GSTM1 genetic polymorphism and susceptibility to bladder, breast and colon cancer. Carcinogenesis 14:1821–1824

Zhong S, Howie AF, Ketterer B, et al (1991) Glutathione S-transferase μ locus: use of genotyping and phenotyping assays to assess association with lung cancer susceptibility. Carcinogenesis 12:1533–1537

8 Human Variability in Susceptibility and Response: Implications for Risk Assessment

David A. Neumann and Carole A. Kimmel

CONTENTS

As part of the review of current and potential uses of information on interindividual variability within the human population in risk assessment, a workshop was convened in Washington, D.C., on October 16–17, 1995. The intent was to bring together scientists from academia, industry, and government with expertise in various scientific disciplines to consider seven specific questions (Table 1). After summaries of the preliminary drafts of several of the background papers were presented, workshop participants formed breakout groups (Table 2) to address the questions relative to neurotoxicity, pulmonary toxicity, reproductive and developmental toxicity, and cancer. Their responses and the general discussion stimulated by their reports are summarized below and reflect the overall conclusions and recommendations for this activity.

The information and ideas presented in this chapter were extracted from a transcript of a tape recording of the final session of the workshop. This chapter should not be construed to reflect a consensus of opinion among workshop participants or even among breakout group members. Rather it is a reflection of the discussion by participants of the current state of scientific understanding and the implications of that understanding for risk assessment. A number of issues and ideas raised during the workshop were subsequently incorporated into the preceding chapters, most of which were drafted before the workshop.

0-8493-2805-5/99/$0.00+$.50
© 1999 by CRC Press LLC

TABLE 1
Focus Issues for the October 16–17, 1995, Workshop on Human Variability, Washington, D.C.

1. Are there suitable quantifiable parameters that allow an assessment of the range of interindividual variability in the human response to carcinogenic processes or to agents that elicit neurotoxic, pulmonary, or reproductive/developmental effects?
2. Can these parameters be used to predict the specificity (target tissue) or severity of the response or process?
3. Are such parameters influenced by age, gender, ethnicity, preexisting disease, diet/nutritional status, socioeconomic status, or other factors?
4. Are the findings and relationships unique to the compound(s), endpoint(s), and process(es) considered during the workshop or are they more broadly applicable?
5. How might this information be used in cancer or non-cancer risk assessment?
6. Do the current default methodologies for cancer risk assessment in the United States adequately account for interindividual variability?
   Does this information enhance or diminish confidence in the currently used 10-fold uncertainty factor in non-cancer risk assessment to account for interindividual variability?
7. Are there specific research needs relative to quantitatively assessing interindividual variability that, if addressed, would improve risk assessment?

ARE THERE SUITABLE QUANTIFIABLE PARAMETERS THAT ALLOW AN ASSESSMENT OF THE RANGE OF INTERINDIVIDUAL VARIABILITY IN THE HUMAN RESPONSE TO AGENTS THAT ELICIT TOXIC/CARCINOGENIC EFFECTS?

NEUROTOXIC EFFECTS

There are unequivocal examples of parameters that can be measured by a variety of techniques, e.g., clinical and neurochemical assays for determining cholinesterase activity (George et al. 1985) either in the blood or in the brain. Lymphocyte neurotoxic esterase activity (Lotti 1987) can be determined repeatedly in humans. Imaging techniques (Ørbaek et al. 1987) can reveal morphological changes in the central nervous system, and newer techniques (Wood et al. 1991) can characterize certain functional changes that occur during the imaging process.

Electrophysiology, e.g., evoked potentials and various factors associated with nerve conduction velocity or other parameters (Johnson 1980), offers a host of endpoints for assessing interindividual variability. Various behavioral parameters (Johnson 1987) also can be used, including measures of performance and motor and sensory function. Parameters such as these can be repeatedly measured on the same population, whereas tissue morphology is unlikely to be determined for the same individuals on multiple occasions.

PULMONARY EFFECTS

This breakout group focused on human variability in response to exposure to ozone as a general model for pulmonary toxicology. For a compound as rich in both human and animal data as ozone (Lippmann 1989), definitive insights into identifying susceptible populations and quantifying variability in responsiveness were elusive. However, numerous techniques are available for studying the effects of ozone on human lungs. Pulmonary function changes (Lebowitz 1991), e.g., forced expiratory volume in the first second (FEV_1) and forced vital capacity, can yield information about different types of effects in the lungs. Bronchoalveolar lavage (Witschi and Last 1996) can be used to evaluate neutrophilic infiltrates in the lungs and characterize the presence of interleukins and other inflammatory mediators in the lavage fluid. Such parameters may aid in interpretation of the significance of lung function changes. Nasal lavage (Koren et al. 1990) can be used to evaluate effects, but observed changes may not be as specific as some of those observed in the bronchoalveolar lavage

TABLE 2
Participants in the October 16–17, 1995, Workshop on Human Variability, Washington, D.C.

Neurotoxic effects

Deborah Cory-Slechta, co-chair	University of Rochester
John O'Donoghue, co-chair	Eastman Kodak Company
Deborah Barsotti	Harding Lawson Associates
David Bellinger	Children's Hospital, Boston
Peter Dews	Harvard Medical School
John Glowa	Uniformed Services University of Health Sciences
Dale Hattis	Clark University
Abby Li	Monsanto/Ceregen
Robert MacPhail	University of North Carolina
David Neumann	ILSI Risk Science Institute
William Slikker	U.S. Food and Drug Administration, National Center for Toxicological Research
Tom Starr	ENVIRON Corporation

Pulmonary effects

Barbara Beck, co-chair	Gradient Corporation
Edward Ohanian, co-chair	U.S. Environmental Protection Agency, Office of Water
Philip Bromberg	University of North Carolina
Mark Frampton	University of Rochester
Jean Grassman	National Institute of Environmental Health Sciences
Les Schwartz	SmithKline Beecham Pharmaceuticals
David Warheit	DuPont Haskell Laboratory

Carcinogenic effects

Penny Fenner-Crisp, co-chair	U.S. Environmental Protection Agency, Office of Pesticide Programs
Clay Frederick, co-chair	Rohm and Haas Company
Neil Caporaso	National Cancer Institute
Rick Corley	Dow Chemical Company
Ronald Estabrook	University of Texas, Southwestern Medical Center
Adam Finkel	Occupational Safety and Health Administration
Herman Gibb	U.S. Environmental Protection Agency, National Center for Environmental Assessment
Nicholas Lang	University of Arkansas for Medical Sciences
Genevieve Matanoski	Johns Hopkins University
John Newton	Sanofi-Winthrop, Inc.
Gerry Raabe	Mobil Oil Corporation
Nathaniel Rothman	National Cancer Institute
Peter Shields	National Cancer Institute

Reproductive/developmental effects

George Daston, co-chair	American Industrial Health Council
Carole Kimmel, co-chair	U.S. Environmental Protection Agency, National Center for Environmental Assessment
Cynthia Bearer	Rainbow Babies and Children's Hospital, Cleveland
Willem Faber	Eastman Kodak Company
Warren Foster	Health Canada
Bryan Hardin	National Institute for Occupational Safety and Health
John Kiely	National Center for Health Statistics
Steve Lewis	Exxon Biomedical Sciences, Inc.
John Rogers	Raleigh, NC
Sherry Selevan	U.S. Environmental Protection Agency, National Center for Environmental Assessment

fluids. Monitoring for changes in the blood (Vogt 1991) may reveal effects that are secondary to the pulmonary effects.

Endobronchial biopsy (Fink 1992) can be performed to identify cellular or histopathologic changes associated with exposure to ozone or other pulmonary toxicants. Cells from the biopsies can be grown in culture to monitor for gene expression and messenger RNA production. However, it is unclear whether the area of the lung accessible for biopsy corresponds to the area of the lungs that is most likely to respond to ozone exposure.

REPRODUCTIVE/DEVELOPMENTAL TOXICITY

Data for a number of variables are routinely collected, e.g., perinatal mortality and malformation incidence, specific malformations, major and minor malformations, birth weight corrected for gestational age (Scialli and Lione, this volume), but are not analyzed to assess for interindividual variability or variability in toxicologic response. However, this database has the potential to provide information on the extent of variability as influenced by intrinsic factors such as age, parity, race, etc.

More specific information can be collected once a hypothesis has been developed to explain an experimental observation. Such information might include the incidence of major and minor malformations that comprise part of a syndrome, postnatal growth parameters, or functional measures such as neonatal indices of neurobehavioral maturation. Various sperm measures can be evaluated when considering male reproductive toxicity (Working 1988). Parameters perceived to be relevant to specific toxicologic responses are unlikely to be routinely reported unless there is an *a priori* reason to do so.

Data for such parameters can be evaluated when a causation hypothesis has been generated and when there is some understanding of the mechanism of action of the toxicant. A good example is the work by Buehler et al. (1994) on susceptibility to diphenylhydantoin-induced fetal hydantoin syndrome and epoxide hydrolase activity in amniocytes collected by amniocentesis. This model is presumed to be a surrogate for the embryo's total capability for metabolizing and, in this case, detoxifying an epoxide intermediate of phenytoin.

CARCINOGENICITY

Although there are acceptable measures of dosimetry, there is uncertainty about whether and how adduct concentrations in the blood, tissues, or urine contribute to response variability (Pitot and Dragan 1996). Thus, it is difficult to determine the relationships between dosimetry and susceptibility and between the expression of dosimetry and susceptibility and disease. A significant problem is that the critical rate-limiting step(s) associated with carcinogenesis is unknown. Currently, there is a lack of even proxy measures of cancer susceptibility.

CAN THESE PARAMETERS BE USED TO PREDICT THE SPECIFICITY (TARGET TISSUE) OR SEVERITY OF THE RESPONSE OR PROCESS?

NEUROTOXIC EFFECTS

The quantifiable parameters available to neurotoxicologists can be used in a compound- or endpoint-specific manner to predict the specificity or severity of the response or effect. This can be done readily with a selective neurotoxicant such as 1-methyl-4-phenyl-1,2,3,6-tetrahydropyridine, which is associated with induction of Parkinson's disease (Kopin and Markey 1988). Similarly, the ability to predict the broad toxic manifestations of exposure to lead are well-known (Needleman et al. 1990). Measurements of brain stem auditory evoked potential (Otto 1986) can dissect the electrophysiology of hearing and can detect changes in particular parts of the auditory system.

Toxicant-induced changes in motor activity (Anger and Johnson 1992), such as grip strength, may reflect altered peripheral nerve function, altered central nervous system function, or interaction between the two.

PULMONARY EFFECTS

Parameters that directly measure changes in the lung, e.g., lung function or bronchoalveolar lavage (Lebowitz 1991, Witschi and Last 1996), can be used to predict effects that are specific to the lungs. Thus, it may be somewhat easier to study certain changes or effects on the lungs than in other organ systems. However, it is unclear whether changes in certain parameters, e.g., specific cytokine concentrations in the blood, that may occur in response to ozone exposure directly reflect a change in the lung or are reflective of secondary effects.

A broad concern is whether the qualitative and quantitative features of an observed response constitute an adverse health effect. Is the significance of a lung function decrement accompanied by inflammation different than a lung function decrement in the absence of inflammation? Does a neurogenic-mediated decline in lung function have a different implication in terms of severity than a decline associated with mechanical changes? Interpretation of the health significance of such results depends on a number of unresolved issues, and their resolution clearly will have implications for risk assessment. Criteria for distinguishing between the lowest-observed-adverse-effect level and a measurable effect that is not adverse, or that perhaps is adaptive or even beneficial, need to be identified (Calabrese 1992). This is especially applicable to ozone.

REPRODUCTIVE/DEVELOPMENTAL TOXICITY

Most available measures of effect are apical, i.e., they do not directly address the specificity of the response to toxic agents. However, certain measures are more specific, e.g., epoxide hydrolase and fetal hydantoin syndrome (Buehler et al. 1994). There may be other examples as well in terms of biomarkers of biological outcome whose variability can be measured. Although their relationship to toxic exposure is unknown, many common prenatal diagnostic tools, such as measurement of serum and amniotic fluid α-fetoprotein concentrations, ultrasonography, and others, potentially could be used to characterize interindividual variability.

CARCINOGENICITY

The currently available parameters cannot be used to predict the tissue specificity or severity of the carcinogenic outcome. This reflects the paucity of data available and lack of understanding of the underlying mechanisms associated with carcinogenesis.

ARE SUCH PARAMETERS INFLUENCED BY AGE, GENDER, ETHNICITY, PREEXISTING DISEASE, DIET AND NUTRITIONAL STATUS, SOCIOECONOMIC STATUS, OR OTHER FACTORS?

NEUROTOXIC EFFECTS

It is likely that these factors affect the distribution and magnitude of the responses, but to different degrees. There is not enough information available about how these various factors interact with each other and with the monitored parameter. Interactions among these various modifying factors may be additive or subtractive, contribute to potentiation, or have other effects. For example, if age and nutrition are factors, then for a young child with nutritional problems those outputs might be additive, although they could result in potentiation.

Genetic polymorphisms may be important in determining individual human susceptibility to neurotoxicants. Information about genetic polymorphisms could facilitate research on specific genes that may be important in the metabolism of particular compounds. There are examples of interindividual differences in enzymatic activity that have a significant impact on neurotoxicity. Fast and slow acetylation plays a role in isoniazid toxicity and influences whether isoniazid produces a neuropathy (Parkinson 1996). Genetic differences are responsible for phenylketonuria (Kaufman 1977) and whether an individual suffers neurologic impairment from exposure to phenyl ketone. Theoretically, polymorphisms in certain solvent-metabolizing enzymes, such as CYP2E1, may be associated with solvent-mediated neurotoxicity (Snyder and Andrews 1996). The toxicity of various chlorinated solvents may be increased or decreased by interaction with glutathione (Monks et al. 1990); this could be associated with their ability to cause cerebellar damage.

Polymorphic genes for molecules other than metabolic enzymes also may contribute to interindividual variability in response. For example, recent studies suggest that aminolevulinic acid desynthase polymorphisms may be involved in mediating the neurotoxic effects of lead (Smith et al. 1995). Other studies have shown associations between various polymorphisms of apolipoprotein E and the risk of developing Alzheimer's disease (Saunders et al. 1993). Some work suggests that the risk of neurological impairment from head trauma may be associated with the expression of certain alleles.

The interaction between genetic polymorphisms and environmental factors is important for understanding human variability in response. Although it may be possible to identify a sensitive subpopulation, perhaps defined by a specific genetic polymorphism, it is important to characterize the variability within that population. Identifying sensitive individuals may not necessarily provide insight into how to measure variability in response.

PULMONARY EFFECTS

The breakout group discussing this topic rearranged the question to ask whether parameters such as lung function, inflammatory changes, and other responses to ozone are influenced by age, gender, ethnicity, disease, diet, etc. In terms of pulmonary function, asthmatic individuals exposed to ozone do not appear to have enhanced responsiveness (Bromberg, this volume). It is not clear whether this is true in terms of more severe effects, because, although studies show that increased hospital admissions for asthma occur on days with elevated ozone concentrations, other environmental parameters, e.g., temperature, sulfate concentrations, are also altered under those conditions (Bromberg, this volume). Although it is possible to correct for such factors, there is residual concern about whether the effect is due to ozone or to the multitude of chemicals and other factors that are associated with high ozone concentrations.

Susceptibility to ozone is inversely related to age; the younger the individual the greater the responsiveness to ozone. Young adults (in the 30- to 35-year age range) are more responsive to ozone than are older adults (McDonnell et al. 1993, 1995), possibly because of an age-related decline in nociceptive receptors. There are no reports of major changes in responsiveness to ozone associated with gender or race, but diet may influence responsiveness. In animal studies, those fed diets deficient in antioxidants, e.g., vitamin E, tend to be more responsive to oxidant gases (Weinstock and Beck 1988). Studies currently in progress may help to elucidate whether diet plays a role in human responsiveness to ozone.

Exercise is also a key factor contributing to ozone susceptibility (McDonnell et al. 1993). Susceptibility may be increased 10-fold with exercise based on studies showing that a nonexercising individual may exhibit little, if any, response to ozone exposures as high as 0.75 parts per million (ppm). With exercise and exposure periods on the order of hours, responses are observed at ozone concentrations of 0.12 ppm or less. Because such studies have been performed by different investigators using different procedures, it is not clear whether there is a true 10-fold difference in susceptibility. In the context of assessing the protectiveness of an uncertainty factor of 10 relative to ozone exposure, if exercisers are considered a susceptible population, the 10-fold uncertainty factor may be consistent with the magnitude of the difference in response between the susceptible and the nonsusceptible populations.

Prior ozone exposure also enhances responsiveness to ozone (Tepper et al. 1989). Exposure to ozone within, perhaps, 4 days or less of a subsequent exposure, is associated with diminished responsiveness to the later exposure. However, such a prior exposure to ozone can enhance an individual's responsiveness to a subsequent allergen exposure (Molfino et al. 1991).

For ozone, susceptible populations include young adults, children, and people who exercise. However, there is no good mechanistic basis to explain these differences in susceptibility, and even among exercisers there is variability in responsiveness. More information is needed before associations between enhanced susceptibility to ozone and the expression of particular gene products or other factors can be understood.

REPRODUCTIVE/DEVELOPMENTAL TOXICITY

Exposure-response relationships are clearly influenced by age, gender, ethnicity, etc. For example, a mother's age, parity, ethnicity, and, quite likely, her nutritional and socioeconomic status influence the likelihood that alcohol abuse will cause fetal alcohol syndrome (Sokol et al. 1986). Another example is dietary intake of folate and the prevalence of neural tube defects (MRC Vitamin Study Research Group 1991, Czeizel and Dudas 1992). Maternal age and the prevalence of Down syndrome, which may have an environmental component, is well-known (Penrose 1933). Evidence strongly suggests that female offspring have more severe responses than male offspring to prenatal exposure to diethylstilbestrol (Bibbo et al. 1975, Kaufman et al. 1977, Senekjian et al. 1988, Wilcox et al. 1995).

CARCINOGENICITY

Age, gender, ethnicity, and various other factors likely influence the carcinogenic process, but the critical question is how? Cancer incidence is known to increase with age (Fraumeni et al. 1993), and there are well-established differences in the distributions of tumors between men and women (Miller et al. 1993) and among ethnic groups (Overfield 1995). Specific dietary factors have been associated with certain types of cancers (Frame et al., this volume). Yet, in spite of these recognized associations, there is little information available on the biological basis for such observations.

ARE THE FINDINGS AND RELATIONSHIPS UNIQUE TO THE COMPOUND(S), ENDPOINT(S), AND PROCESS(ES) CONSIDERED DURING THE WORKSHOP OR ARE THEY MORE BROADLY APPLICABLE?

NEUROTOXIC EFFECTS

These relationships generally appear to be broadly applicable, although there is insufficient information to definitively address the question. In certain cases, there may be chemical-specific factors that would necessitate the measurement of chemical-specific parameters, endpoints, and interactions that would not be readily predictable of the response(s) to exposure to other compounds.

PULMONARY EFFECTS

Information on interindividual variability in the response of humans to ozone exposure offers insight into how to generate and evaluate information that can be more useful for risk assessment. The basis for susceptibility to ozone exposure appears to be age- and exercise-related, and the observed variation in responsiveness between the least and the most susceptible individuals may be within one order of magnitude. However, these findings cannot be generalized to apply to other compounds, including other oxidant gases and certain classic pneumotoxicants such as silica or asbestos. This reflects the different mechanisms associated with different toxic agents and, likely, the different biological features of populations susceptible to the other toxicants.

Some of the critical issues described above for neurotoxicants probably apply to compounds that affect other target organs, including the lungs. It is important to consider the endpoint to be used when trying to assess or characterize susceptibility because, as illustrated by the ozone model, the use of different endpoints may reveal different susceptible populations. Individuals who are more responsive to ozone in terms of changes in pulmonary function are distinct from those who are more responsive in terms of inflammatory changes (Koren et al. 1991). Thus, for a single compound and the same target organ, it is important to consider the endpoint being used, because that will influence the definition of the susceptible population.

Ozone concentration and duration of exposure are important (Bromberg, this volume). Except, perhaps, within a very narrow range, Haber's law [i.e., the product of concentration and time is equal to the product of concentration and time $(C_1T_1 = C_2T_2)$] is probably not correct. Exposure to 1 ppm for 10 hours is not the same as exposure to 10 ppm for 1 hour. For ozone, as well as for other compounds, it is useful to design studies that address tissue dosimetry (Bromberg, this volume). Quantitative differences in responsiveness to ozone may reflect differences in inhalation rate or other factors associated with certain individuals that result in a higher dose of ozone at the target tissue. Conversely, different individuals given the same tissue dose may respond differently.

REPRODUCTIVE/DEVELOPMENTAL TOXICITY

Based on our current knowledge, such findings and relationships appear to be unique to certain compounds. Importantly, there are critical periods of susceptibility during development that change rapidly over time and that influence both the probability of any outcome occurring as well as the nature of the outcome.

CARCINOGENICITY

The available examples may be broadly applicable, but they are too few and too poorly understood to allow sweeping generalizations. Examination of the available aflatoxin and arsenic data may contribute to better understanding of interindividual variability, although the applicability of the results of such an analysis to other chemicals is unclear. Data are available for a stable, homogeneous human population in China that has been exposed to high concentrations of aflatoxin and to other synergistic factors (Qian et al. 1994). There also are considerable data available on aflatoxin dosimetry and effects in both animals and humans (Strickland and Groopman 1995). Data for humans exposed to arsenic (Landrigan 1992) may be particularly useful for evaluating the adequacy of the 10-fold uncertainty factor. Valuable insights into interindividual variability may result from examination of data for pharmacologically active agents, e.g., tamoxifen and cyclophosphamide, where human exposure can be accurately assessed.

Changes in cancer incidence are of concern and will affect approaches to characterizing and quantifying interindividual variability. For example, the rapid increase of ~3% per year in the incidence of human liver cancer may be related to coding procedures used in the cancer registry, increases in viral hepatitis transmission, an increased prevalence of liver cancer among AIDS patients, or other causes (Kosary and Diamondstone 1993). Factors that contribute to interindividual variability are complex and likely operate through various mechanisms. Clinical and epidemiologic studies are intended to identify additive risk from environmental agents and, although there are often good mechanistic data on the effects of such agents on rodents, there is frequently much uncertainty about how such agents operate in humans.

HOW MIGHT THIS INFORMATION BE USED IN NON-CANCER AND CANCER RISK ASSESSMENT?

NEUROTOXIC EFFECTS

The risk assessment process could be refined by identifying additional sources of variability. Currently, a default value of 10 is used to account for interindividual variability for noncancer endpoints (Dourson et al. 1996). If additional data are available on the sources of variability in response, it may be possible to replace the default value. The opportunity to potentially replace the 10-fold default uncertainty factor with a chemical- or exposure-specific model would encourage research and yield important new information on neurotoxicity.

A number of important items need to be considered to convince the public and regulatory agencies that the default assumption could be replaced. One point is to more clearly identify the source(s) of the variability. Information about why individuals vary in a particular response parameter is critical to any reevaluation of the adequacy of the uncertainty factor. The slope and the shape of the dose–response curve are additional concerns (Faustman and Omenn 1996). If the slope is very shallow, there is less confidence in replacing the 10-fold uncertainty factor with a smaller number. Conversely, if the slope is relatively steep, it would be easier to predict the impact of changing the uncertainty factor.

Another concern is the method for determining the point of extrapolation, i.e., the number to which the uncertainty factor is applied, and the degree of confidence associated with that value. Information that enhances the level of confidence one has in that value may encourage the use of a nondefault uncertainty factor. For example, if the benchmark dose (Crump 1984), which is derived from curve fitting, provides a more relevant value on which to base the extrapolation, the appropriate uncertainty factor might be less than the one that would be applied to a no-observable-adverse-effects-level (NOAEL) value (Barnes et al. 1995). The availability of such information is contingent on improved understanding of the mechanisms of action of neurotoxic processes. Importantly, altering the uncertainty factor would require demonstrating that such a change would not place the health of an individual or the population as a whole at increased risk.

PULMONARY EFFECTS

Risk assessors must apportion variability into at least two components: that reflecting exposure-to-dose relationships, and that reflecting dose–response relationships (Pierson et al. 1991). As an extreme example, if dosimetry were fully (100%) responsible for observed variations in response, there would be no need to use uncertainty factors. Simply modeling the dosimetry and perhaps including some estimate of variability would describe the response variability as a function of the percentage of individuals in the population receiving a certain target organ dose for a given concentration in the air.

Another important issue is that of extrapolating from animals to humans (D'Amato et al. 1992), recognizing that human exposure dosimetry does not necessarily equate to the tissue dosimetry of, e.g., a rat. Rats and humans exposed to an external ozone concentration of 0.12 ppm may experience very different internal doses. Knowing the target tissue dosimetry is essential for determining whether the rats received an exposure that is relevant to the human exposure.

REPRODUCTIVE/DEVELOPMENTAL TOXICITY

Currently, when information is available on a particular susceptible subpopulation, i.e., children, adults of reproductive age, or other groups, it is incorporated into the risk assessment. In most cases, such information is broad in scope and generally nonspecific.

The breakout group for this topic reviewed the extrapolation of animal data to humans and how those data might relate to understanding human variability. For example, when a value from an animal dose–response curve is used as the basis for extrapolation to humans, it is unclear where the extrapolated value falls in relationship to the human dose-response curve. It could represent the median, mean, or a value in either tail of the distribution. Such concerns are confounded by recognition that developmental studies are typically performed with two species.

Variability observed in animal studies represents only the variability associated with the specific population of the specific strain under study. It does not address variability among all the various strains of the species under study or within the human population. Currently, there is little information from animal models that can facilitate understanding or characterization of human variability. The development of animal models to study specific factors known to be important in susceptibility to particular agents ultimately may contribute to better understanding of human variability.

CARCINOGENICITY

Appropriate data on interindividual variability can and should be used in human health risk assessment. The critical problems are determining what data are appropriate and how they should be collected. As noted previously, an important issue is derivation of the value to which the 10-fold uncertainty factor is applied and what it predicts relative to the human dose-response curve. Although the uncertainty factor approach is not used in cancer risk assessment (Pitot and Dragan 1996), understanding how to measure and characterize interindividual variability will likely inform the cancer risk assessment process.

DO THE CURRENT DEFAULT METHODOLOGIES FOR NON-CANCER AND CANCER RISK ASSESSMENT IN THE UNITED STATES ADEQUATELY ACCOUNT FOR INTERINDIVIDUAL VARIABILITY?

NEUROTOXIC EFFECTS

One approach is to assess whether the 10-fold uncertainty factor is protective of human health with regard to neurotoxicity. Although the breakout group was unable to identify a case where the default uncertainty factor had failed when applied to a neurotoxic agent, such an exercise is confounded by the use of multiple uncertainty factors in the risk assessment process (Dourson et al. 1996). Because of this compounded conservatism, if the 10-fold uncertainty factor for interindividual variability is not sufficient to account for the range of variability in response, the other uncertainty factors, e.g., the 10-fold factor reflecting interspecies differences, may compensate, in part, for possible inadequacies. These types of evaluations are confounded by uncertainty about the derivation of the value on which the extrapolation is based. For example, permissible lead exposures for children have declined significantly over the years, yet this reduction is more reflective of changes in the NOAEL value than of recognition that specific uncertainty factors are inappropriate. As noted previously and by other breakout groups, the derivation of the value (NOAEL or benchmark dose) to which the uncertainty factor is applied and the confidence one has in that value is critical to evaluating the adequacy of the uncertainty factor. Increased confidence in this value might also provide the opportunity to reduce some of the other default uncertainty factors, recognizing that such changes must reflect the scientific community's confidence that the public is being protected.

Are there cases where the default uncertainty factor is overprotective? Use of the 10-fold uncertainty factor may be unnecessarily costly in terms of the effort, costs, equipment, and people required to control exposure to a particular substance. If a 2- or 3-fold uncertainty factor is adequate, society as a whole might experience considerable cost savings and other benefits.

For most of the materials addressed in the accompanying report on neurotoxicants (Eckerman et al., this volume), the 10-fold uncertainty factor appears to provide an adequate level of protection. However, this uncertainty factor may not be protective for cisplatin (Hattis 1995) and mercury

(Hattis and Silver 1994) because of questions about the quality of the exposure data and their impact on the analysis of interindividual variability. A further concern is that where there are human data that are applicable to the risk assessment of neurotoxicants, e.g., ethanol, such data are often obtained primarily from healthy young males. It is unclear whether a 10-fold uncertainty factor would be sufficient if the population of concern included a subpopulation of individuals of greater susceptibility.

PULMONARY EFFECTS

It is clear that most studies are not designed to evaluate the adequacy of the 10-fold uncertainty factor. An exception is a recent study by McDonnell et al. (1995) in which individuals were exposed to ozone after exercising so that their FEV_1 values declined by 15%, 10%, or 5%. McDonnell et al. examined the resulting probability distribution curves to determine what ozone exposure (ppm × hours) elicited similar responses among members of the study population. For example, among those exhibiting a 10% decline, some respond at just below 0.1 ppm × hours with perhaps one-half of the population responding at ~0.35 ppm × hours. Under the conditions of the study, barely 50% of the individuals responded.

This is important when considering the protectiveness of the 10-fold uncertainty factor with respect to ozone. If the average person corresponds to an individual in the 50th percentile, that person would respond at 0.7 ppm × hours. The most sensitive person may be in the 5th percentile and would respond at ~0.1 ppm × hours. Given this endpoint, i.e., a 10% decline in FEV_1, there is perhaps a 7-fold variation in ozone exposure (ppm × hours) required to elicit the same response. For this group of healthy, exercising young adult males, the 10-fold uncertainty factor is certainly protective or even overprotective.

Rather than asking whether the 10-fold uncertainty factor is or is not protective, it may be more appropriate to consider what proportion of the population is protected by a specific uncertainty factor (e.g., Dourson and Stara, 1983). For example, an uncertainty factor of 10 might protect 95% of the population from ozone; an uncertainty factor of 3 might protect 50%; and an uncertainty factor of 2 might protect 25%. Instead of thinking of 10 as safe or unsafe, the question becomes how safe should it be. This would make the job of the risk manager more difficult, particularly with respect to criteria pollutants, but it would provide for fuller and more accurate use of available information. Such an assessment would require fitting distributions to data such as those reported by McDonnell et al. (1995). Scientists conducting such studies should be encouraged to interact with risk assessors and risk managers to ensure that resulting data are appropriate for use in quantitative risk assessment.

REPRODUCTIVE/DEVELOPMENTAL TOXICITY

The degree of confidence scientists have in the 10-fold uncertainty factor varies considerably. Often the level of confidence reflects the degree to which the various components of variability have been considered. For example, in the case of toxicity where there is both prenatal and lactational exposure, the exposure is indirect and thus interindividual variability has some components of exposure variability to both mothers and their offspring as well as inherent response variability. These different components of variability need to be considered when the uncertainty factor(s) used in risk assessment are evaluated.

Because of the variety of outcomes possible from either reproductive or developmental toxicity studies in animals, it is often unclear whether the most sensitive endpoints have been measured. A study designed to assess whether developmental exposure might result in toxicities up to the time of birth addresses only some aspects of developmental toxicity. Measuring other parameters later in life, e.g., functional endpoints or growth, may reveal a more sensitive endpoint. It is unclear whether additional uncertainty should be factored into the risk assessment process when there is limited information available on the range of possible outcomes or when there is little or no information on either reproductive or developmental effects.

CARCINOGENICITY

It is unclear whether the current default methodologies for cancer risk assessment in the United States adequately account for individual variability (National Research Council 1994). There may be certain subpopulations of particular concern, e.g., children, women, or others, yet there are not enough data available to be able to evaluate the adequacy of the default methodology.

One issue underlying such an evaluation is that of appropriate measures of dosimetry. Although adduct quantitation and the use of urine and plasma dosimeters as internal dose measures may facilitate a shift from measures of external exposure to measures of internal susceptibility, there is considerable uncertainty about the relevance of adducts for risk assessment (Sutherland and Woodhead 1990). For example, the presence of adducts in the heart is not associated with tumor formation, yet urinary aflatoxin adducts may be useful composite indicators of risk from complex exposures (Groopman et al. 1992).

Metabolism and the relationships between metabolism, gene polymorphisms, and enzyme activity were considered with respect to tumor formation. Although one can demonstrate 100-fold or greater differences in the concentrations and activities of certain enzymes (Nebert 1991), such differences do not correspond to increased cancer risk because of the complexity of metabolic pathways and over-lapping substrate specificity. The distinction between constitutive and inducible enzymes is important, yet little is known about the role of inducible enzymes in human variability. A controversial issue is whether people who have diminished detoxification capacity are exposed to toxicants at sufficiently high concentrations that their ability to detoxify the material is truly compromised.

ARE THERE SPECIFIC RESEARCH NEEDS RELATIVE TO QUANTITATIVELY ASSESSING INTERINDIVIDUAL VARIABILITY THAT, IF ADDRESSED, WOULD IMPROVE RISK ASSESSMENT?

NEUROTOXIC EFFECTS

There is a need for comparative analysis of currently used risk assessment techniques to determine how they would be affected by changes in measures of individual variability. Are there instances in which the risk assessment scenarios change as a result of changes in such measures? Similarly, it would be useful to analyze the current reference dose (RfD) and acceptable daily intake (ADI) values to determine whether they are under- or overprotective. Because a number of both RfDs and ADIs have been in place for some time, it may be possible to determine whether they have been protective. Are there individuals who remain susceptible to neurotoxic diseases, even though these RfDs and ADIs represent relatively conservative values?

Further research is needed to define and evaluate sensitive biomarkers of effect and suscepti-bility. Neurotoxicity studies should be designed to identify sensitive subgroups as well as to characterize average responses. Most neurotoxicity studies report mean values with standard deviations and provide relatively little data about the distribution of effects within the population. Investigators should be encouraged to report on the interindividual variability observed within study groups in addition to reporting average values.

Neurotoxicity studies could focus on sensitive subpopulations or be stratified for susceptible individuals to examine variability within that subpopulation. Data evaluation and analysis should recognize that sensitive subpopulations could be present within the study group. Data from such individuals should be considered and possibly analyzed separately to characterize the variability between individuals and within the group as a whole.

Data from both animal and human studies occasionally include individual values that are considered outliers. However, such outliers may represent the susceptible portion of the population. It is important to confirm that those data are considered in the risk assessment and are not simply discarded as outliers. For example, within any fairly large population with average susceptibility to lead, there usually is a small group of individuals who apparently have an increased susceptibility to the effects of lead. One needs to determine whether this group is biologically different by collecting data specifically on that subpopulation.

There is a need to develop databases of quantitative information about human variability. Such databases would foster additional research on human variability and susceptibility and should be accessible to the broad scientific community. Better mechanisms are needed to facilitate sharing of data, particularly epidemiologic and human effects data, while protecting the confidentiality of the patient and the intellectual property rights of the original investigators.

More data are needed on the relative contributions of pharmacokinetics and pharmacodynamics to susceptibility. This would facilitate partitioning the source of variability within an individual. Is the variability due to absorption of the material, its distribution, metabolism, or excretion? Is it due to differences in the rate of interaction between the target tissue and the chemical or to properties that are associated with development of the pathologic process? There is a need for better understanding of the mechanisms by which neurotoxicity occurs.

International collection and analysis of data on human responses to neurotoxic agents need to be encouraged. It is clear that exposures to chemicals in both the workplace and the environment have tended to decrease in recent years. As individual exposures decrease, it becomes necessary to study larger groups of individuals to detect specific effects. Similarly, because of changing industrial processes the number of people exposed to materials in the workplace is decreasing, making it more difficult to perform epidemiologic studies in an occupational environment. In the future, multicountry rather than multicompany investigations may be necessary to obtain data from enough exposed individuals to perform a meaningful analysis.

It is necessary to perform repetitive tests on the same individuals to determine how much of the observed variability is due to inherent interindividual variability and how much is due to variability associated with the methodology. Similarly, there is a need to distinguish inter- and intraindividual variability from measurement errors.

PULMONARY EFFECTS

It is important to better understand inflammation, not only with respect to ozone but also for sulfur dioxide and other oxidant gases. What is the significance of inflammatory changes as far as risk for chronic effects, particularly for individuals who may be susceptible? The relationship(s) between genetic differences and inflammatory changes warrants further characterization. For example, neutrophils from different animal strains have very different responses to ozone, which need to be examined in terms of their relevance to humans.

Ozone may enhance the response to allergens, which is important in terms of understanding the secondary effects of ozone exposure. In chamber studies, asthmatic individuals are no more sensitive to ozone than the general population, yet epidemiologic studies indicate an association between days with high ozone concentrations and hospital admissions for asthmatic patients. As noted earlier, a number of caveats associated with these generalizations call into question the use of ozone concentrations as surrogate environmental variables for such studies.

It is important to know how much of the observed variability is attributable to dosimetric effects and how much is due to variability in response to identical tissue doses. If there is an innate difference in responsiveness, those differences may vary depending on the parameter under consideration, e.g., pulmonary function or inflammation. Presumably, different mechanisms are associated with different responses./

The biological basis for the variability observed among young, healthy individuals exposed to ozone remains unclear. Neurogenic differences, differences in nociceptor function and distribution, and other genetic and nongenetic factors, e.g., dietary status, may contribute to this variability. Further investigation of the basic biology of nociceptors and their response to ozone exposure is warranted. The mechanistic basis for age-related variability in the response to ozone exposure is unknown, and it is not clear why 10% of all exercisers are particularly susceptible to ozone exposure. To extrapolate from chamber studies to the broader population, it would be valuable to know the basis for that susceptibility. If the correct interpretation of the epidemiologic studies is that hospital admissions are associated with ozone exposure, then the disparity between those results and the results of chamber studies must be reconciled.

Developing optimal study designs for assessing pulmonary responses to ozone and other respiratory toxicants is important. For example, a well-designed study may later be disputed if new studies of daily variations in ozone exposure suggest that the animals may have adapted to the variable exposures, thus diminishing their responses. Can a strain of rat be created that does not experience the pain (?) that is felt on inspiration of ozone, and would that alter their responses to long-term ozone exposure?

It may be useful to reanalyze the results of some of the earlier ozone studies with probability distributions similar to those reported by McDonnell et al. (1995). This would afford an opportunity to better understand human variability and the implications of such variability for risk assessment.

REPRODUCTIVE/DEVELOPMENTAL TOXICITY

There are several existing large epidemiologic databases [e.g., Centers for Disease Control (1980)] that could be analyzed to specifically assess interindividual variability, recognizing that as the data are segregated into various components and the number of individuals per subpopulation decreases, the power of the analysis may be compromised. It also might be possible to pool data from various sources and perform meta-analyses to identify and characterize factors influencing variability in response.

As noted previously, the critical periods of exposure or timing of exposure are important with respect to developmental toxicity and must be considered in the risk assessment. There remains much uncertainty about the relationships between critical periods of exposure and outcome, especially preconceptional exposures, lactational exposures, and exposures during other periods of childhood.

There is little information available on mechanisms of action of agents that alter development or reproduction. This is a promising area of research because the rapidly accruing knowledge about the basic molecular aspects of developmental biology are beginning to be applied to developmental toxicology. As understanding about underlying mechanisms improves, the development of animal models of susceptibility would be extremely useful.

There is the potential for discovery and characterization of birth defect susceptibility genes analogous to cancer susceptibility genes, e.g., the uncommon allele for transforming growth factor α is associated with a higher incidence of cleft lip and palate in children of women who smoke (Hwang et al. 1995, Shaw et al. 1996). Another promising research area is that of genetic imprinting. There is some evidence that gene function can be altered either pre- or periconceptionally, raising the possibility of altered gene expression during development (Stewart et al. 1997).

CARCINOGENICITY

There are four areas in which research that could improve cancer risk assessment is urgently needed. Epidemiologic studies of the role of mechanistically plausible, polymorphic cancer susceptibility genes need to be undertaken. Critical intermediate endpoints, e.g., adducts or urinary biomarkers, for determining human susceptibility to cancer need to be identified. The relationships among human susceptibility, dosimetry, and disease need to be determined for a variety of representative compounds. Research to identify susceptibility factors for children, women, and specific ethnic groups should be encouraged. Better understanding of exposures during childhood is critical to identifying susceptibility factors. For example, the diets of children may differ considerably from those of adults for a variety of reasons, with children on soy-based formula potentially being exposed to very high levels of phytoestrogens. Such exposures may have ramifications for subsequent biological responses.

GENERAL DISCUSSION

After the reports of the breakout groups (summarized above), workshop participants broadly addressed the application of information on human interindividual variability to the risk assessment process. The discussion considered not only existing data but also data that might be collected during future studies that could enhance understanding of interindividual variability. Because this issue has not been emphasized in the past, examination of existing databases and resources likely will yield a high return for the effort.

The volume, diversity, and complexity of information available on the effects of toxicants and carcinogens confounds the identification and analysis of information relevant to understanding interindividual variability. Human response variability is clearly confounded by variability in exposure (National Research Council 1994). For the purpose of the workshop, participants were asked to address variability in response to a constant level of exposure, recognizing that exposures are complex and that internal dose likely requires participation of certain genetically defined pathways. Exposures can have multiple effects manifested by multiple different endpoints. Understanding the genetic basis for variability in internal dose will provide insight into response variability and its application to risk assessment.

Information about exposure variability is a principal source of uncertainty in epidemiologic studies of susceptibility and risk (LaGoy 1994). Studies incorporating accurate exposure and genetic information generally are not large enough to yield definitive results. Workplace studies potentially can identify specific exposures and associate them with specific responses, thus providing insight into interindividual variability. When such studies are conducted, archiving saliva, blood, and urine specimens can provide the opportunity for retrospective analysis as new, more sensitive, and more specific analytical and interpretive methods become available.

The relationship between exposure and the observed toxicologic or carcinogenic responses is particularly complex when responses to prenatal and lactational exposure to chemicals are considered. In this case, two individuals are exposed, each having a distinct genome. The maternal genome will determine, in part, the dose of the parent compound and its metabolites that is delivered to the embryo or young. Physiological factors associated with both the mother and the offspring, e.g., perfusion rates, also might influence the response. The involvement of two genomes complicates understanding of both the exposure and the response variability.

In the context of a constant exposure, two factors influence the response: target tissue dosimetry and the intrinsic biological response of the individual to that dosimetry (Rodricks and Turnbull 1994). Although there appear to be suitable bases for measuring dosimetry, incomplete understanding of the mechanisms associated with observed responses confound determination of the critical indicator of the effective dose. Moreover, the relationship between that critical dosimeter and the pharmacodynamic factors that ultimately contribute to the manifestation of the response is not well-understood. This type of information is essential to understanding interindividual variability in response to chemical exposures.

Data for certain pharmaceutical agents may be useful for addressing target tissue dosimetry, because pharmacokinetic data as well as biological effects data for those types of chemicals often are available for both rodents and humans. Concentrations of the active agent in the blood of both animals and humans are used to establish the range of human variability; this value may be as high as 30 (Gibson and Skett 1986). Because some of these compounds are known to be carcinogenic but have been in use for relatively long periods of time, they may provide valuable insight into the variability of human responses. Two examples of data sets that might facilitate better understanding of interindividual variability were suggested. Data from studies of secondary myelogenous leukemia associated with cyclophosphamide (Swerdlow et al. 1992) may soon lead to identification of critical rate-limiting steps in the disease process. Similarly, ongoing studies of animals and humans exposed to tamoxifen (White and Smith 1996) likely will yield new insights into interindividual variability of response.

These or other approaches potentially could be applied to nonpharmaceutical chemicals. However, data from animals suggest that in some cases different mechanisms of response may be elicited by different doses (Faustman and Omenn 1996). It is unclear whether this occurs in humans, particularly over the likely range of environmental exposures. Studies intended to address this question for nonpharmaceutical agents likely would be fraught with ethical and social, as well as scientific, concerns.

Studies of internal radiation dosimetry suggest that the actual dose to tissues is a function of the rate of cell turnover in those tissues. For example, the concentration of radium or other bone-seeking radioisotopes in humans is a function of age; concentrations are greatest during infancy and puberty. Because organs may undergo multiple spurts of growth from infancy to maturity there may be multiple periods of heightened or diminished susceptibility to toxicants or carcinogens. Age-related, episodic changes in cellular turnover rates reflect genetic regulation of aging and development and suggest that organ growth kinetics should be considered during the risk assessment process.

A critical concern is determining the normal range of response and whether an observed response represents a 3-fold or a 100-fold change in either direction relative to the normal response. Data, if they exist, to address this question most likely will be associated with pharmaceutical compounds for which there are large clinical databases.

Workshop participants developed a conceptual framework (Figure 1) reflecting the idea that, for a given exposure, various factors related to pharmacokinetics will influence interindividual variability in target tissue dose. Internal dose will interact in various ways with different pharmacodynamic factors to yield various biological outcomes. These factors might include enzymes critical to DNA repair, cell proliferation, or other processes essential to the observed biological response. Changes after the initial biological response (pathogenesis) will determine the final outcome or disease.

Insight into the variation associated with the initial step(s) in toxic and carcinogenic processes is rapidly accruing, at least with respect to understanding the influence of external dose. There are many candidate gene polymorphisms that may act to modify the effect of an exposure, regardless of whether the exposure is measured externally or as the biologically effective dose. Despite these emerging insights, understanding remains elusive. For example, although DNA adducts may arise as a consequence of exposure to a chemical (Sutherland and Woodhead 1990), some may be repaired, whereas others are not. Similarly, some adducts may form in critical genes and others may form elsewhere in the genome. According to the multistage model of carcinogenicity (Faustman and Omenn 1996), one or more additional insults and the ensuing response to each must occur to complete neoplastic transformation. During this process, some of the cells that have acquired mutations may be destroyed by apoptosis. Each step in this complex process is likely to be marked by considerable interindividual variability.

Until the basic biological mechanisms and processes leading from the absorption and activation of a carcinogen to tumor formation are more fully understood, understanding the variation in those processes will be incomplete. Although understanding of the early events in the carcinogenic process is improving, the critical rate-limiting steps remain unidentified. Simply summing individual measures of variability associated with each of the early components of the process is unlikely to yield

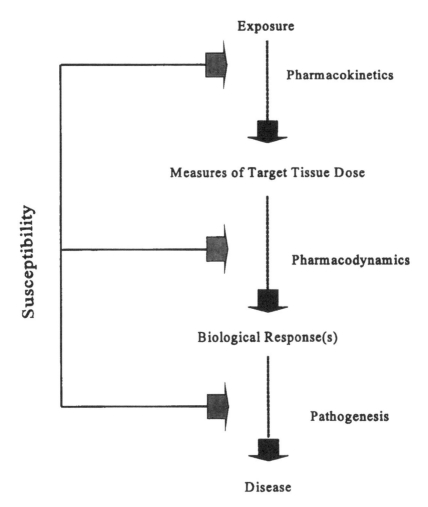

FIGURE 1   Schematic representation of where susceptibility factors impinge on exposure-effect relationships.

meaningful estimates of the range of human variability. Such estimates likely would vary over several orders of magnitude but would be unlikely to reflect correspondingly large changes in relative risk. Similar concerns permeated the discussion of human variability with respect to neurotoxic, pulmonary, and reproductive and developmental effects of exposure to toxicants.

Much of the general discussion focused on the issue of susceptibility genes and the use of information about such genes in the risk assessment process. Although the discussion was couched primarily in terms of cancer susceptibility genes, the issues raised were perceived to be broadly applicable to other effects because of emerging information suggesting that gene polymorphisms contribute to both the likelihood and magnitude of a variety of responses to toxicants.

Workshop participants were uncertain how to apply such information to risk assessment. This hesitancy reflected, in part, the notion that detection of cancer-associated genes implies that such genes equate to heightened sensitivity when, in fact, genetic epidemiology also can lead to the identification of populations that are insensitive or resistant to certain chemical exposures. Current methods of data collection and analysis, particularly the use of two-tailed statistical tests, can identify individuals who are more resistant as well as those who are more sensitive. Thus, risk assessors need to determine how to protect populations that include both more sensitive and more resistant individuals.

The implications of this type of information for risk assessors and managers was illustrated by the following example. Consider two factories, one in Boston and the other in Los Angeles, that use the same toxic material in the manufacturing process. Genetic screening of the workers at the two factories reveals that workers in Boston possess a gene associated with resistance to the effects of the chemical, whereas workers in Los Angeles do not. Does this mean that the Los Angeles workers are more sensitive or are they representative of the average sensitivity for the population at large? Does this mean that workers in the Boston factory could be exposed to higher concentrations of the chemical than their counterparts in Los Angeles? From a risk assessment perspective, it would be prudent to understand the ethnic or other characteristics of both sets of workers that contribute to this apparent difference in susceptibility. It remains to be seen whether the scientific community will achieve sufficient understanding to tailor exposures to specific populations based on their genetic composition. Such understanding, even if attained, would raise the host of ethical, moral, and social issues already associated with genetic screening for heritable disease susceptibility.

A second factor contributing to the hesitancy of the workshop participants to address the potential use of information on cancer susceptibility genes in risk assessment was recognition of the complex, protracted mechanisms and processes associated with carcinogenesis. Outcomes may not be observed for 30 years or more after the initiating or promoting event, and often the precipitating event is unknown or speculative at best (National Research Council 1992). From a population perspective, investigators typically must begin by identifying a tumor and work backward to infer the critical events and processes leading to tumorigenesis. In the case of noncancer outcomes, manifestation of the effects of exposure to a toxicant may not be observed until long after the exposure (Eaton and Klaassen 1996); yet in some instances, e.g., asthma, the precipitating events and disease progression are relatively well-appreciated, if not understood.

Identification of cancer (and other) susceptibility genes in the absence of understanding of the disease process (or outcome) limits the use of such information in risk assessment. Much of the cancer research effort is devoted to identifying agents that slow or halt cell turnover, yet identification of such rate-limiting steps generally provides little insight into carcinogenic processes. Understanding cancer initiating and promoting events in the context of cancer susceptibility genes is critical to understanding how to use such information in risk assessment.

The current approach to understanding the genetic basis for susceptibility is deterministic, i.e., attempting to identify specific genes associated with individuals expressing specific phenotypes [e.g., breast cancer (Langston et al. 1996), ataxia-telangiectasia (Savitsky et al. 1995)]. With the accrual of more information, it is likely that a probability-based understanding of the role of susceptibility genes will emerge. Such an approach would recognize that a variety of subtle alterations in various genes will slightly alter the probability of a specific effect or outcome based on a specific exposure. The interaction between various genes and exposure factors likely will be so complex as to preclude the identification of meaningful single or even multiple gene associations as is done currently. Indeed, it is widely perceived that the human genome project will result in identification of individual genes that confer susceptibility to specific diseases; this likely is an overly simplistic expectation.

Nevertheless, the human genome project has the potential to significantly affect the understanding of interindividual variability in susceptibility and response. The emerging information likely will challenge how risk assessors evaluate the toxic and carcinogenic potential of chemicals. It is critical to develop thoughtful, well-reasoned approaches to incorporating this information into the risk assessment process. However, it is not clear that the resources available to investigate the basic biological processes associated with important toxic effects will be sufficient to keep pace with the information emerging from the human genome project. Absent comparable increments in understanding, the risk assessment community will be unable to effectively use the human genome data in a timely fashion.

Workshop participants also discussed the adequacy of the currently used 10-fold uncertainty factor to account for human variability when performing non-cancer risk assessments. As noted in several of the breakout group reports (summarized above), there is little evidence to suggest that risk assessment as currently practiced has failed to protect the public at large. Conversely, the neurotoxicity breakout group report (summarized above) suggested that there may be cases in which use of the 10-fold uncertainty factor leads to overprotection. This point was illustrated during the general discussion by the example of kidney tumors in male rats. Risk assessments based on such findings could be overly protective if performed before the determination by the U.S. Environmental Protection Agency (EPA) (Baetcke et al. 1991) that kidney tumors associated with exposure to chemicals that induce α_{2u}-globulin accumulation in the male rat kidney likely are not relevant for humans exposed to the same chemical. When there is evidence that the mode of carcinogenic activity in rodents is irrelevant for humans, the use of uncertainty factors becomes a moot point.

There have been few, if any, attempts to rigorously evaluate the adequacy of the uncertainty factor for human variability, recognizing that the use of multiple uncertainty factors for non-cancer risk assessment (and the conservative assumptions inherent to cancer risk assessment) confound one's ability to perform such evaluations. During the general discussion it became clear that different approaches could be used to examine this issue.

One approach might be to reassess previously performed risk assessments or previously derived RfDs, recommended daily allowances (RDAs), or ADIs to determine whether the population is adequately protected. Because many of these values were established a number of years ago, it also may be possible to recalculate them by incorporating information on human variability that has become available in the interim. Changes in the newly derived values relative to the original values may provide insight into the adequacy of the uncertainty factor.

Another approach to evaluating the adequacy of the uncertainty factor that was suggested during the general discussion was to compare the responses of environmentally exposed with those of occupationally exposed populations. Individuals exposed to a chemical in the workplace likely experience greater exposures than people exposed to the same chemical in the environment. The presence or absence of effects among occupationally exposed individuals could provide a benchmark against which to evaluate the adequacy of the environmental standard. For example, the amount of methylene chloride that can be released into the environment (EPA 1992) is well below the concentration to which workers may be exposed in the workplace (Doull 1996). Because methylene chloride-associated cancers have not been observed among a large cohort of occupationally exposed individuals that has been followed for up to 30 years, the environmental standard may be lower than is necessary to protect the public. Conversely, if the worker population is healthier than the general population [e.g., the healthy worker effect (Rockette 1992)], susceptible groups may not be represented and a standard based on the occupationally exposed population may be underprotective. Although the example given does not directly address the adequacy of the uncertainty factor, it does illustrate points raised earlier about the derivation of the value to which the uncertainty factor is applied and about the need to make better use of human data.

A more prospective approach suggested during the general discussion was to identify individuals who express specific susceptibility genes and to determine whether they are at greater risk than the general population for developing cancer or for experiencing specific toxic effects (e.g., neurotoxicity). For similar exposures to known toxicants or carcinogens, if the effect or outcome is more common among the genetically susceptible individuals, the uncertainty factor for human variability may not be sufficient. If the incidences are similar, the uncertainty factor may be adequate or more than adequate.

The workshop participants also considered the distribution of risk as outlined in the report of the breakout group reviewing pulmonary effects. Using the methylene chloride emissions example, it would be instructive to know what percentage of individuals within the exposed population would have defined risks for a specific standard, e.g., 1 ppm. Estimates should reflect available information on variability in inhalation rates and metabolism as well as pharmacodynamic models of responses to methylene chloride. Such an analysis may reveal that 99.999% of the population is being protected from an adverse effect. Although this may be too conservative with respect to the resources needed to achieve this level of protection, such an approach offers a new perspective for evaluating the cost-effectiveness of risk assessments. Better quantitation of interindividual variability should lead to better understanding of the level of protection afforded by regulations based on risk assessments and facilitate determining whether the resources available to reduce human risk are adequate.

Distribution of risk for potential high-risk groups is important, even though little is known about the distribution of risk for children relative to older individuals or for women relative to men (National Research Council 1994). Risk assessment may benefit more by addressing such issues than by attempting to incorporate mechanistic information when critical and rate-limiting steps and processes are largely unknown. Summing known or suspected susceptibility factors across high-risk ethnic or age groups is of dubious value and likely will provide little insight into whether that group is being adequately protected. By considering risk distributions and other information, it may be possible to better account for interindividual variability even in the absence of full understanding of the genetics of susceptibility.

The general discussion reinforced the observation that a plethora of factors compound the uncertainty associated with understanding human variability. It is important that these sources of variability and other limitations be recognized in the risk assessment process. To use perhaps an overly simple example, dietary RDAs recognize that each individual has different nutritional requirements for specific elements, vitamins, etc. Consequently, RDAs are set at two standard deviations above the mean value for each dietary component to account for interindividual variability. The estimated RDA is excessive for some members of the population and inadequate for others (perhaps 5% of the population). By stating these limitations in the risk assessment and risk communication, risk assessors will acknowledge the uncertainty in the process while characterizing the most susceptible and most resistant members of the population. Such information is essential to the risk managers who must decide whether the concept of protecting the most susceptible individual is attainable.

REFERENCES

Anger WK, Johnson BL (1992) Human behavioral neurotoxicology: workplace and community assessments. In Rom WN (ed), Environmental and occupational medicine, 2nd ed. Little, Brown, Boston, pp 573–592

Baetcke KP, Hard GC, Rodgers IS, et al (1991) Alpha$_{2u}$-globulin: association with chemically induced renal toxicity and neoplasia in the male rat. EPA, Risk Assessment Forum, Washington, D.C.

Barnes DG, Datson GP, Evans JS, et al (1995) Benchmark dose workshop: criteria for use of a benchmark dose to estimate a reference dose. Regul Toxicol Pharmacol 21:296–306

Bibbo M, Al-Naqeeb M, Baccarini I, et al (1975) Follow-up study of male and female offspring of DES-treated mothers, a preliminary report. J Reprod Med 15:29–32

Buehler BA, Rao V, Finnell RH (1994) Biochemical and molecular teratology of fetal hydantoin syndrome. Neurol Clin 12:741–748

Calabrese EJ, ed (1992) Biological effects of low level exposures to chemicals and radiation. Lewis Publishers, Ann Arbor, MI

Centers for Disease Control (1980) Congenital malformations surveillance report, January–December 1980. Centers for Disease Control, Atlanta, GA

Crump KS (1984) A new method for determining allowable daily intakes. Fundam Appl Toxicol 4:854–871

Czeizel A, Dudas I (1992) Prevention of the first occurrence of neural-tube defects by periconceptional vitamin supplementation. N Engl J Med 327:1832–1835

D'Amato R, Slaga TJ, Farland WH, et al, eds (1992) Relevance of animal studies to the evaluation of human cancer risk. Wiley-Liss, Washington, D.C.

Doull J (1996) Recommended limits for exposure to chemicals. In Klaassen CD, Amdur MO, Doull J (eds), Casarett and Doull's toxicology: the basic science of poisons, 5th ed. McGraw-Hill, New York, pp 1025–1049

Dourson ML, Stara JF (1983) Regulatory history and experimental support of uncertainty (safety) factors. Regul Toxicol Pharmacol 3:224–238

Dourson ML, Felter SP, Robinson D (1996) Evolution of science-based uncertainty factors in noncancer risk assessment. Regul Toxicol Pharmacol 24:108–120

Eaton DL, Klaassen CD (1996) Principles of toxicology. In Klaassen CD, Amdur MO, Doull J (eds), Casarett and Doull's toxicology: the basic science of poisons, 5th ed. McGraw-Hill, New York, pp 13–33

Faustman EM, Omenn GS (1996) Risk assessment. In Klaassen CD, Amdur MO, Doull J (eds), Casarett and Doull's toxicology: the basic science of poisons, 5th ed. McGraw-Hill, New York, pp 75–88

Fink JN (1992) Hypersensitivity pneumonitis. In Rom WN (ed), Environmental and occupational medicine, 2nd ed. Little, Brown, Boston, pp 367–372

Fraumeni JF, Hoover RN, Devesa SS, et al (1993) Epidemiology of cancer. In De Vita VT, Hellman S, Rosenberg SA (eds), Cancer: principle and practice of oncology, 4th ed. Lippincott, Philadelphia, pp 150–181

George ST, Varghese M, John L, et al (1985) Aryl acylamidase activity in human erythrocyte, plasma and blood in pesticide (organophosphates and carbamates) poisoning. Clin Chem Acta 145:1–8

Gibson GG, Skett P (1986) Introduction to drug metabolism. Chapman and Hall, London

Groopman JD, Roebuck BD, Kensler TW (1992) Molecular dosimetry of aflatoxin DNA adducts in humans and experimental rat models. In D'Amato R, Slaga TJ, Farland WH, et al (eds), Relevance of animal studies to the evaluation of human cancer risk. Wiley-Liss, New York, pp 139–155

Hattis D (1995) Variability in susceptibility — how big, how often, for what responses to what agents? Proceedings WHO/IPCS Workshop, Southampton, England, 26 pp

Hattis D, Silver K (1994) Human interindividual variability — a major source of uncertainty in assessing risks for noncancer health effects. Risk Anal 14:421–431

Hwang S-J, Beaty TH, Panny SR, et al (1995) Association study of transforming growth factor alpha (TGFα) TaqI polymorphism and oral clefts: indication of gene-environment interaction in a population-based sample of infants with birth defects. Am J Epidemiol 141:629–636

Johnson BL, ed (1987) Prevention of neurotoxic illness in working populations. Wiley, New York

Johnson BL (1980) Electrophysiological methods in neurotoxicity testing. In Spencer PS, Schaumburg HH (eds), Experimental and clinical neurotoxicology. Williams and Wilkins, Baltimore, MD, pp 726–742

Kaufman RH, Binder GL, Gray PM, et al (1977) Upper genital tract changes associated with exposure in utero to diethylstilbestrol. Obst Gynecol Surv 32:611–613

Kopin IJ, Markey SP (1988) MDTP toxicity implication for research in Parkinson's disease. Annu Rev Neurosci 11:81–96

Koren HS, Devlin RB, Becker S, et al (1991) Time-dependent changes of markers associated with inflammation in the lungs of humans exposed to ambient levels of ozone. Toxicol Pathol 19:406–411

Koren HS, Hatch GE, Grahara DE (1990) Nasal lavage as a tool in assessing acute inflammation in response to inhaled pollutants. Toxicology 60:15–25

Kosary CL, Diamondstone LS (1993) Liver and intrahepatic bile duct. In Miller BA, Ries LAG, Hankey BF, et al (eds), SEER cancer statistics review: 1973–1990. National Cancer Institute, Bethesda, MD, pp XIV.1–XIV.17

LaGoy PK (1994) Risk assessment: principles and applications for hazardous waste and related sites. Noyes Publications, Park Ridge, NJ

Landrigan PJ (1992) Ethylene oxide. In Rom WN (ed), Environmental and occupational medicine, 2nd ed. Little, Brown, Boston, pp 1033–1039

Langston AA, Malone KE, Thompson JD, et al (1996) BRCA1 mutations in a population-based sample of young women with breast cancer. N Engl J Med 334:137–142

Lebowitz MD (1991) Methods to assess respiratory effects of complex mixtures. Environ Health Perspect 95:75–80

Lippmann M (1989) Health effects of ozone: a critical review. J Air Pollut Cont Assoc 39:67–96

Lotti M (1987) Organophosphate-induced delayed polyneuropathy in man and perspectives for biomonitoring. Trends Pharmacol Sci 8:176–177

McDonnell WF, Andreoni S, Smith MV (1995) Proportion of moderately exercising individuals responding to low-level, multi-hour ozone exposure. Am J Respir Crit Care Med 152:589–596

McDonnell WF, Muller KE, Bromberg PA, et al (1993) Predictors of individual differences in acute response to ozone exposure. Am Rev Respir Dis 147:818–825

Miller BA, Ries LAG, Hankey BF, et al, eds (1993) SEER cancer statistics review: 1973–1990. NIH Publication No. 93-2789. National Cancer Institute, Bethesda, MD

Molfino NA, Wright SC, Katz I, Tarlo S, et al (1991) Effect of low concentrations of ozone on inhaled allergen responses in asthmatic subjects. Lancet 338:199–203

Monks TJ, Anders MW, Dekant W, et al (1990) Glutathione conjugate mediated toxicities. Toxicol Appl Pharmacol 106:1–19

MRC Vitamin Study Research Group (1991) Prevention of neural tube defects: results of the Medical Research Council vitamin study. Lancet 338:131–137

National Research Council (1994) Science and judgment in risk assessment. National Academy Press, Washington, D.C.

National Research Council (1992) Multiple chemical sensitivities — an addendum to biologic markers in immunotoxicology. National Academy Press, Washington, D.C.

Nebert DW (1991) Role of genetics and drug metabolism in human cancer risk. Mutat Res 247:267–281

Needleman HL, Schell A, Bellinger D, et al (1990) Long-term effects of childhood exposure to lead at low dose; an eleven-year follow-up report. N Engl J Med 322:83–88

Ørbaek P, Lindgren M, Olivecrona H, et al (1987) Computed tomography and psychometric test performances in patients with solvent induced chronic toxic encephalopathy and healthy controls. Br J Ind Med 44:175–179

Otto D (1986) The use of sensory evoked potentials in neurotoxicity testing of workers. Sem Occup Med 1:175–183

Overfield T (1995) Biologic variation in health and illness: race, age, and sex differences, 2nd ed. CRC Press, Boca Raton, FL

Parkinson A (1996) Biotransformation of xenobiotics. In Klaassen CD, Amdur MO, Doull J (eds), Casarett and Doull's toxicology: the basic science of poisons, 5th ed. McGraw-Hill, New York, pp 113–186

Penrose LS (1933) The relative effects of paternal and maternal age in mongolism. J Genet 27:219–224

Pierson TK, Hetes RG, Naugle DF (1991) Risk characterization framework for noncancer end points. Environ Health Perspect 95:121–129

Pitot HC III, Dragan YP (1996) Chemical carcinogenesis. In Klaassen CD, Amdur MO, Doull J (eds), Casarett and Doull's toxicology: the basic science of poisons, 5th ed. McGraw-Hill, New York, pp 201–267

Qian G, Ross RK, Yu MC, et al (1994) A follow-up study of urinary markers of aflatoxin exposure and liver cancer risk in Shanghai, P.R.C. Cancer Epidemiol Biomarkers Prev 3:3–11

Rockette HE (1992) Occupational biostatistics. In Rom WN (ed), Environmental and occupational medicine, 2nd ed. Little, Brown, Boston, pp 51–58

Rodricks JV, Turnbull D (1994) Risk assessment: biological considerations. In Milman HA, Weisburger EK (eds), Handbook of carcinogen testing, 2nd ed. Noyes Publications, Park Ridge, NJ, pp 630–650

Saunders AM, Strittmatter WJ, Schmechel D, et al (1993) Association of apolipoprotein E allele epsilon 4 with late-onset familial and sporadic Alzheimer's disease. Neurology 43:1467–1472

Savitsky K, Bar-Shira A, Gilad S, et al (1995) A single ataxia telangiectasia gene with a product similar to PI-3 kinase. Science 268:1749–1753

Senekjian EK, Potkul RK, Frey K, et al (1988) Infertility among daughters either exposed or not exposed to diethylstilbestrol. Am J Obstet Gynecol 158:493–498

Shaw GM, Wasserman CR, Lammer EJ, et al (1996) Orofacial clefts, parental cigarette smoking, and transforming growth factor-alpha gene variants. Am J Hum Genet 58:551–561

Smith CM, Wang X, Hu H, et al (1995) A polymorphism in the δ-aminolevulinic acid dehydratase gene may modify the pharmacokinetics and toxicity of lead. Environ Health Perspect 103:248–253

Snyder R, Andrews LS (1996) Toxic effects of solvents and vapors. In Klaassen CD, Amdur MO, Doull J (eds), Casarett and Doull's toxicology: the basic science of poisons, 5th ed. McGraw-Hill, New York, pp 737–771

Sokol RJ, Ager J, Martier S, et al (1986) Significant determinants of susceptibility to alcohol teratogenicity. Ann NY Acad Sci 477:87–102

Stewart CL, Pedersen R, Rotwein P, et al (1997) Genomic imprinting. Reprod Toxicol 11:309–316

Strickland PT, Groopman JD (1995) Biomarkers for assessing environmental exposure to carcinogens in the diet. Am J Clin Nutr 61:710S–720S

Sutherland BM, Woodhead AD, eds (1990) DNA damage and repair in human tissues. Plenum, New York

Swerdlow AJ, Douglas AJ, Hudson GV, et al (1992) Risk of second primary cancers after Hodgkin's disease by type of treatment: analysis of 2846 patients in the British National Lymphoma Investigation. Br Med J 304:1137–1143

Tepper JS, Costa DL, Lehmann JR, et al (1989) Unattenuated structural and biochemical alterations in the rat during functional adaption to ozone. Am Rev Respir Dis 140:493–501

U.S. Environmental Protection Agency (1992) Methylene chloride. Integrated Risk Information System (IRIS) Record No. 68, EPA, Washington, D.C.

Vogt RF Jr (1991) Use of laboratory tests for immune biomarkers in environmental health studies concerned with exposure to indoor air pollutants. Environ Health Perspect 95:85–91

Weinstock S, Beck BD (1988) Age and nutrition. In Brain JD, Beck BB, Warren AJ, et al (eds), Variations in susceptibility to inhaled pollutants: identification, mechanisms, and policy implications. Johns Hopkins Univ Press, Baltimore, MD, pp 104–126

White INH, Smith LL (1996) Mechanisms of tamoxifen-induced genotoxicity and carcinogenicity. In Li JJ, Li SA, Gustafsson L, et al (eds), Hormonal carcinogenesis II. Proceedings of the Second International Symposium. Springer-Verlag, New York, pp 228–239

Wilcox AJ, Baird DD, Weinberg CR, et al (1995) Fertility in men exposed prenatally to diethylstilbestrol. N Engl J Med 332:1411–1416

Witschi HR, Last JA (1996) Toxic responses of the respiratory system. In Klaassen CD, Amdur MO, Doull J (eds), Casarett and Doull's toxicology: the basic science of poisons, 5th ed. McGraw-Hill, New York, pp 443–462

Wood FB, Flowers DL, Naylor CE (1991) Cerebral laterality in functional neuroimaging. In Kitterle FL (ed), Cerebral laterality: theory and research. Erlbaum, Hillsdale, NJ, pp 103–115

Working PK (1988) Male reproductive toxicology: comparison of the human to animal models. Environ Health Perspect 77:37–44

Index

S